Renormalization and Invariance in Quantum Field Theory

NATO ADVANCED STUDY INSTITUTES SERIES

A series of edited volumes comprising multifaceted studies of contemporary scientific issues by some of the best scientific minds in the world, assembled in cooperation with NATO Scientific Affairs Division.

Series B: Physics

Volume 1 – Superconducting Machines and Devices
edited by S. Foner and B. B. Schwartz

Volume 2 – Elementary Excitations in Solids, Molecules, and Atoms
(Parts A and B)
edited by J. Devreese, A. B. Kunz, and T. C. Collins

Volume 3 – Photon Correlation and Light Beating Spectroscopy
edited by H. Z. Cummins and E. R. Pike

Volume 4 – Particle Interactions at Very High Energies
(Parts A and B)
edited by David Speiser, Francis Halzen, and Jacques Weyers

Volume 5 – Renormalization and Invariance in Quantum Field Theory
edited by Eduardo R. Caianiello

The series is published by an international board of publishers in conjunction with NATO Scientific Affairs Division

A	**Life Sciences**	Plenum Publishing Corporation
B	**Physics**	New York and London
C	**Mathematical and Physical Sciences**	D. Reidel Publishing Company Dordrecht and Boston
D	**Behavioral and Social Sciences**	Sijthoff International Publishing Company Leiden
E	**Applied Sciences**	Noordhoff International Publishing Leiden

Renormalization and Invariance in Quantum Field Theory

Edited by

Eduardo R. Caianiello

Cybernetics Laboratory
Consiglio Nazionale delle Ricerche
Naples, Italy

PLENUM PRESS • NEW YORK AND LONDON
Published in cooperation with NATO Scientific Affairs Division

Library of Congress Cataloging in Publication Data

NATO Advance Study Institute, Capri, 1973.
Renormalization and invariance in quantum field theory.

(The NATO Advanced Study Institutes series. Section B: Physics, v. 5)
1. Quantum field theory—Congresses. 2. Renormalization (Physics)—Congresses. 3. Symmetry (Physics)—Congresses. I. Caianiello, Eduardo R., 1921- ed. II. Title. III. Series: NATO Advanced Study Institutes series. Series B: Physics, v. 5.
QC174.45.A1N67 1973 530.1'43 74-8902
ISBN 0-306-35705-4

Lectures presented at the NATO Advanced Study Institute, Capri, Italy, July 1-14, 1973

A Division of Plenum Publishing Corporation
227 West 17th Street, New York, N.Y. 10011

United Kingdom edition published by Plenum Press, London
A Division of Plenum Publishing Company, Ltd.
4a Lower John Street, London, W1R 3PD, England

Printed in the United States of America

Foreword

The subject matter of this Advanced Study Institute, which has been rendered possible by the generous support of NATO, gratefully acknowledged here, is of central importance to quantum field theory today. The problems involved are both deep and complicated, to a point that perhaps does not find easily a parallel in other branches of theoretical physics. From the first rule-of-thumb prescriptions on how to perform renormalization to the most recent advances--which brighten our hopes to find, eventually, in field theory indications as to the ultimate composition of matter--a great amount of progress has certainly been made. We were fortunate to have among us many of the people who have most contributed to past and recent developments. Although clues that point to what to do next are clearer now than might have been expected only a few years ago, much hard work is still ahead. It is hoped that both our meeting, which brought together leaders in the field as well as bright and eager beginners, and the present volume, based on the NATO meeting, may be useful as a mise-à-point and as an up-to-date reference book for researchers interested in the field.

Eduardo R. Caianiello

Contents

DIMENSIONAL RENORMALIZATION

P.Butera, G.M.Cicuta, and E.Montaldi

Istituto di Scienze Fisiche dell'Universita'

and I.N.F.N. - Sezione di Milano -

INTRODUCTION

Dimensional renormalization, a renormalization involving an analytic interpolation in the space-time dimension, is a recently devised method [4,6,10,21] for renormalization of quantum field theories. Although being equivalent to the conventional Bogoliubov-Parasiuk-Hepp and analytic renormalization theories, it is simpler than both. This is a decisive advantage for theories where it is convenient to keep gauge symmetries in all steps of the renormalization (quantum electrodynamics and particularly gauge theories of weak interactions and gravitation); furthermore, asymptotic behaviour and scale transformations appear to be much simpler.

All the technicalities of the method pertain to one of the two conventional methods, of which this may be described as a happy combination.

The basic idea is that the space-time dimension being four is the feature responsible for ultraviolet divergences (e.g., in bidimensional scalar models, Feynman integrals of polynomial interactions are convergent).

As an oversimplified case, let us consider the model integral

$$I_{(n)} = \int g(p^2)\, d^n p$$

where p^2 is the square of the length of the euclidean n-component vector p. We have

$$I_{(n)} = \Omega_n \int_0^\infty p^{n-1} g(p^2)\, dp$$

where

$$\Omega_\lambda = 2\pi^{\frac{\lambda}{2}} / \Gamma(\tfrac{\lambda}{2})$$

is the surface of the unit sphere in λ dimensions. It is clear that if the integrand has an inverse power behaviour

$$g(p^2) \underset{p^2 \to \infty}{\sim} (p^2)^{-K}$$

independent of n, $I_{(n)}$ is divergent for all $n \geq 2K$. Let us now consider the integral

$$I_{(\lambda)} = \Omega_\lambda \int_0^\infty p^{\lambda-1} g(p^2)\, dp$$

for complex values of λ, which is an obvious interpolating function for $I_{(n)}$. One easily sees that $I_{(\lambda)}$ is a meromorphic function for $\lambda \geq 2K$ and regular for $\lambda < 2K$.

Before discussing the general case, it is useful to examine the lowest order Feynman graph for the selfenergy in a ϕ^4 theory. Its Feynman integral is

$$I_{(\lambda)}(p^2) = \int_0^\infty [C(\alpha_i)]^{-\frac{\lambda}{2}} \exp\left[i \frac{D(p^2, \alpha_i, m^2)}{C(\alpha_i)}\right] d\alpha_1 d\alpha_2 d\alpha_3 \tag{1}$$

where

$$C(\alpha_i) = \alpha_1\alpha_2 + \alpha_1\alpha_3 + \alpha_2\alpha_3$$

$$D(p^2, \alpha_i, m^2) = \alpha_1\alpha_2\alpha_3\, p^2 - m^2 C(\alpha_i) \sum_1^3 \alpha_i + i\varepsilon$$

This is an obvious interpolating function for the Feynman integral $I_{(n)}(p^2)$ associated with the Feynman graph in Fig.1 in an n-dimensional space-time. As shown in [9], the bad asymptotic behaviour of $I_{(n)}(p^2)$ for large negative n does not allow us to define the interpolating function in a unique way. Under suitable restrictions, we have characterized the class of analytically interpolating functions which we here denote by $\{I_{(\lambda)}(p^2)\}$.

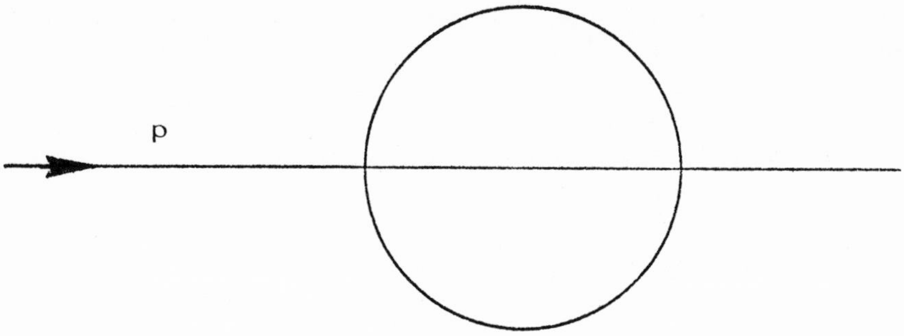

Fig.1.The second order selfenergy graph

By investigating the analyticity properties of $I_{(\lambda)}(p^2)$ as a function of the complex space-time dimension λ, for values of λ close to four, we find the decomposition

$$I_{(\lambda)}(p^2) = \Gamma(3-\lambda)\,\Gamma\left(2-\frac{\lambda}{2}\right)R_1(\lambda) + \Gamma(4-\lambda)\,p^2 R_2(\lambda) + R_3(\lambda) \qquad (2)$$

where the $R_i(\lambda)$ are regular functions at $\lambda = 4$.

We now introduce an operator $\mathcal{J}$ which acts on the class $\{I_{(\lambda)}(p^2)\}$ of interpolating functions in such a way that

$$\mathcal{J}\{I_{(\lambda)}(p^2)\} = \{f_{(\lambda)}(p^2)\},$$

where $f_{(\lambda)}(p^2)|_{\lambda=4}$ is a finite function of p^2 ; moreover, the class $\{f_{(\lambda)}(p^2)|_{\lambda=4}\}$ is an equivalence class of functions modulo $c + bp^2$, where c and b are finite arbitrary constants. We shall fix them by always choosing the subtraction point at the origin. We give here two explicit realizations of the operator $\mathcal{J}$, namely

$$\mathcal{J}\{I_{(\lambda)}(p^2)\} = \frac{1}{2\pi i}\oint \frac{dz}{z-\lambda}\{I_{(z)}(p^2)\} = (1-\mathcal{M}_{\mu,a})\{I_{(\lambda)}(p^2)\} \qquad (3)$$

where a is arbitrary and $\mathcal{M}_{K,a}\, g(x)$ is the Taylor expansion of $g(x)$ around $x = a$, truncated at order K, and μ is the superficial divergence.

The realizations (3) of the operator $\mathcal{J}$ are very

instructive, because the first one is an analytic evaluator, which is a basic tool of the analytic renormalization theory, while the second realization is just the familiar subtraction operator of the BPH theory.

The reader however should be warned against a straightforward generalization of the operator $\mathfrak{Y}$ in the case of an arbitrary Feynman graph. As we show in the next section, the simple prescription (3) is correct only for a special class of divergent Feynman graphs; the general prescription must take into account the standard analysis of the divergences associated with the subgraphs. On the other hand, this warning agrees with general arguments indicating the impossibility of performing the analytic renormalization by using a single complex parameter.

Let us return to the previously considered Feynman integral. By applying the operator $\mathfrak{Y}$ given in (3) to the equivalence class of interpolating functions $\{I_{(\lambda)}(p^2)\}$, where $I_{(\lambda)}(p^2)$ is given in (1), we obtain for $\lambda = 4$ and with the above described choice of the subtraction constants

$$I_{ren}(p^2) = R_3(4) = i\int_0^1 d\alpha_1 d\alpha_2 d\alpha_3 \frac{\delta(1-\sum_1^3 \alpha_i)}{[C(\alpha_i)]^3}\Big\{\alpha_1\alpha_2\alpha_3 p^2 \cdot$$

$$\cdot\Big[1-\log\Big(1-\frac{\alpha_1\alpha_2\alpha_3 p^2}{C(\alpha_i)m^2}\Big)\Big] + C(\alpha_i)m^2 \log\Big(1-\frac{\alpha_1\alpha_2\alpha_3 p^2}{C(\alpha_i)m^2}\Big)\Big\} \tag{4}$$

Before closing this section, we would like to add a remark concerning the asymptotic behaviour of dimensionally regularized Feynman integrals. It is traditional to study the asymptotic behaviour, $\sigma \equiv -p^2 \to \infty$, by Mellin transforms. The Mellin transform of $I_{(\lambda)}(\sigma)$ is

$$F_{(\lambda)}(\beta) = \int_0^\infty I_{(\lambda)}(\sigma)\,\sigma^{-\beta-1} d\sigma = \Gamma(-\beta)\, e^{i\frac{\pi}{2}\beta} A_{(\lambda)}(\beta) \tag{5}$$

where

$$A_{(\lambda)}(\beta) = \int_0^\infty (\alpha_1\alpha_2\alpha_3)^\beta [C(\alpha_i)]^{-(\beta+\frac{\lambda}{2})} e^{-im^2\sum_1^3\alpha_i} \prod_1^3 d\alpha_i$$

One easily finds that $A_{(\lambda)}(\beta) = \Gamma(3+\beta-\lambda)$ times a function which has double poles at $\beta = -1, -2, \ldots.$

According to the usual analysis, the rightmost poles of $A_{(\lambda)}(\beta)$ in the β plane determine the asymptotic behaviour of $I_{(\lambda)}(\sigma)$. For $\lambda = 4$, the dominant pole is in $\beta = 1$. We have

$$F_{(4)}(\beta) \underset{\beta \sim 1}{\approx} i\,\Gamma(-\beta)\Gamma(\beta-1)\int_0^1 \alpha_1\alpha_2\alpha_3 \frac{\delta(1-\sum_1^3 \alpha_i)}{[C(\alpha_i)]^3} \prod_1^3 d\alpha_i \tag{6}$$

By inverting the Mellin transform one obtains the asymptotic behaviour of the renormalized integral

$$I_{(4)}(\sigma) \sim i\,\sigma \log\frac{\sigma}{m^2} \int_0^1 \alpha_1\alpha_2\alpha_3 \frac{\delta(1-\sum_1^3 \alpha_i)}{[C(\alpha_i)]^3} \prod_1^3 d\alpha_i \tag{7}$$

which is checked by looking at (4).

We like to remark that the poles of $A_{(\lambda)}(\beta)$ in the β plane which are independent of the space-time dimension λ are those which can be predicted by the usual considerations about shortest paths.

RENORMALIZATION OF A GENERAL SCALAR GRAPH

We now describe a suitable extension of the techniques sketched in the introduction, in order to renormalize the generic Feynman integral of a renormalizable scalar field theory.

Let G be a general graph. A basic notion is the non-recursive characterization of the class $\{S(G)\}$ of the dominant (or complete) divergent subgraphs S_i. These are the irreducible subgraphs S_i of the graph G which :

1) are superficially divergent, $\mu_i \geq 0$;
2) cannot be formed from another superficially divergent graph by simply opening one line.

$\{S(G)\}$ consists of those graphs of the class singled out by the recursive B.P.H. procedure which are associated with Taylor subtractions.

We define the $\mathcal{A}$ class of Feynman graphs as the class of graphs such that $\{S(G)\}$ contains only one subgraph, which is the graph itself. An example of graph that belongs to the $\mathcal{A}$ class is that discussed in the introduction. The renormalization of a graph of the $\mathcal{A}$ class is performed by simply applying the operator $\mathcal{J}$ of equation (3).

The renormalization of a general Feynman integral is performed by applying the procedure of the previous section to each subgraph S_i of the class $\{S(G)\}$.

The regularized integral has the form

$$I_{(\lambda)}(p_i) = \int_0^\infty [C(\alpha_i)]^{-\frac{\lambda}{2}} \exp\left[i \frac{D(p_i, \alpha_i, m^2)}{C(\alpha_i)}\right] \prod_1^{\ell} d\alpha_i \qquad (8)$$

where $C(\alpha_i)$ and $D(\alpha_i, p_i, m^2) = W(\alpha_i, p_i) - m^2 C(\alpha_i) \sum_1^{\ell} \alpha_i + i\varepsilon$ are the familiar parametric functions. The formal, possibly divergent Feynman integral is recovered by letting the complex variable λ be equal to four.

As a consequence of a Speer's theorem, $I_{(\lambda)}(p_i)$ is a meromorphic function in the complex λ plane. More precisely,

$$I_{(\lambda)} \left\{ \sum_P \prod_{j=2}^{\ell} \Gamma\left(j - \frac{\lambda}{2} N_j(P)\right) \right\}^{-1}$$

is an entire function of λ. Here, $\sum_P$ means sum over all permutations of the labels of the ℓ lines of the graph and, for a given permutation P, $N_j(P)$ is the number of loops of the graph consisting of lines 1 through j with their vertices.

Beside the obvious interpolating function given by (8), it is necessary to define the general class of interpolating functions, which we shall denote by $\{I_{(\lambda)}(p_i)\}$.

The elements of this class are obtained by substituting, in the evaluation of the Feynman integral, the obvious interpolating function associated to the subgraph $S_i \in \{S(G)\}$ with its equivalence class of interpolating functions, as discussed in the introduction.

It will be convenient to use the subtractive realization of the operator $\mathcal{J}$. The renormalization procedure we are going to describe is equivalent to the conventional one, where for each subgraph $S_i \in \{S(G)\}$ the familiar Taylor expansion subtraction is performed around

$k_i = 0$, k_i being the external momenta of S_i (of course, these momenta may be either internal or external for the graph G itself).

Let us choose a subgraph $S_k \in \{S(G)\}$. By performing a scale transformation, see Appendix, on the variables α_j associated with the lines belonging to S_k we obtain

$$I_{(\lambda)}(p_i) = \int_0^\infty \prod_{\alpha_i \notin S_k} d\alpha_i \int_0^1 \prod_{\alpha_j \in S_k} d\alpha_j \, \delta(1 - \sum_{\alpha_j \in S_k} \alpha_j) \cdot \int_0^\infty \rho^{-r_k - 1} \hat{C}^{-\frac{\lambda}{2}} \exp(i \hat{D}/\hat{C}) \, d\rho . \tag{9}$$

Here $r_k = -n_k + \frac{\lambda}{2} \ell_k$, n_k being the number of lines of the subgraph S_k and ℓ_k the order of the zero of $C(\rho\alpha_j)$ for $\rho = 0$, that is also the number of loops of S_k; furthermore

$$\hat{C}(\alpha_i, \alpha_j, \rho) = \rho^{-\ell_k} C(\alpha_i, \alpha_j \to \rho\alpha_j) \tag{10}$$

$$\hat{D}(\alpha_i, \alpha_j, \rho, p_i, m^2) = \rho^{-\ell_k} D(\alpha_i, \alpha_j \to \rho\alpha_j, p_i, m^2) = \\ = \rho^{-\ell_k} \{ W(\alpha_i, \alpha_j \to \rho\alpha_j, p_i) - m^2 C(\alpha_i, \alpha_j \to \rho\alpha_j)[\Sigma\alpha_i + \rho\Sigma\alpha_j] \}$$

Let us now introduce an operator $\mathcal{J}_{(S_k)}$ to remove the divergence from the ρ integration in (9).

$$\mathcal{J}_{(S_k)} I_{(\lambda)}(p_i) = \int_0^\infty \prod_{\alpha_i \notin S_k} d\alpha_i \int_0^1 \prod_{\alpha_j \in S_k} d\alpha_j \, \delta(1 - \sum_{\alpha_j \in S_k} \alpha_j) \cdot \\ \cdot \int_0^\infty \rho^{-r_k - 1} e^{-im^2(\Sigma\alpha_i + \rho\Sigma\alpha_j)} \left[1 - \mathcal{M}^{(\rho)}_{(r_k)}\right] \{ \hat{C}^{-\frac{\lambda}{2}} \cdot \\ \cdot \exp\left[i W(\alpha_i, \alpha_j \to \rho\alpha_j, p_i)/\hat{C}\right] \} d\rho \tag{11}$$

$\mathcal{M}^{(\rho)}_{(r_k)} f(\rho)$ being the Taylor expansion of $f(\rho)$ around the point $\rho = 0$, truncated at order $r_k|_{\lambda=n}$. The regularized amplitude $I_{(\lambda)}(p_i)$ may be written as

$$I_{(\lambda)}(p_i) = I_{(\lambda)}^{(G)} \left[I_{(\lambda)}^{(S_k)} (q_1 \ldots q_r) \right]$$

where the functional dependence of $I_{(\lambda)}^{(G)}$ upon $I_{(\lambda)}^{(S_k)}$ is denoted by the square brackets and $q_1, \ldots, q_r$ are the external momenta of S_k.

Let $\mathcal{J} = \prod \mathcal{J}(S_k)$, where the product is the successive application of $\mathcal{J}_{(S_k)}$ for each element S_k of the class $\{S(G)\}$ in a given order.

Our procedure is described by the following theorem, which is essentially a consequence of Appelquist's work, see also [9].

Theorem. $\mathcal{J} I_{(\lambda)}(p_i)$ is finite at $\lambda = 4$ and equals the renormalized Feynman integral

$$\mathcal{J} I_{(\lambda)}(p_i) \Big|_{\lambda=4} = I_{ren}(p_i) , \qquad (12)$$

and the result does not depend on the ordering of the operators $\mathcal{J}_{(S_k)}$.

As a final remark, we recall that the choice of the origin in momentum space as a subtraction point poses no restriction on this formalism. Subtraction in a different point, say $p_i = a$, may be performed by applying the convenient integral representation for the operator $\mathcal{M}_{(r_k, a)}$ given in [10] to each subgraph of the class $\{S(G)\}$.

GAUGE SYMMETRIES AND OTHER THINGS

As it is well known, every regularization procedure (Lorentz-invariant cutoff, Pauli–Villars, cutting the origin in the parametric representation, analytic renormalization, and so on) modifies in some way the formal properties of Feynman integrals. While it is possible to devise a regularization such that, for special choices of some parameters, gauge symmetry is preserved (e.g. Pauli-Villars), the superiority of the dimensional regularization is exhibited by the preservation of the gauge symmetry for any value, even complex, of the unique regularizing parameter. This advantage is such that this technique has become a standard tool for the renormalization of non-abelian gauge theories. We shall sketch this property in the simpler case of quantum electrodynamics. To this aim, we have to define a proper interpolation of Dirac γ matrices in a n-dimensional space-time, n being an integer different from four. We shall consider the set of n Dirac matrices γ_i satisfying $\{\gamma_i, \gamma_j\} = 2g_{ij} I$ where $g_{oo}=1$, $g_{ii}=-1$, if $i = 1,..,n-1$, $g_{ij}=0$ if $i \neq j$.

The computation of traces and products follows then trivially, for instance

$$\sum_{\ell} g^{\ell\ell} \gamma^{\ell} \gamma^{k} \gamma^{\ell} = (2-n)\gamma^{k} \tag{13}$$

$$\sum_{\ell} g^{\ell\ell} \gamma^{\ell} \gamma^{a} \gamma^{b} \gamma^{\ell} = (n-4)\gamma^{a} \gamma^{b} + 4g^{ab}$$

$$\sum_{\ell} g^{\ell\ell} \gamma^{\ell} \gamma^{a} \gamma^{b} \gamma^{c} \gamma^{\ell} = (4-n)\gamma^{a} \gamma^{b} \gamma^{c} - 2\gamma^{c} \gamma^{b} \gamma^{a}$$

The interpolation to complex values of the dimension has to be done after products or traces of Dirac matrices have been performed.

Let us consider the second order photon polarization tensor

$$\Pi^{(2)}_{\mu\nu}(K) = -\frac{\alpha}{4\pi^3 i} \int dp \, \frac{\mathrm{Tr}\{\gamma_\mu [\gamma\cdot(p+K)+m]\gamma_\nu (\gamma\cdot p+m)\}}{[(p+K)^2 - m^2 + i\varepsilon](p^2 - m^2 + i\varepsilon)} \tag{14}$$

Its obvious interpolating function, according to the rules described above, is

$$\Pi^{(2)}_{\mu\nu\,(\lambda)}(K) = -\frac{\alpha}{2\pi^3} e^{-i\pi\frac{\lambda}{2}} \cdot \lambda \pi^{\frac{\lambda}{2}} \, \Gamma\left(2-\frac{\lambda}{2}\right)(g_{\mu\nu} K^2 - K_\mu K_\nu) \cdot$$

$$\cdot \int_0^1 d\alpha \, \frac{\alpha(1-\alpha)}{[\alpha(1-\alpha)K^2 - m^2]^{2-\frac{\lambda}{2}}} \tag{15}$$

The gauge symmetry factor does not depend on λ.

As a further example of the same property, we consider the lowest non trivial order for the Ward identity.

The second order electron self-energy is

$$\Sigma(p) = -\frac{4\pi\alpha}{(2\pi)^n} i \sum_\ell g^{\ell\ell} \int dq \frac{\gamma^\ell [\gamma\cdot(p-q)+m]\gamma^\ell}{[(p-q)^2-m^2](q^2-\mu^2)} \tag{16}$$

and its obvious interpolating function is

$$\Sigma_{(\lambda)}(p) = -i\frac{4\pi\alpha}{(2\pi)^\lambda} e^{-i\pi\frac{\lambda}{4}} \pi^{\frac{\lambda}{2}} e^{i\frac{\pi}{2}(2-\frac{\lambda}{2})} \cdot$$

$$\cdot \Gamma(2-\tfrac{\lambda}{2}) \int_0^1 [(2-\lambda)\alpha\gamma\cdot p + \lambda m] B^{-(2-\frac{\lambda}{2})} d\alpha$$

where $B = \alpha(1-\alpha)p^2 - \alpha\mu^2 - (1-\alpha)m^2$, (17)

and the photon has been given mass μ.

The insertion of an external photon, carrying zero momentum, gives the fermion vertex function at third order

$$\Gamma^\mu(p,0) = -i\frac{4\pi\alpha}{(2\pi)^n}\sum_\ell g^{\ell\ell}\int dq \frac{\gamma^\ell[\gamma\cdot(p-q)+m]\gamma^\mu[\gamma\cdot(p-q)+m]\gamma^\ell}{[(p-q)^2-m^2]^2(q^2-\mu^2)}$$

whose obvious interpolating function is

$$\Gamma^\mu_{(\lambda)}(p,0) = -i\frac{4\pi\alpha}{(2\pi)^\lambda} e^{-i\frac{\pi}{4}\lambda} \pi^{\frac{\lambda}{2}} \Big\{ -e^{i\frac{\pi}{2}(2-\frac{\lambda}{2})}\Gamma(2-\tfrac{\lambda}{2})\cdot$$

$$\cdot\frac{(2-\lambda)^2}{2}\gamma^\mu\int_0^1 \frac{d\alpha}{B^{2-\frac{\lambda}{2}}} + i e^{i\frac{\pi}{2}(3-\frac{\lambda}{2})}\Gamma(3-\tfrac{\lambda}{2})\int_0^1\frac{d\alpha}{B^{3-\frac{\lambda}{2}}} Y^\mu\Big\}$$

where $Y^\mu = (2-\lambda)\alpha^2\gamma\cdot p\,\gamma^\mu\gamma\cdot p + 2m\lambda\alpha p^\mu + (2-\lambda)m^2\gamma^\mu$. (18)

By using the identity, proved in [9]

$$\Gamma(3-\tfrac{\lambda}{2})\int_0^1 \frac{(1-\alpha)(m^2-\alpha^2p^2)}{B^{3-\frac{\lambda}{2}}} d\alpha = \Gamma(2-\tfrac{\lambda}{2})\int_0^1\frac{(\frac{\lambda}{2}-1)(1-\alpha)-\alpha}{B^{2-\frac{\lambda}{2}}} d\alpha \tag{19}$$

valid for arbitrary λ , the Ward identity obtains

$$\Gamma^{\mu}_{(\lambda)}(p,0) = -\frac{\partial}{\partial p_{\mu}} \Sigma_{(\lambda)}(p) \tag{20}$$

The simplicity of the dimensional regularization has made it a convenient tool in the investigation of a number of theoretical problems. For the interested reader, we list some recent pertinent works.

Anomalous Ward identities and a suitable definition for γ_5 have been considered in [1-3,11].

Dimensionally regularized Feynman integrals behave in a simple way under scale transformations $x \to \eta x$, $p \to p/\eta$, and this property simplifies the investigation of the short distance behaviour and of the renormalization group [22,23].

The quantum theory of gravitation presents formidable problems, among them are the non polynomial character of the Lagrangian and the non abelian gauge symmetry. Both these features are usefully dealt with by dimensional regularization techniques [8,12,14,16,19,24,25].

Infrared divergences are also regularized by changing the space-time dimension. The technique can then be used to handle the infrared divergences associated with real and virtual radiative corrections [18].

Finally, we mention that renormalization procedures in some way analogous to that one here described, have been developed by other authors, in the form of analytic renormalization [5] and in the form of a subtraction scheme [17].

APPENDIX

In dealing with parametric functions it is often convenient to perform a scale transformation. It is a transformation from a set of n variables α_i to a set of n+1 variables ρ, $\bar{\alpha}_i$ with the constraint $\sum_1^n \bar{\alpha}_i = 1$.

One has

$$\int_0^\infty \left(\prod_1^n d\alpha_i\right) f(\alpha_i) = \int_0^1 \left(\prod_1^n d\bar{\alpha}_i\right) \delta\left(1-\sum_1^n \bar{\alpha}_i\right) \int_0^\infty f(\rho\bar{\alpha}_i)\rho^{n-1} d\rho \tag{21}$$

To perform the inverse transformation one uses the following relation

$$\int_0^1 \left(\prod_1^n dx_i\right) \delta\left(1-\sum_1^n x_i\right) \int_0^\infty d\rho \, g(\rho, x_i) =$$

$$= \int_0^\infty \left(\prod_1^n dy_i\right) \left(\sum_1^n y_i\right)^{1-n} g\left(\sum_1^n y_i, \frac{y_i}{\sum_1^n y_i}\right) . \qquad (22)$$

REFERENCES

The traditional subtractive renormalization theory is a result accomplished by many people over a period of many years. Standard references are the books : N.N.Bogoliubov and D.V.Shirkov,Introduction to the theory of quantized fields,Interscience Pub.1959,New York, and K.Hepp,Theorie de la renormalisation,Springer Verlag 1969. The analytic renormalization has been developped by C.G. Bollini,J.J.Giambiagi,W.Guttinger,E.R.Caianiello and coworkers,E.Speer and others.Speer's book,Generalized Feynman Amplitudes,Princeton Univ.Press,1969, is the basic reference. More recent developments are F.Guerra,Nuovo Cimento 1A,523 (1971),H.Saller and H.P.Dűrr,Nuovo Cimento 64A,145 (1969),E.Speer,Lectures on Analytic Renormalization,Univ.of Maryland Tech.Rep.73-067 (1972),M.Marinaro, L.Mercaldo,G.Scarpetta,to appear in Nuovo Cimento A.

As we mentioned before,the most explored use of dimensional renormalization is in dealing with gauge theories. We confine ourselves to quote the report given by B.W.Lee,XVI Int.Conf.on High Energy Nuclear Physics,Batavia 1972 and many contributions to this volume.

Our formulation uses many results of T.Appelquist, Ann.of Phys. 54,27 (1969).Furthermore,it should be remarked that Feynman parametric representation, with arbitrary space-time dimension,explicitly appears in E.Speer and M.J.Westwater,Ann.Inst.Henri Poincare' A14,1 (1971) and N.Nakanishi,Graph Theory and Feynman Integral,Gordon and Breach Pub.1971. K.Wilson's work in recent years,stressing the relevance of models with non integer space-time dimension has certainly influenced the development of the dimensional renormalization.

We proceed now to list for the interested reader a number of recent (1972 or after) papers that deal with the dimensional renormalization, its applications (other than gauge theories) or closely related approaches.

[1,2,3] D.A.Akyeampong,R.Delbourgo,Imperial College preprints,ICTP 72/22, 72/32, 72/33.

[4] J.F.Ashmore,Lett.Nuovo Cimento 4, 289 (1972).

[5] J.F.Ashmore,Comm.Math.Phys.29, 177 (1973).

[6] C.G.Bollini,J.J.Giambiagi,Phys.Lett. 40B, 566 , (1972)

[7] C.G.Bollini,J.J.Giambiagi,Nuovo Cimento 12B , 20, (1972)

[8] M.Brown,Nucl.Phys.B56, 194 (1973).

[9] P.Butera,G.M.Cicuta,E.Montaldi, INFN / AE 73 / 2, Frascati 1973.

[10] G.M.Cicuta,E.Montaldi, Lett.Nuovo Cimento 4, 329 (1972).

[11] G.M.Cicuta, SLAC-PUB-1076 (TH) Stanford 1972.

[12,13] D.M.Capper,M.Ramon Medrano, Int.Cen.for Theor. Phys.Trieste preprints IC/72/138, IC/73/32.

[14,15] D.M.Capper,G.Leibbrandt, Int.Cen.for Theor. Phys.Trieste preprints IC/72/145, IC/72/158.

[16] D.M.Capper,G.Leibbrandt,M.Ramon Medrano,Int.Cen. for Theor.Phys. preprint IC/73/26.

[17] H.J.de Vega,F.A.Shaposnik,Univ.Nacional de la Plata,Argentina, preprint Sept.1972.

[18] R.Gastmans,R.Meuldermans, Leuven preprint 1973.

[19] G.Geist,H.Kuhnelt,W.Lang, Karlsruhe preprint 1972.

[20] P.K.Mitter, Phys.Rev. D 7, 2927 (1973).

[21] G.t'Hooft,M.Veltman,Nucl.Phys. B44, 189 (1972).

[22] G.t'Hooft, CERN preprint Th.1666 (1973).

[23] G.t'Hooft, CERN preprint Th.1692 (1973).

[24] G.t'Hooft,M.Veltman, CERN preprint Th.1723 (1973).

[25] H.Vucetich,Lett.Nuovo Cimento 6, 314 (1973).

[26] K.G.Wilson, Phys.Rev. D 7, 2911 (1973).

A NEW APPROACH TO DIVERGENT INTEGRALS

Eduardo R. Caianiello

Istituto di Fisica, Università di Salerno, Italy

Laboratorio di Cibernetica del CNR
Arco Felice, Napoli, Italy

1. - Introduction

Divergent integrals are of course what we are all concerned about here; wherever we go with fields or particles, there we are dead sure to meet them. They may appear disguised as distributions, as bearers of ambiguities, or simply as functions that refuse to be integrated. They plague general relativity even more fundamentally than the kind of field theory we are discussing here, because "singularities" ought to mean "particles" in relativity, and no one is thus far quite clear about how this concept can be meaningfully expressed in mathematical terms.

I wish to present a fresh approach to this problem, which stems from my very naïve attitude whenever confronted with questions of a like nature: must we cure a disease that unavoidably exists, or did we not perhaps cause it ourselves because of basic definitions which we took for granted, and were posed at a time when troubles did not yet appear? To play this game, I shall show that if the original definition of the Riemann integral through Darboux sums is suitably modified, correspondingly to a specified class of divergent integrands, then the new integral is convergent and yields the same result which one would expect from renormalization. I shall do so for one-dimensional functions, and will freely borrow from a lengthier paper which was just published [1]. It will be apparent that a mathematically complete theory of the proposed integration procedure would appear as a "quasi-local" measure theory, as contrasted with the standard "local" theory.

2. - Generalization of Darboux's sums

Consider, for simplicity, integrals on the closed interval [0,1], and only the upper Darboux sum:

$$(1) \quad \int_0^1 f(x)dx = \lim_{\substack{N\to\infty \\ \max h_i \to 0}} \sum_{i=1}^{N} h_i^{(N)} f(x_i^{(N)}) \quad \text{(when everything goes well)}$$

where

$$(2) \quad \sum_{i=1}^{N} h_i^{(N)} = 1 \quad ; \quad 0 < h_i^{(N)} < \varepsilon(N) \qquad \forall N$$

$$\lim_{N\to\infty} \varepsilon(N) = 0$$

The partitions (2) of the interval [0,1] have, obviously, remarkable properties, typified by

$$(3) \quad \lim_{N\to\infty} \sum_{i=1}^{N} \left(h_i^{(N)}\right)^{\alpha} = 0 \qquad , \alpha > 1,$$

or more generally, for any smooth functions $u(x)$ such that

$$\lim_{x\to 0} \frac{u(x)}{x} = 0 \qquad , \text{by:}$$

$$\lim_{N\to\infty} \sum_{i=1}^{N} u(h_i^{(N)}) = 0 \qquad \text{when (2) holds.}$$

We want to replace the definition (1) with a more general one, the specific form of which shall be typical for each class of singular functions to be integrated, so as to achieve the objectives outlined in Section 1.

We start with some preliminary, intuitive, remarks. Suppose the interval [0,1] contains a finite number of points a_p, $p=1,2,\ldots,M$ and, as an example, that $f(x)$ has at these points singularities of type:

$$\underset{x \sim a_p}{f(x)} \stackrel{\sim}{=} \frac{g(x)}{(x-a_p)^\rho} \quad , |g(x)| < \infty, \ g(x) \text{ as smooth as required,}$$

with ρ fixed, e.g., $1 < \rho < 2$.

Search for differential operators which have $1/x^\rho$ as an eigen-function:

$$(5) \qquad (x \frac{d}{dx})^h \frac{1}{x^\rho} = (-\rho)^h \frac{1}{x^\rho} \quad ,$$

so that

$$x \frac{d}{dx} [g(x) \frac{1}{x^\rho}] = -\rho \frac{1}{x^\rho} g(x) + \frac{1}{x^{\rho-1}} g'(x) \quad ,$$

$$(x \frac{d}{dx})^2 [g(x) \frac{1}{x^\rho}] = \rho^2 \frac{1}{x^\rho} g(x) - (2\rho-1) \frac{1}{x^{\rho-1}} g'(x) + \frac{1}{x^{\rho-2}} g''(x),$$

etc.

Search now for numbers λ and μ such that

$$(6) \ [1 + \lambda x \frac{d}{dx} + \mu (x \frac{d}{dx})^2] [g(x) \frac{1}{x^\rho}] =$$

$$= (1 - \lambda\rho + \mu\rho^2) \frac{g(x)}{x^\rho} + [\lambda - \mu(2\rho-1)] \frac{g'(x)}{x^{\rho-1}} + \mu \frac{g''(x)}{x^{\rho-2}}$$

is finite at $x = 0$. Clearly, numbers λ and μ can be found, that cancel the first two terms on the r.h.s. of (6).

It is also clear that similar devices can be found for far more general classes of singular functions. The essential point is that linear combinations of operators such as (5) (or of more general types) in expressions like (6) bring about, for suitable choices of the coefficients, the cancellation of divergent terms at $x = 0$.

These remarks provide the intuitive background for introducing our procedure, if we observe that in the neighborhood of <u>each</u> singular point the corresponding interval $h_i^{(N)}$ and its powers (or other suitable functions of it) behave like infinitesimal quantities when $N \to \infty$ and can play, therefore, a rôle similar to that of x in $(x \frac{d}{dx})^h$ acting on $\frac{1}{x^\rho}$ when $x \to 0$.

As an illustration, which should suffice to understand the whole procedure, we begin the next section with a discussion of the simple example just quoted.

The ordinary integral being divergent for $1 < \rho < 2$, we shall denote henceforth, conventionally, the finite value obtained with our method with the symbol $⨍$ instead of $\int$ (a notation for finite-part integrals which we adopted long ago in field theory[2]).

$$(7) \quad ⨍_0^1 f(x)dx = \lim_{\substack{N\to\infty \\ \max h_i^{(N)} \to 0}} \sum_{i=1}^{N} h_1^{(N)} [f(x_i^{(N)}) + \lambda_1 h_i^{(N)} f'(x_i^{(N)}) + \lambda_2 h_i^{(N)2} f''(x_i^{(N)})].$$

If the interval of integration is restricted to a sub-interval <u>not</u> containing singular points a_p, because of (3) (applied to that sub-interval), it is apparent that our definition reduces to (1), since the terms with λ_1 and λ_2 give no contribution. Consider, then, a sub-interval containing only one singular point, a_p, as an internal or a boundary point; we must show that the values of λ_1 and λ_2 can be determined so that (7) is convergent; such values depend on the partitioning rule adopted and on ρ (different partitions may give different λ_1 and λ_2, but must lead to the same value for (7), to within "legitimate" ambiguities, which are excluded, however, by physics in this example). Given any class of singular functions, the writing of a formula like (7) by using (4) and (3) and the technique indicated here is straightforward.

3. - <u>Computation</u>

1. We discuss here the example reported in the preceding section, with functions of type (4); to simplify matters further, we assume that there is only one singular point, at x = 0. For

simplicity, we consider only the simplest possible partition:

$$\begin{cases} x_k^{(N)} = \frac{k}{N}, \text{ so that:} \\ h_k^{(N)} = \frac{1}{N}, \quad \forall k . \end{cases}$$

Then:

$$(8) \qquad \sum_{k=1}^{N} \frac{h_k^{(N)\,p}}{x_k^{(N)\,s}} = N^{s-p} \sum_{k=1}^{N} \frac{1}{k^s} .$$

For N large, a straightforward use of the Riemann ζ-function and of the Laplace transformation yields for $s > 1$:

$$\sum_{k=1}^{N} \frac{1}{k^s} \cong \zeta(s) - \frac{1}{\Gamma(s)} \int_0^1 \frac{e^{-N\nu}\,\nu^{s-1}}{e^{\nu}-1}\, d\nu \cong \zeta(s) - \frac{1}{s-1} \frac{1}{N^{s-1}}$$

so that:

$$(9) \qquad \sum_{k=1}^{N} \frac{h_k^{(N)\,p}}{x_k^{(N)\,s}} \cong N^{s-p} \quad \zeta(s) - \frac{1}{s-1} \quad N^{1-p} \qquad (s>1,\ p\geq 1).$$

For $s < 1$ $(p \geq 1)$, since $\sum_{k=1}^{N} \frac{1}{k^s} = \int_1^N \frac{dx}{x^s} + O(1)$,

$$(10) \qquad \sum_{k=1}^{N} \frac{h_k^{(N)\,p}}{x_k^{(N)\,s}} \cong \frac{1}{1-s} \quad N^{1-p}$$

(for s = 1 the expression (7) would not be adequate in either treatment; such and more general instances need a more general prescription, which is mentioned later, and correspond to what one expects from physics (i.e. logarithmic ultraviolet divergences).

2. We wish to compute, with (7), the expression:

$$\underline{\int}_0^1 \frac{g(x)}{x^\rho}\,dx = \lim_{\substack{N\to\infty \\ \max h_i^{(N)}\to 0}} \sum_{k=1}^{N} h_k^{(N)} \left[\frac{g(x_k^{(N)})}{x_k^{(N)\rho}} + \lambda_1 h_k^{(N)} \left(\frac{g(x)}{x^\rho}\right)'_{x=x_k^{(N)}} + \lambda_2 h_k^{(N)2} \left(\frac{g(x)}{x^\rho}\right)''_{x=x_k^{(N)}} \right],$$

that is

$$\underline{\int}_0^{1\,'} \frac{g(x)}{x^\rho}\,dx = \lim_{\substack{N\to\infty \\ \max h_k^{(N)}\to 0}} \sum_{k=1}^{N} \left\{ g(x_k^{(N)}) \left[\frac{h_k^{(N)}}{x_k^{(N)\rho}} - \lambda_1 \rho \frac{h_k^{(N)2}}{x_k^{(N)\rho+1}} + \lambda_2 \rho(\rho+1) \frac{h_k^{(N)3}}{x_k^{(N)\rho+2}} \right] + g'(x_k^{(N)}) \left[\lambda_1 \frac{h_k^{(N)2}}{x_k^{(N)\rho}} - \lambda_2 \rho \frac{h_k^{(N)3}}{x_k^{(N)\rho+1}} \right] + \lambda_2 g''(x_k^{(N)}) \frac{h_k^{(N)3}}{x_k^{(N)3}} \right\} \tag{11}$$

We know already that $\underline{\int}_\varepsilon^1 \equiv \int_\varepsilon^1$; our only concern must be, therefore, with determining λ_1 and λ_2 in such a way that $\lim \Sigma$ stays finite also when $\varepsilon\to 0$. To avoid a more cumbersome proof, we assume (which is not clearly necessary) that $g(x) = g_0 + g_1 x + g_2 x^2 + \ldots$ within [0,1]; we can thus restrict our attention to each square bracket in (11), that is to:

$$(12)\ \lim_{N\to\infty} \sum_{k=1}^{N} \left[\frac{h_k^{(N)}}{x_k^{(N)\,\rho}} - \lambda_1 \rho \frac{h_k^{(N)\,2}}{x_k^{(N)\,\rho+1}} + \lambda_2 \rho(\rho+1) \frac{h_k^{(N)\,3}}{x_k^{(N)\,\rho+2}} \right] = \int_o^1 \frac{dx}{x^\rho}$$

$$(13)\ \lim_{N\to\infty} \sum_{k=1}^{N} \left[\lambda_1 \frac{h_k^{(N)\,2}}{x_k^{(N)\,\rho}} - \lambda_2 \rho \frac{h_k^{(N)\,3}}{x_k^{(N)\,\rho+1}} \right],$$

$$(14)\ \lim_{N\to\infty} \lambda_2 \sum_{k=1}^{N} \frac{h_k^{(N)\,3}}{x_k^{(N)\,\rho}}$$

If (11) ÷ (14) converge, then it would trivially follow that, for any smooth g(x), also (11) converges. We find:

$$(12') = \lim_{N\to\infty} \left\{ \left[N^{\rho-1} \zeta(\rho) - \frac{1}{\rho-1} \right] - \lambda_1 \rho \, [N^{\rho-1} \zeta(\rho+1) - \frac{1}{\rho} \frac{1}{N}] + \right.$$

$$\left. + \lambda_2 \rho(\rho+1) \, [N^{\rho-1} \zeta(\rho+2) - \frac{1}{\rho+1} \frac{1}{N^2}] \right\} = -\frac{1}{\rho-1} +$$

$$+ \lim_{N\to\infty} N^{\rho-1} [\zeta(\rho) - \lambda_1 \rho \, \zeta(\rho+1) + \lambda_2 \rho \, (\rho+1) \, \zeta(\rho+2)],$$

$$(13') = \lim_{N\to\infty} \left\{ \lambda_1 \, [N^{\rho-2} \zeta(\rho) - \frac{1}{\rho-1} \frac{1}{N}] - \lambda_2 \rho \, [N^{\rho-2} \zeta(\rho+1) - \frac{1}{\rho} \frac{1}{N^2}] \right\} =$$

$$= \lim_{N\to\infty} N^{\rho-2} [\lambda_1 \zeta(\rho) - \lambda_2 \rho \, \zeta(\rho+1)],$$

$$(14') = \lim_{N\to\infty} \lambda_2 \, [N^{\rho-3} \zeta(\rho) - \frac{1}{\rho-1} \frac{1}{N^2}] \, .$$

3. We discuss in detail as an example, for simplicity, only the case $1 < \rho < 2$; then

$$\int_0^1 \frac{dx}{x^\rho} = -\frac{1}{\rho-1} \quad ,$$

provided λ_1 and λ_2 are taken so as to satisfy the condition:

$$\zeta(\rho) - \lambda_1 \rho \zeta(\rho+1) + \lambda_2 \rho(\rho+1)\,\zeta\,(\rho+2) = 0 \quad .$$

For g(x) smooth enough one finds thus:

$$\int_0^1 \frac{g(x)}{x^{3/2}} = \frac{1}{\rho-1}\, g(0) + \int_0^1 \frac{g(x) - g(0)}{x^{3/2}}\, dx$$

which exhibits with full clarity the "quasi-local" character of our integration method.

It is instructive to compare it with the historic example of Hadamard; in his notation 3) :

$$\text{P.F.} \int_0^1 \frac{dx}{x^{3/2}} = \left| \int_0^1 \frac{dx}{x^{3/2}} \equiv \int_0^1 \frac{dx}{x^{3/2}} \overset{def}{=} -2 \lim_{\varepsilon \to 0} \left[\frac{1}{\sqrt{x}} \Big|_\varepsilon^1 + \frac{1}{\sqrt{\varepsilon}} \right] = -2 \right. .$$

Analytic continuation would give:

$$\int_0^1 \frac{dx}{x^{3/2}} = \lim_{\lambda \to 0} (\text{regular part of}) \int_0^1 \frac{dx}{x^{3/2}} \cdot x^\lambda = -2 \quad .$$

Our technique works, therefore, as expected. The same method can be applied to integrable singular integrands; take e.g., ad abundantiam formula (12) to evaluate:

$$\int_0^1 \frac{dx}{\sqrt{x}} \equiv \int_0^1 \frac{dx}{\sqrt{x}} \qquad ;$$

we find

$$\int_0^1 \frac{dx}{\sqrt{x}} = \lim_{N\to\infty} \left[2 + \lambda_1 \frac{1}{N} - \frac{1}{2} \lambda_2 \frac{1}{N^2}\right] = +2$$

regardless of the choice of λ_1 and λ_2, which does not affect the result.

4. - More General Cases

The extension of formulas like (7) to singular integrands of a more general nature, such as do arise in concrete physical problems, is an appropriate subject for a systematic investigation. We merely indicate here on some examples, at the heuristic level used in formulas (5) - (9), ways to extend formula (7) to a few more cases of interest. We shall use the trick expressed by (5); as was seen before, this is adequate to fix a convenient expression for the definition (7), but the parameters $\lambda_1, \lambda_2, \ldots$ will depend upon the specific partitioning rule adopted. It will suffice for this purpose, as before, to consider only values around $x = 0$. We omit discussing actual computational procedures.

a. For (integrable) integrands which behave, at each point a_p, like:

$$f(x) \sim \sum_{s=1}^{k} g_s(a_p) \quad (\ell g|x-a_p|)^s \quad , k \leq K$$

we can use

$$(x\ell g|x|\frac{d}{dx})^r \quad (\ell g|x|)^s = s^r \quad (\ell g|x|)^s \quad ,$$

whence:

$$\sum_{s=1}^{k} g_s(x) \quad [(\ell g|x|)^s + \sum_{r=1}^{s} \mu_r \quad (x\ell g|x| \frac{d}{dx})^r (\sum_{s=1}^{k} g_x(x) \quad (\ell g|x|)^s)] =$$

$$= \sum_{s=1}^{k} g_s(x) \quad (\ell g|x|)^s \quad [1 + \sum_{r=1}^{k} \mu_r \quad s^r] \quad + \ldots ;$$

what corresponds to the term $[1 + \sum_{r=1}^{k} \mu_r s^r]$, in a concrete computation, making use of this rule, must be made to vanish. For example, for k = 2 we would find:

$$\int_0^1 f(x)dx = \lim_{\substack{N\to\infty \\ \max h_i^{(N)}\to 0}} \sum_{h_i=1}^{N} h_i^{(N)} \Big\{ f(\bar{x}_i^{(N)}) + h_i^{(N)} [\lambda_1 \ell g h_i^{(N)} + \lambda_2 (\ell g h_i^{(N)})^2] f'(\bar{x}_i^{(N)}) + h_i^{(N)^2} \lambda_2 (\ell g h_i^{(N)})^2 f''(\bar{x}_i^{(N)}) \Big\} .$$

b. Singularities up to $(\ell g|x|/x)^k$ can be handled by noting that:

$$\left(\frac{x\ell g|x|}{1-\ell g|x|} \frac{d}{dx}\right)^r \left(\frac{\ell g|x|}{x}\right)^s = s^r \left(\frac{\ell g|x|}{x}\right)^s .$$

c. For singularities up to $(\ell g|x|)^s/x$:

$$[(1 + x \frac{d}{dx})^{r+i} - 1] \frac{\ell g|x|^s}{x} = - \frac{(\ell g|x|)^s}{x} \qquad (r \geq s)$$

d. Finally, it is instructive to consider the singular integrands $g(x)/x$, on which, as was seen, formula (7) fails. The origin of the failure of (7) is clear: logarithmic terms must be compensated, and (7) cannot do so; but then, a simple way out is first to compensate for $1/x$, then for $\ell g|x|$. By looking at (5) and at a. above, we see immediately that our starting point must be:

$$(1 + \lambda_1 \ell g|x| \frac{d}{dx}) (1 + \lambda_2 x \frac{d}{dx}) = 1 + [\lambda_1 (1 + \lambda_2) x \ell g|x| + \lambda_2 x] \frac{d}{dx} + \lambda_1\lambda_2 x^2 \ell g|x| \frac{d^2}{dx^2}$$

This yields:

$$\int_0^1 \frac{g(x)}{x}\,dx = \lim_{N\to\infty} \sum_{k=1}^{N} \Big\{ g\left(\frac{k}{N}\right) \left[\frac{1}{k} + \lambda_1 (1+\lambda_2)\frac{\ell gN}{k^2} - \lambda_2 \frac{1}{k^2} +\right.$$

$$\left. -2\lambda_1\lambda_2 \frac{\ell gN}{k^3}\right] + g'\left(\frac{k}{N}\right)\left[-\lambda_1 (1+\lambda_2)\frac{\ell gN}{N}\frac{1}{k} + \lambda_2 \frac{1}{N}\frac{1}{k} + 2\lambda_1\lambda_2 \frac{\ell gN}{N}\frac{1}{k^2}\right] +$$

$$-g''\left(\frac{k}{N}\right)\lambda_1\lambda_2 \frac{\ell gN}{N}\frac{1}{k^2}\Big\} = \lim_{N\to\infty}\sum_{k=1}^{N} g\left(\frac{k}{N}\right)\left[\frac{1}{k} + \lambda_1 (1+\lambda_2)\,\zeta(2)\,\ell gN +\right.$$

$$\left. -\lambda_2\,\zeta(2) - 2\lambda_1\lambda_2\,\ell gN\,\zeta(3)\right] = \lim_{N\to\infty}\sum_{k=1}^{N} g\left(\frac{k}{N}\right)\Big\{\frac{1}{k} +$$

$$+ [\lambda_1 (1+\lambda_2)\,\zeta(2) - 2\lambda_1\lambda_2\,\zeta(3)]\,\ell gN - \lambda_2\,\zeta(2)\Big\}.$$

Consider now, as before:

$$\int_0^1 \frac{dx}{x} = \lim_{N\to\infty}\Big\{\sum_{k=1}^{N}\frac{1}{k} + \ell gN \cdot [\lambda_1 (1+\lambda_2)\,\zeta(2) - 2\lambda_1\lambda_2\,\zeta(3)]\Big\} - \lambda_2\zeta(2)$$

and note that, if

$$\lambda_1 [(1+\lambda_2)\zeta(2) - 2\lambda_2\zeta(3)] = -1$$

and $\mathcal{C}$ is the Euler-Mascheroni constant, calling $\mathcal{C} - \lambda_2\,\zeta(2) = \alpha$, we find

$$\int_0^1 \frac{g(x)}{x}\,dx = \alpha g(0) + \int_0^1 \frac{g(x) - g(0)}{x}\,dx ,$$

which contains the <u>indeterminate</u> quantity α , just as one expects from physics [3), 4)].

The Cauchy P.V. integral implies an additional ad hoc prescription, which is valid only for intervals including the point $x = 0$ and gives $\alpha = 0$; our procedure exhibits the ambiguity, which is more interesting, e.g., in renormalization theory.

Conclusion

The previous discussions purport to be merely indicative as regards the problems which will arise in a formally complete study of the integration procedure introduced here; they already suffice for the practical computation of integral of the types just seen and, with obvious extensions, of type

$$\frac{(\ell g|x|)^s}{x^k} \qquad (\underline{e.g.}\ k \leq 2 \text{ or } 4)$$

which is all that is required for renormalizable field theories. Use of this method implies then ipso facto, as is shown in references 3) and 4), that the correct renormalization has been carried out.

Can we hope that the subject of renormalization, which is fully "solved" theoretically in many ways, e.g. through analytic continuation techniques, be in this way "solved" fully as well from the computational aspect (for which analytic continuation and most other techniques are grossly inadequate or expensive)?

Among other topics from physics, applications to general relativity should prove of particular interest; for instance, the necessity of introducing ad hoc the so-called "good" δ functions in the study of the motion of particles as singularities of gravitational fields [5] is completely bypassed if integrals are computed with our generalized definition.

References

1) E. R. Caianiello - "Generalized Integration Procedure for Divergent Integrals" Nuovo Cimento Vol.15/A 145, (1973).

2) E. R. Caianiello - "Colloque sur les Problèmes Mathématiques de la Théorie Quantique des Champs" Lille, June 1957.

3) J. Hadamard - "Lectures on Cauchy's Problem in Linear Partial Differential Equations", London (1922)

4) E. R. Caianiello - "Combinatorics and Renormalization in Quantum Field Theory", Beljamin Inc., Mass. In press, and references cited therein.

5) L. Infeld and J. Plebanski - "Motion and Relativity" Pergamon Press, Warsawa (1960).

BRANCHING EQUATIONS OF I AND II TYPE AND THEIR CONNECTIONS WITH FINITE EQUATIONS OF MOTION AND THE CALLAN-SYMANZIK EQUATION

Eduardo R. Caianiello
Istituto di Fisica, University of Salerno
and Laboratorio di Cibernetica of CNR, Naples

and
Maria Marinaro
Istituto di Fisica, University of Salerno

1. - Introduction

The aim of this paper is to recall, and present in the context of today's quantum field theory, some results which we found in a sequel of past works. Most of them were obtained many years ago[1]; our main objective was at the time to discriminate, by using appropriate techniques, combinatoric from analytic problems. The first were brought into a naturally compact form by introducing some algorithms, which need not be mentioned here; the second were treated, at the beginning of our research, by defining "finite parts" of all quantities of interest, together with the formal properties which had to be satisfied by them in order that combinatorics could apply. Specific definitions were also given for the computation of finite part integrals, but no special emphasis was then placed on this aspect of the problem.

In more recent works [2] we have introduced a particular realization of the finite part integral, based on the method of the analytic regularization [3];this realization, besides being convenient for actual computation, permits an easy comparison between our approach to renormalization and those used by other authors. In the following, for the sake of concreteness and of clarity, we shall restrict ourselves throughout to this realization. Besides, we shall limit our study to the model of field theory that obtains by considering a scalar field $\phi(x)$ with $\phi^4(x)$ coupling.

The plan of this paper is the following. In Section 2 we define the finite parts of the vacuum expectation values of the time ordered products of the field and current operators. In Section 3 we write the equations of motion in finite form. In Section 4 we define the derivatives of the Green functions with respect to the parameters which appear in the lagrangian as well as their variations under an infinitesimal change of the regularization rule. In the last Section we show that the Callan-Symanzik (C.S.) equations can be derived quite naturally with our formalism, that gives also additional information on the structure of the coefficients which appear in them.

2. - Basic definitions

We consider a scalar field $\phi(x)$ with $\phi^4(x)$ coupling, and define the finite parts of the vacuum expectation values of the time ordered products of field operators by means of the following prescriptions:

$$G_N(x_1 \ldots x_N) = \langle o|T(\phi(x_1)\ldots\phi(x_N))|o\rangle = \frac{K_N(x_1\ldots x_N)}{K_o} \tag{1}$$

$$\begin{aligned} K_N(x_1\ldots x_N) &= \\ &= \sum_p \frac{(i)^p}{p!}\,\Theta(\underline{\lambda}_1)\ldots\,\Theta(\underline{\lambda}_p)\int d\xi_1\ldots d\xi_p\ {}_o\langle o|T(\phi_o(x_1)\ldots\phi_o(x_N)\cdot \\ &\quad\cdot\hat{L}_I(\xi_1)\ldots\hat{L}_I(\xi_p))|o\rangle_o \end{aligned} \tag{1'}$$

where $\phi_o(x_i)$ denotes the free field of mass m, $|o\rangle$ is the Fock-space vacuum state for ϕ_o , T is the chronological ordering.

$$\hat{L}_I(x) = g\int \phi_o(x^1)\,\phi_o(x^2)\,\phi_o(x^3)\,\phi_o(x^4)\ \Gamma_x^{(4)}\ dx^1\,dx^2\,dx^3\,dx^4 \tag{2}$$

with

$$\Gamma_x^{(4)} = \hat{\delta}(x^1-x;\lambda^1)\ \ldots\ \hat{\delta}(x^4-x;\lambda^4) \tag{3}$$

in (3) the quantity $\hat{\delta}(x,\lambda)$ is defined by the expression

$$(3') \qquad \hat{\delta}(x;\lambda) = \frac{i}{\pi^2} \, e^{i\pi\lambda} \, (M^2)^{\lambda} \, \frac{\Gamma(2-\lambda)}{\Gamma(\lambda)} \, (x^2 - i\overset{+}{o})^{\lambda-2}$$

M is an arbitrary dimensional parameter,

$$\lim_{\lambda\to o} \int \delta\,(x;\lambda)\, F(x) = \int \hat{\delta}\,(x)\, F(x) = F(o) \quad \forall\, F \text{ regular at } x=0$$

$\Theta(\underline{\lambda}_i)$ denotes the limiting procedure

$$(4) \qquad \Theta(\underline{\lambda}_i) = \Theta\,(\lambda_i^1) \ldots \Theta(\lambda_i^4)$$

$$(5)\ \Theta(\lambda)\ f(\lambda\ldots) = \lim_{\lambda\to 0} \text{ of the regular part in } \lambda \text{ of the Laurent expansion of } f(\lambda\ldots)$$

$f(\lambda_i^j\ldots)$ is a meromorphic function in the complex plane λ_i^j. Equation (1) can be written more conveniently by remembering that the vacuum expectation value of time ordered products of a scalar field can be expressed through a hafnian [1], namely we have:

$$(6) \qquad K_N(x_1\ldots x_N) = \sum_{p=o}^{\infty} \Theta(\underline{\lambda}_1) \ldots \Theta(\underline{\lambda}_p)\ K^p(x_1\ldots x_N)$$

where

$$(7)\ K^p(x_1\ldots x_N) = \frac{(ig)^p}{p!} \int dR_p\ \Gamma^{(4)}_{\xi_1} \ldots \Gamma^{(4)}_{\xi_p} [x_1\ldots x_N\ \xi_1^1\ldots\xi_1^4\ldots\xi_p^1\ldots\xi_p^4]$$

$$\int dR_p = \prod_{j=1}^{p} \int d\xi_j \, d\xi_j^1 \ldots d\xi_j^4$$

(8) $[x_1 \ldots \xi_p^4]$ denotes the hafnian defined by

$$[z_1 \ldots z_{2s}] = \sum_{j=2}^{2s} [z_1 \; z_j] \, [z_2 \ldots \not{z}_j \ldots z_{2s}]$$

where $[z_2 \ldots \not{z}_j \ldots z_{2s}]$ is the hafnian of 2s-2 variables obtained from $[z_1 \ldots z_{2s}]$ by dropping the variables z_1 and z_j, and $[z_1 \; z_j] = \Delta_c(z_i - z_j; m)$; $(\, (\Box_x - m^2) \, \Delta_c(x;m) = i\delta(x) \,)$

K^P contains all the regularized Feynman graphs of order 1...P. It is worth noting that (6) defines a tempered distribution at each perturbative order P because it yields meromorphic distributions in the complex plane of $\underline{\lambda}_1 \ldots \underline{\lambda}_p$ and the coefficients of the Laurent expansion of this distribution are tempered distributions [4]. The vacuum expectation value of the time ordered products of space and time derivatives of field operators can be also defined. If we denote with D_x a differential operator containing spatial derivatives only we have:

$$\begin{aligned} &\langle o | T(D_{x_1} \phi(x_1) \ldots D_{x_k} \phi(x_k) \, \phi(y_1) \ldots \phi(y_n) \,) | o \rangle = \\ (9) \qquad & \\ &= D_{x_1} \ldots D_{x_k} \langle o | T(\phi(x_1) \ldots \phi(x_k) \, \phi(y_1) \ldots \phi(y_n)) | o \rangle \end{aligned}$$

The time derivatives are given instead by the recursive formula

$$\langle o|T(\partial_{t_1}\,\phi(x_1)\ldots\partial_{t_k}\,\phi(x_k)\,\phi(y_1)\ldots\,\phi(y_n))|o\rangle =$$

$$=\partial_{t_1}\,\langle o|T(\phi(x_1)\,\partial_{t_2}\,\phi(x_2)\ldots\partial_{t_k}\,\phi(x_k)\,\phi(y_1)\ldots\phi(y_n))|\,o\rangle\,+ \tag{10}$$

$$+\sum_{i=2}^{k}\,\delta(x_1-x_i)\,\langle o|T(\partial_{t_2}\,\phi(x_2)\ldots\cancel{\partial_{t_i}\,\phi(x_i)}\ldots\partial_{t_k}\,\cdot$$

$$\cdot\,\phi(x_k)\,\phi(y_1)\ldots\,\phi(y_n))|\,o\rangle$$

From (6), (9) and (10), setting

$$D^{n_i}_{x_i}\;=\;\partial_{x_i^{\mu_{i_1}}}\ldots\partial_{x_i^{\mu_{i_{n_i}}}}$$

we obtain the final expression:

$$\langle o|T(D^{n_1}_{x_1}\phi(x_1)\ldots D^{n_k}_{x_k}\,\phi(x_k)\,\phi(y_1)\ldots\phi(y_n))|\,o\rangle\;=$$

$$=\frac{1}{K_o}\,\sum_{p}\,\Theta(\underline{\lambda}_1)\ldots\Theta(\underline{\lambda}_p)\,\Big\{\,D^{n_1}_{x_1}\ldots D^{n_k}_{x_k}\;K^p(x_1\ldots x_k\;y_1\ldots y_n)\;+ \tag{11}$$

$$+\;A^p(x_1\ldots x_k\;y_1\ldots y_n)\Big\}$$

A^P is the sum of the contact terms which are produced on iteration of the recursive formula (10) by the last term on the r.h.s. A simple generalization of (6) and (11) allows us to define the vacuum expectation value of time ordered produced of field and current operators. Denote with $O(x)$ the quantity

$$O(x)\;=\;D^{n_1}_{x}\;\;\phi(x)\ldots\;D^{n_r}_{x}\;\;\phi(x)$$

and define

$$<o|T(O(x)\ \phi(y_1)\dots\phi(y_n))|\ o> \equiv \frac{1}{K_o}\sum_{p=o}^{\infty} S(\Theta(\underline{\lambda}_1)\dots\Theta(\underline{\lambda}_p)\ \Theta(\underline{\mu}))\cdot$$

(12)
$$\cdot\left\{\int D_x^{n_1}\dots D_{x^r}^{n_r}\ K^p(x^1\dots x^r\ y_1\dots y_n)\Gamma_x^{(r)}dx^1\dots dx^r\ +\right.$$

$$\left.+\ A^p(x^1\dots x^r\ y_1\dots y_n)\Gamma_x^{(r)}\ dx^1\dots dx^r\right\}$$

$$\Theta(\underline{\mu}) = \Theta(\mu^1)\dots\Theta(\mu^r)\ ; \qquad \Gamma_x^{(r)} = \hat{\delta}(x^1-x;\ \mu^1)\dots\hat{\delta}\ (x^r-x;\mu^r)$$

S denotes the symmetrization operator. In particular if $D_x^{n_1}=\dots=D_x^{n_r}=1$, $O(x) = \phi^r(x)$ and eq. (12) becomes

$$<o|T(O(x)\ \phi(y_1)\dots\phi(y_n))|o> =$$

(13)
$$= \frac{1}{K_o}\sum_{p=o}^{\infty}\ S(\Theta(\underline{\lambda}_1)\dots\Theta(\underline{\lambda}\ )\ \Theta(\underline{\mu}))\int K^p(x^1\dots x^r y_1\dots y_n)\Gamma_x^{(r)}dx^1\dots dx^r$$

In the following we denote the previous expression with

$$K_{n+r}^{(x)}\ (\underbrace{x\dots x}_{r}\ y_1\dots y_n)$$

Eq. (13) can be easily generalized

$$<o|T(O(x_1)\dots O(x_k)\ \phi(y_1)\dots\phi(y_n))|o> \equiv$$

(14)
$$\frac{1}{K_o}\sum_{p=o}^{\infty}\ S(\Theta(\underline{\lambda}_1)\dots\Theta(\underline{\lambda}_p)\ \Theta(\underline{\mu}_1)\dots\Theta(\underline{\mu}_k))$$

$$\int K(x_1^1\dots x_1^{r_1}\dots x_k^1\dots x_k^{r_k}\ y_1\dots y_n)\Gamma_{x_1}^{(r_1)}\dots\Gamma_{x_k}^{(r_k)}dx_1^1\dots dx_k^{r_k}$$

(note that it is easy to prove that eq.'s(12), (13), (14) are well defined distributions in S').

In conclusion, we observe that it is possible in the frame of analytic regularization to define vacuum expectation values of field and current operators that are finite to every order of perturbation theory; as for the standard open questions of locality, causality and unitarity, work is being done at present, and what indications can be gathered seem to justify some optimism. Some properties relating to the commutation and divergence of current operators have been proved to hold in the context of the σ-model [5] theory.

3. - The equation of motion

The classical expression of the equation of motion in the ϕ^4 model is

$$(\square_x - m^2)\,\phi(x) = -4\,g\phi^3(x) \tag{15}$$

According to the definitions (11) and (13) we can easily write the quantized form of equation (15)

$$\begin{aligned}(\square_x - m^2)\,K_n(xy_1\ldots y_{n-1}) &= i\sum_{j=1}^{n=1}\delta(x-y_j)K_{n-2}(y_1\ldots \not{y}_j\ldots y_{n-1}) \\ &\quad - 4g\,K^{(x)}_{n+2}(xxx\,y_1\ldots y_{n-1})\end{aligned} \tag{16}$$

we note that equation (16) in our previous works was obtained directly from eq. (6) by expanding the hafnian with respect to the variables x_i [1),6)]. These equations were called branching equations of the I type by us; they were independently studied by many authors [7)].

It may be convenient to write the equation of motion by using instead of the above the propagator (NS ≡ Non Symmetric):

$$
{}^{NS}K^{(x)}_{n+2}(xxx\, y_1 \ldots y_{n-1}) = \Theta(\underline{\mu}) \int dx^1 \ldots dx^3 \;\; K(x^1\, x^2\, x^3\, y_1 \ldots y_{n-1}) \Gamma^{(3)}_x =
$$

$$
(17) \qquad = \sum_{p=o}^{\infty} \Theta(\underline{\mu})\Theta(\underline{\lambda}_1) \ldots \Theta(\underline{\lambda}_p) \int K^p(x^1 \ldots x^3 y_1 \ldots y_{n-1}) \Gamma^{(3)}_x \; dx^1 \ldots dx^3
$$

which is obtained from $K(x_1^1\, x_1^2\, x_1^3\, y_1 \ldots y_n)$ through some suitable limiting procedure on $x_1^1\, x_1^2\, x_1^3$.

In terms of ${}^{NS}K^{x_1}(x_1 x_1 x_1 y_1 \ldots y_n)$ the equation of motion becomes

$$
(\Box_x - m^2)\, K_n(x\, y_1 \ldots y_{n-1}) = i \sum_{j=1}^{n-1} \delta(x-y_j) K_{n-2}(y_1 \ldots \not{y}_j \ldots y_{n-1}) \; +
$$

$$
(18) \qquad -4g(\bar{a} - \frac{\bar{c}a}{c}) K_n(x\, y_1 \ldots y_{n-1}) + 4g(\bar{b} - \frac{\bar{c}b}{c}) \Box_x K_x(x\, y_1 \ldots y_{n-1}) \; +
$$

$$
-4g\, \frac{\bar{c}}{c}\; {}^{NS}K^{(x)}_{n+2}(x\, x\, x\;\; y_1 \ldots y_{n-1})
$$

where a, b, c are power series in g and $\bar{a}$, $\bar{b}$, $\bar{c}$ are related to a, b, c through the relations

$$
\frac{\partial(\bar{a}\, g)}{\partial\, g} = a \quad ; \quad \frac{\partial(\bar{b}\, g)}{\partial\, g} = b \quad ; \quad \frac{\partial(\bar{c}\, g)}{\partial\, g} = c
$$

The proof of these results will be published shortly. The explicit expressions of the quantities a, b, c, $\bar{a}$, $\bar{b}$, $\bar{c}$ were given, with a different notation, in ref. 6).

4. - Differentiation with respect to the parameters g, m and M

In this Section we report some relations which connect the Green function and their derivatives. These relations were found in our previous papers [1) 6)] and called branching equations of the II type. From equation (6) we have immediately

$$\frac{\partial K_N(x_1 \dots x_N)}{\partial g} =$$

$$(19) \quad = i \sum_{p=1}^{\infty} \Theta(\underline{\lambda}_1)\dots\Theta(\underline{\lambda}_p) \int d\xi_p \; d\xi_p^1 \dots d\xi_p^4 \; K^{p-1}(x_1 \dots x_N \; \xi_p^1 \dots \xi_p^4) \Gamma^{(4)}_{\xi_p}$$

$$= i \int d\xi_1 \;\; K^{(\xi_1)}_{N+4} \; (\xi_1 \xi_1 \xi_1 \xi_1 \; x_1 \dots x_N)$$

$$(20) \quad \frac{\partial K_N(x_1 \dots x_N)}{\partial m^2} = - \frac{i}{2} \sum_{p=o}^{\infty} \Theta(\underline{\lambda}_1)\dots\Theta(\underline{\lambda}_p) \int d\xi_1 \; K^p(x_1 \dots x_N \; \overset{\circ}{\xi}_1 \; \overset{\circ}{\xi}_1)$$

The last equation is obtained by reminding that

$$\frac{\partial}{\partial m^2} [x \; y] = - i \int [x \, \xi] \, [\xi \, y] \; d\xi = - \frac{i}{2} \int [x \; y \; \overset{\circ}{\xi} \; \overset{\circ}{\xi}] \; d\xi$$

the zeros on the ξ's indicate that in the expansion of the hafnian the term $[\overset{\circ}{\xi} \, \overset{\circ}{\xi}]$ must be put equal to zero. The connection between $\frac{\partial K}{\partial g}$ and $\frac{\partial K}{\partial m}$ with the operation called Δ_1 and Δ_3 by Lowenstein[7)] is immediate.

Finally from (6), (3) and (3') we can calculate the variation of $K \, (x_1 \dots x_N)$ under an infinitesimal change of the regularizing parameter M:

$$\frac{\partial K_N(x_1 \dots x_N)}{\partial M} = \sum_{p=o}^{\infty} \sum_{i=1}^{p} \frac{(i\,g)^p}{p!} \Theta(\underline{\lambda}_1) \dots \Theta(\lambda_p) \int dR_p \tag{21}$$

$$\Gamma^{(4)}_{\xi_1} \dots (\partial_M \; \Gamma^{(4)}_{\xi_i}) \dots \Gamma^{(4)}_{\xi_p} \; [x_1 \dots x_n \xi_1^1 \dots \xi_1^4 \dots \xi_p^1 \dots \xi_p^4]$$

After some manipulations [8] we obtain

$$\frac{\partial K_N(x_1 \dots x_N)}{\partial M} = (R_o + \frac{N}{2} R_2^{(2)} - 3R_4 \; B^2 \Omega) \; K_N(x_1 \dots x_N) + \tag{22}$$

$$+(-6BR_4 + R_2^{(1)}) \; 2i \; \frac{\partial}{\partial m^2} K_N(x_1 \dots x_N) + (-iR_4 + 2gR_2^{(2)}) \; \frac{\partial}{\partial g} K_N(x_1 \dots x_N)$$

where

$$B = \Theta(\lambda) \int d\,y \quad \Delta_c \; (x - y) \;\; \hat{\delta}(x - y; \lambda \;)$$

and R_o, $R_2^{(1)}$, $R_2^{(2)}$, R_4 are expressed by series in g as follows:

$$R_o = \sum_{n=1}^{\infty} \frac{(i\,g)^n}{n!} \Theta_{\lambda_1,\lambda_2\ldots\lambda_n} \int d\,R_n\,(\partial_M\,\Gamma^{(4)}_{\xi_1}\,)\Gamma^{(4)}_{\xi_2}\,\ldots\Gamma^{(4)}_{\xi_n}$$

(23a)

$$[\xi^1_1\ldots\xi^4_1\ldots\xi^1_n\ldots\xi^4_n]$$

$$R_4 = \sum_{n=1}^{\infty} \frac{(i\,g)^n}{n!} \Theta_{\lambda_1,\lambda_2\ldots\lambda_n} \int dR_{n-1} \int d\xi^1_1\ldots d\xi^4_1\,(\partial_M\,\Gamma^{(4)}_{\xi_1}\,)\Gamma^{(4)}_{\xi_2}\ldots\Gamma^{(4)}_{\xi_n}$$

(23b)

$$\sum_{(4)} [\xi^1_1\ldots\xi^4_1\ldots\xi^1_n\ldots\xi^4_n]$$

$$R^{(1)}_2 = \sum_{n=1}^{\infty} \frac{(i\,g)^n}{n!} \Theta_{\lambda_1,\lambda_2\ldots\lambda} \int dR_{n-1} \int d\xi^1_1\ldots d\xi^4_1(\partial_M\,\Gamma^{(4)}_{\xi_1})\Gamma^{(4)}_{\xi_2}\ldots\Gamma^{(4)}_{\xi}$$

(23c)

$$(A_n + \frac{m^2}{2}\,B_n)$$

$$R^{(2)}_2 = i\sum_{n=1}^{\infty} \frac{(i\,g)^n}{n!} \Theta_{\lambda_1,\lambda_2\ldots\lambda_n} \int dR_{n-1} \int d\xi^1_1\ldots d\xi^4_1\,(\partial_M\,\Gamma^{(4)}_{\xi_1})\Gamma^{(4)}_{\xi_2}\ldots\Gamma^{(4)}_{\xi}\,B_n$$

(23d)

$$A_n = \sum_{(2)} [\xi^1_1\ldots\xi^4_1\ldots\xi^1_n\ldots\xi^4_n\,]$$

$$B_n = \frac{1}{2} \sum_c \left\{ (p_1-\xi_1)\cdot(p_2-\xi_1) - (p_1-\xi_1)^2 - (p_2-\xi_1)^2 \right\} \; [z_1\ldots z_{4n-2}]$$

where $\sum_{(q)}$ denotes summations over all the hafnians obtainable from $[\xi_1^1 \ldots \xi_n^4]$ by deleting any q variables from it. $\sum_c$ denotes summations over all the ways of extracting two variables, P_1, P_2, from the sequences $\xi_1^1 \ldots \xi_n^4$, and $z_1 \ldots z_{4n-2}$ denote the remaining variables $\Theta_{\lambda_1,\lambda_2\ldots\lambda_n}$ is defined by the recursive formula

$$\Theta_{\lambda_1,\lambda_2\ldots\lambda_n} = \left\{ \Theta_{\lambda_1,\lambda_2\ldots\lambda_{n-1}}, \Theta(\underline{\lambda}) \right\}$$

$\{A,B\}$ denotes the commutator.

It is important to stress that g,m and M are finite arbitrary parameters devoid of physical meaning. In order to introduce the physical parameters we must of course impose the usual normalization conditions. We have only two independent parameters g and m; these parameters can be fixed by requiring that:

1) The pole of the momentum space two-point Green functions coincides with the physical mass m_{ph}.

2) The four-point Green function for some fixed values of external momenta coincides with the physical coupling constant g_{ph}.

These conditions allow us to express

$$m = m\,(m_{ph}\; g_{ph}\; M)$$

$$g = g\,(m_{ph}\; g_{ph}\; M)$$

Note that with these normalizing conditions the residue of G at $p^2=m^2_{ph}$ is different from 1: it takes a value Z which depends on m, g, M.

5. - Renormalization group and Callan-Symanzik equation[9),10)]

In this Section we show that a very simple derivation of the C.S. Equations can be given if one makes use of equation (22).

Let us start from the n-point Green function defined by

$$G_n(\underline{x}, m, g, M) = \frac{K_n(\underline{x}\ m, g, M)}{K_o} \tag{24}$$

In (24) we denote with $\underline{x}$ the external variables $(x_1, \ldots x_N)$ and write explicitly the dependence on the parameters g, m, M. K_n and K_o are obtained from (6) by setting N = n,0 respectively. For dimensional reasons we have:

$$\begin{aligned} G_n(\underline{x}, m, g, M) &= M^n F_n(\underline{s}, g, z) \\ s = M\underline{x} \qquad & z = \frac{m}{M} \end{aligned} \tag{25}$$

From (25) it follows:

$$\frac{\partial G_n}{\partial M} = M^{n-1} \left(n F_n + \underline{s} \frac{\partial F_n}{\partial \underline{s}} - z \frac{\partial F_n}{\partial z}\right) \tag{26}$$

Then, keeping in mind eq. (22), (23), (24), and (25) equation (26) can be written

$$\left(-\bar{\beta}(g) \frac{\partial}{\partial g} + \underline{x} \frac{\partial}{\partial \underline{x}} + n(1-\bar{\gamma}(g))\right) G_n = m \left(\bar{\alpha}(g, \frac{m}{M}) + 1\right) \frac{\partial G_n}{\partial m} \tag{27}$$

where we have put

$$i\left(-6BR_4 + R_2^{(1)}\right) = \frac{m^2}{M} \bar{\alpha}(g, \frac{m}{M})$$

$$R_2^{(2)} = \frac{2}{M} \bar{\gamma}(g)$$

$$-iR_4 + 2\bar{g} R_2^{(2)} = \frac{1}{M} \bar{\beta}(g)$$

Relations quite similar to the above are given in ref.6).

The usual considerations[9] allow us to write eq. (27) in the asymptotic region as follows:

$$(28) \qquad \left(- \bar{\beta}(g) \frac{\partial}{\partial g} + \underline{x} \frac{\partial}{\partial \underline{x}} + n(1 - \bar{\gamma}(g))\right) G_n = 0$$

therefore if g_∞ is a zero of $\bar{\beta}(g)$ we recover the asymptotic scaling behaviour of the Green functions

$$G_n = r^{-(\bar{\gamma}(g_\infty)-1)n} \qquad\qquad r = \sqrt{\sum_{h=1}^{n} (x_h)^2}$$

It is worth noting that equations (28) differ from the asymptotic form of the C.S. equations for the fact that in (28) one finds the parameter g instead of the physical coupling charge and mass. It presents the advantage that a compact form of the coefficients $\bar{\beta}(g)$ and $\bar{\gamma}(g)$ is given by formulas (23).

From equation (28) we can obtain the asymptotic form of the C.S. equations by the following procedure:

By using the normalizing conditions (see last Section) we write the Green functions in terms of $m_{ph}(m,g,\mu)$ and $g_{ph}(m,g,\mu)$:

$$(29) \qquad \begin{aligned} Z^{-n} G_n(\underline{x}\, m, g, M) &= Z^{-n} M^n F_n(xM, g, \frac{m}{M}) = H_n(\underline{x}, m_{ph}, g_{ph}) = \\ &= m_{ph}^n \mathcal{H}_n(x m_{ph}, g_{ph}) \end{aligned}$$

Note that the factor $Z = Z(m,g,M)$ has been introduced in order to normalize to 1 the residue of H at the point $p^2 = m_{ph}^2$. By substituting (29) into (28) we find the equation for $\mathcal{H}(\underline{x}, m_{ph}\, g_{ph})$:

(30) $$m_{ph} \frac{\partial \mathcal{H}_n}{\partial m_{ph}} + \beta(g_{ph}) \frac{\partial \mathcal{H}_n}{\partial g_{ph}} + n(-\gamma(g_{ph}) + 1) \mathcal{H}_n = 0$$

where:

(31) $$\beta(g_{ph}) = - \bar{\beta} \frac{\partial g_{ph}}{\partial g} \left(1 - \frac{\bar{\beta}}{m_{ph}} \frac{\partial m_{ph}}{\partial g} \right)^{-1}$$

$$\gamma(g_{ph}) = \bar{\gamma} + \bar{\beta} \frac{\partial \ell z_g}{\partial g} \left(1 - \frac{\bar{\beta}}{m_{ph}} \frac{\partial m_{ph}}{\partial g} \right)^{-1}$$

It is interesting to compare (31) with the relations which connect the coefficients of the C.S. equations to the zero-mass theory vertex-functions [11].

REFERENCES

1) E.R.Caianiello - Nuovo Cimento 10, 1634 (1953); 11,492 (1954); 13, 637 (1959), 14, 185 (1959).

2) F.Guerra,M.Marinaro - Nuovo Cimento 60A, 756 (1969)
F. Guerra - Nuovo Cimento 1A, 523 (1971)
M. Marinaro - Nuovo Cimento 9A, 62 (1972)

3) E. R. Speer - Journ.Math. Phys. 9, 1404 (1968)

4) I.M. Gel'fand and G.E.Shilov - Generalized Functions Vol. I, (New York 1964)

5) M. Marinaro, L. Mercaldo, G. Vilasi - Analytic Renormalization of the model - Proceedings of NATO Capri School (July 1973)

6) E. R. Caianiello,F.Guerra, M.Marinaro - Nuovo Cimento 60A, 713 (1969)

7) J. H. Lowenstein, University of Maryland Lecture Notes (1972)
W. Zimmermann Lectures on E.P. and Q.T. Brandeis (1970)

8) E. R. Caianiello - Combinatorics and Renormalization in Quantum Field Theory (Benjamin Inc. 1974)

9) K. Symanzik - Comm. Math. Phys. 18, 227 (1970); 23,49 (1971)

10) C. Callan - Phys. Rev. D2, 1541 (1970)

11) K. Symanzik - Desy 73/6.

NELSON'S SYMMETRY AT WORK: THE INFINITE VOLUME BEHAVIOR OF THE VACUUM FOR TWO-DIMENSIONAL SELF-COUPLED BOSE FIELDS(1)

FRANCESCO GUERRA

Institute of Physics (2)

University of Salerno, Italy

1. Introduction

We consider a self-coupled scalar field in two-dimensional space-time in the framework of the constructive quantum field theory program [1-11], successfully advanced by Glimm and Jaffe and their followers in the last years.

Let H_0 be the Hamiltonian for a time zero free field $\phi(x)$ of positive mass m in the Fock representation, with Hilbert space $\mathfrak{F}$ and bare vacuum Ω_0. The interaction is specified by a polynomial $P(x)$ with real coefficients, bounded below and, without loss of generality, normalized to $P(0)=0$.

Since we work in a one dimensional space all ultraviolet divergences are eliminated by the Wick ordering prescription in the interaction [1], but in order to avoid difficulties connected with Haag theorem [1] it is necessary to introduce a space cut off in

---

(1) Expanded version of a talk given at the Summer School on "Renormalization and Invariance in Quantum Field Theory", held in Capri, July 2-14, 1973

(2) Postal address: Istituto di Fisica dell'Università, Via Vernieri, 52, 84100, Salerno, Italy

the interaction and try to recover the complete theory at the end by removing the cutoff.

Therefore we start from a space cutoff interacting Hamiltonian of the type

$$H_\ell = H_0 + V_\ell \quad , \quad V_\ell = \int_{-\frac{\ell}{2}}^{\frac{\ell}{2}} : P(\phi(x)) : dx$$

Roughly speaking the physical meaning of the interaction H_ℓ is that the bare particles extended all over the physical (one dimensional) space but are allowed to interact only in the region of length ℓ centered at the origin.

It is well known [3, 4, 7] that the Hamiltonian H_ℓ can be rigorously defined as an unbounded operator on $\mathfrak{F}$, essentially selfadjoint on the intersection of the domains of H_0 and V_ℓ . It turns out that H_ℓ has a unique eigenstate Ω_ℓ of lowest energy E_ℓ . The state Ω_ℓ plays the role of approximate vacuum state for the system, the true vacuum state Ω should be recovered in the limit $\ell \to \infty$.

In the following we are interested in properties of Ω_ℓ , and E_ℓ as $\ell \to \infty$. Besides other known results we make use of a remarkable symmetry discovered by Nelson, who stated it implicitly in [12] and used it for a simple proof of the Glimm-Jaffe linear lower bound for the vacuum energy [13] . In order to state Nelson's symmetry let us remark that, since H_ℓ is bounded below then the positive operator e^{-tH_ℓ} is bounded above for $t \geq 0$ and for its norm we have $\| e^{-tH_\ell} \| \leq e^{-tE_\ell}$.
Nelson's symmetry states that the bare vacuum average of e^{-tH_ℓ} is a symmetric function in t and ℓ

$$<\Omega_0, \ e^{-tH_\ell} \ \Omega_0> \ = \ <\Omega_0, \ e^{-\ell H_t} \ \Omega_0> .$$

This relation looks very surprising because the parameters t and ℓ play a completely different role in e^{-tH_ℓ} , since ℓ is the size of the interaction region and t is a semigroup parameter. On the other hand the symmetry is very powerful because it allows to use the control of e^{-tH_ℓ} with respect to t to get informations about the large ℓ behavior of Ω_ℓ , and E_ℓ , as will be explained in the following.

The power of Nelson's symmetry was firstly appreciated in [14] and then further exploited in joint work with Rosen and Simon [15, 16].

The most natural proof of the symmetry can be obtained in the framework of the Euclidean-Markov methods [12, 17, 18, 10, 11] recently developed, mainly by Nelson, for the study of the constructive program.

In this talk we present some of the results about the infinite volume behavior of the vacuum, established in [14] and [15].

2. Preliminary considerations

In the last years most of the technical progress in constructive quantum field theory has been based on the systematic use of the Q representation [2, 3, 4, 6, 7]. Roughly speaking in this representation the time zero field operators $\phi(x)$ are simultaneously diagonalized, in the same way as the configuration variables $q_i(t)$ are diagonalized in the Schrödinger representation of usual quantum mechanics.

We refer to the mentioned literature for a complete mathematical treatment and limit ourselves here to state the basic facts.

In the Q representation the Fock space $\mathfrak{F}$ is represented as the Hilbert space $L^2(Q, \mu)$ on some probability space (Q, μ), in such a way that each element u of $\mathfrak{F}$ is described by a square integrable function $u(q)$ on Q with respect to the normalized measure μ, so that

$$\langle u, u' \rangle = \int_Q u^*(q)u'(q)d\mu(q)$$

where $\langle \, , \, \rangle$ is the scalar product in $\mathfrak{F}$.

The bare vacuum (no-particle state) Ω_0 is represented by the function identically equal to one on Q. Fields and functions of fields are represented by Q space functions, in particular it turns out that the interaction part V_ℓ of the Hamiltonian is an L^p function on Q for $1 \leq p < \infty$.

With a convenient choice of normalization and phase factor, we have $\|\Omega_\ell\|_2 = 1$ and $\Omega_\ell > 0$ almost everywhere on Q (no nodes for the ground state!). Moreover $\Omega_\ell \in L^p(Q,\mu)$ for $1 \leq p < \infty$.

For non trivial interactions and $\ell > 0$ we have $\Omega_\ell \neq \Omega_o$ and $E_\ell < 0$ because $<\Omega_o, H_\ell \Omega_o>$, moreover

$$\| \Omega_\ell \|_1 = \int_Q \Omega_\ell d\mu = <\Omega_\ell, \Omega_o> < 1$$

We introduce the energy density $\alpha(\ell) = -E_\ell/\ell$ and the overlapping factor $\eta(\ell)$ by $\| \Omega_\ell \|_1^2 = \exp[-\ell\eta(\ell)]$.
Obviously $\alpha(\ell) \geq 0$ and $\eta(\ell) \geq 0$.

3. The main results with some comments

Theorem 1. The energy density $\alpha(\ell)$ is monotone non decreasing and bounded above, therefore if we define

$$\alpha_\infty = \sup_\ell \alpha(\ell) \qquad \text{we have}$$

$$\lim_{\ell \to \infty} \alpha(\ell) = \alpha_\infty .$$

This theorem states the convergence of the energy density in the infinite volume limit established in [14, 15] . The upper bound for $\alpha(\ell)$ is nothing but the Glimm-Jaffe linear lower bound for the ground state energy [13] . For generalizations of theorem 1 see [16] and [19] .

Corollary 2. E_ℓ is monotone decreasing in ℓ .

Theorem 3. There are positive constants c, d, $\bar{\ell}$, such that $c\ell \leq \alpha(\ell) \leq d\ell$, for $0 \leq \ell \leq \bar{\ell}$, in particular $\alpha(\ell)$ is no constant in ℓ.

Theorem 4. There are positive constants η_1, η_2 such that $\eta_1 \leq \eta(\ell) \leq \eta_2$ for large ℓ. As a consequence

$$\|\Omega_\ell\|_p \to 0 \quad \text{for} \quad 1 \leq p < 2$$

$$\to \infty \quad \text{for} \quad p > 2, \text{ as } \ell \to \infty$$

Theorem 5. As $\ell \to \infty$ Ω_ℓ tends weakly to zero in the Fock space, that means $\langle u, \Omega_\ell \rangle \to 0$ for any fixed $u \in \mathfrak{F}$.

This is the famous Van Hove phenomenon [1]. In some sense Ω_ℓ disappears from the Fock space $\mathfrak{F}$ as $\ell \to \infty$. Therefore the limit $\ell \to \infty$ of the theory involves a very delicate procedure and the limiting theory lives outside Fock space. For recent results see [9, 10, 20, 21, 22].

Theorem 6. For any $\ell, \ell' \geq 0$ the following bound holds

$$|E_\ell - E_{\ell'}| \leq \alpha_\infty \, |\ell - \ell'|$$

and E_ℓ is concave in ℓ.

Corollary 7. The energy density $\alpha(\ell)$ is increasing in ℓ for a non trivial interaction.

Theorem 8. Define $\beta(\ell) \equiv E_\ell + \ell\alpha_\infty$. Then $\beta(\ell)$ is concave, monotone non decreasing and bounded above. Therefore putting

$$\beta_\infty = \sup_\ell \beta(\ell) \qquad \text{we have}$$

$$\lim_{\ell\to\infty} \beta(\ell) = \beta_\infty .$$

This theorem gives us the beginning of an asymptotic expansion of E_ℓ in ℓ of the type

$$E_\ell = -\ell\alpha_\infty + \beta_\infty - \gamma_\infty/\ell + \ldots,$$

unfortunately up to now it has not been possible to go beyond β_∞. So the existence of γ_∞ as

$$\lim_{\ell\to\infty} \ell\,[\,-E_\ell - \ell\alpha_\infty + \beta_\infty\,]$$

has not been proved yet.

4. Three useful technical lemmas

We collect here some useful results which are used in the next section for the proof of the statements made in section 3.

Lemma 1. Nelson's symmetry. See Section 1.

Lemma 2. For ℓ, $t \geq 0$, $r, s, \beta \geq 1$, $\beta^{-1} + \beta'^{-1} = 1$, the following estimate holds

$$|\langle u, e^{-t(H_0+V_\ell)}v\rangle| \leq \| u \|_{\beta' r} \| v \|_{\beta' s} \langle \Omega_0, e^{-t(H_0+\beta V_\ell)}\Omega_0\rangle^{\frac{1}{\beta}},$$

where $(r-1)(s-1) = e^{-2tm}$ and $\|\cdot\|_p$ denotes the norm in $L^p(Q,\mu)$ spaces.

Lemma 2 is a consequence of the representation of e^{-tH_ℓ} through the Feynman-Kaĉ path integral formula and the hypercontractivity of the free Hamiltonian. We cannot enlarge our considerations to include a full proof of lemma 2 but refer to [2, 3, 6, 7, 10, II, I2] for the basic ideas about Feynman-Kaĉ formula and hypercontractivity. For the statement of lemma 2 we have used the best Nelson hypercontractive estimate [18].

Lemma 3. For $t, a, b, c \geq 0$ and $k \geq 1$ the following inequality holds

$$-tE_{a+b+c} - (a+b+c)\,\eta\,(a+b+c) \leq -\frac{t}{2k}[E_{2ka} + E_{2kb}] - c\,E_t$$

Proof. First of all note that for any $k \geq 1$ we have, using Nelson symmetry

$$\| e^{-aH_t}\,\Omega_o \|_2^2 = \langle\Omega_o, e^{-2aH_t}\,\Omega_o\rangle \leq \langle\Omega_o, e^{-2akH_t}\,\Omega_o\rangle^{\frac{1}{k}} =$$

$$= \langle\Omega_o, e^{-tH_{2ak}}\,\Omega_o\rangle^{\frac{1}{k}} \leq e^{-\frac{t}{k}E_{2ak}} .$$

Therefore we can write in general

$$e^{-tE_{a+b+c}} \|\Omega_{a+b+c}\|_1^2 = <\Omega_{a+b+c}, e^{-\frac{t}{2}H_{a+b+c}}\Omega_o>^2 \leq$$

$$\leq \| e^{-\frac{t}{2}H_{a+b+c}}\Omega_o\|_2^2 = <\Omega_o, e^{-tH_{a+b+c}}\Omega_o> =$$

$$= <\Omega_o, e^{-(a+b+c)H_t}\Omega_o> = <e^{-aH_t}\Omega_o, e^{-cH_t} e^{-bH_t}\Omega_o> \leq$$

$$\leq \| e^{-aH_t}\Omega_o\|_2 \| e^{-cH_t}\| \| e^{-bH_t}\Omega_o\|_2 \leq e^{-\frac{t}{2k}[E_{2ak}+E_{2bk}]} e^{-cE_t}.$$

Recalling the definition $\|\Omega_\ell\|_1^2 = e^{-\ell\eta(\ell)}$ and taking the logarithm we have the inequality of the lemma.

5. Proof of the results of Section 3

Proof of theorem 1. Put $c = 0$, $a=b=\frac{\ell}{2}$, $\ell'=k\ell$, with $\ell'\geq\ell$, in lemma 3, divide by t and let $t\to\infty$, then we have

$$-E_\ell \leq (\ell/\ell')(-E_{\ell'}),$$

therefore $\alpha(\ell) \leq \alpha(\ell')$ for $\ell\leq\ell'$ and monotonicity is proved. Corollary 2 is then obvious. To prove boundedness of $\alpha(\ell)$ we use lemma 2 with $\beta' r = \beta' s = 2$, then for t fixed there is β such that

$$e^{-tE_\ell} = \|e^{-tH_\ell}\| \le \langle \Omega_0, \; e^{-\ell(H_0+\beta V_t)}\Omega_0\rangle^{\frac{1}{\beta}} \le e^{-\frac{\ell}{\beta}E_t(\beta)} ,$$

where Nelson symmetry has been used and $E_t(\beta)$ is the ground state energy of $H_0 + \beta E_t$. Therefore we have

$$\alpha(\ell) \le -E_t(\beta)/t\beta$$

and the theorem is established.

Proof of Theorem 3. As in the proof of theorem 1 we have

$$\langle \Omega_0, e^{-\ell H_t}\Omega_0\rangle \le e^{-tE_\ell} \le \langle \Omega_0, e^{-(H_0+\beta V_t)}\Omega_0\rangle^{\frac{1}{\beta}} .$$

Using the bounds

$$(x-1)/2 \le \log x \le x-1$$

for x small enough, we have

$$\frac{1}{2t}\frac{1}{\ell^2}(\langle\Omega_0, e^{-\ell H_t}\Omega_0\rangle - 1) \le \frac{\alpha(\ell)}{\ell} \le \frac{1}{\beta t}\frac{1}{\ell^2}(\langle\Omega_0, e^{-\ell(H_0+\beta V_t)}\Omega_0\rangle - 1)$$

Since $\langle\Omega_0, H_t\Omega_0\rangle = 0$ we have by the spectral theorem

$$\lim_{\ell\to 0}\frac{1}{\ell^2}(\langle\Omega_0, e^{-\ell H_t}\Omega_0\rangle - 1) = \frac{1}{2}\langle\Omega_0, H_t^2\,\Omega_0\rangle$$

therefore there is a constant C such that

$$C \leq \alpha(\ell)/\ell$$

In the same way we can find a constant d such that

$$d \geq \alpha(\ell)/\ell$$

and the theorem is proved. In particular we have $\lim_{\ell\to 0} \alpha(\ell) = 0$.

Proof of Theorem 4. Put $a = b = 0$, $c=\ell$ in lemma 3, then we have

$$\eta(\ell) \geq t(\alpha(\ell) - \alpha(t)).$$

By theorem 3 we can find ℓ_o, t such that $\alpha(\ell_o) > \alpha(t)$, therefore for $\ell \geq \ell_o$ we have

$$\eta(\ell) \geq \eta_1 \qquad \text{where}$$

$$\eta_1 \equiv t(\alpha(\ell_o) - \alpha(t)) > 0 .$$

Consider now lemma 2 and put $u=v=\Omega_\ell$ and $\beta'r=\beta's=p$ with $1<p<2$, then we find immediately

$$\| \Omega_\ell \|_p^2 \geq e^{-\eta'\ell}$$

for some positive constant η'. Using convexity of $\log \| \Omega_\ell \|_p^p$ in the interval $1 \leq p \leq 2$ we find

$$\| \Omega_\ell \|_1^2 \geq e^{-\eta_2 \ell}$$

for some positive constant η_2. Therefore we have established the bounds

$$0 < \eta_1 \leq \eta(\ell) \leq \eta_2 .$$

The rest of the theorem follows again for the convexity of $\log \|\Omega_\ell\|_p^p$.

<u>Proof of Theorem 5</u>. By theorem 4 we have $\|\Omega_\ell\|_1 \to 0$ as $\ell \to \infty$, therefore $\Omega_\ell \to 0$ weakly in L^∞ . Since L^∞ is dense in L^2 in the L^2 norm and $\|\Omega_\ell\|_2 = 1$ we have also that $\Omega_\ell \to 0$ weakly in L^2.

<u>Proof of Theorem 6</u>. Put k=1 , $a=b=\ell/2$, $c=\ell'-\ell$, $a+b+c=\ell'$, $\ell' \geq \ell$ in lemma 3. We have

$$-tE_{\ell'} - \ell'\eta(\ell') \leq -tE_\ell - (\ell'-\ell)E_t ,$$

divide by t and take $t \to \infty$, then

$$-E_{\ell'} \leq -E_\ell + (\ell'-\ell)\, \alpha_\infty .$$

Since $\ell' \geq \ell$ and $E_{\ell'} \leq E_\ell$ by Corollary 2, the bound

$$|E_\ell - E_{\ell'}| \leq \alpha_\infty\, |\ell - \ell'|$$

is established.

To prove concavity of E_ℓ in ℓ , put c=0, k=1, $2a=\ell_1$, $2b=\ell_2$

in lemma 3, then

$$E_{(\ell_1+\ell_2)/2} \geq (E_{\ell_1} + E_{\ell_2})/2$$

Proof of Corollary 7. We already know that $\alpha(\ell)$ is non decreasing. Since $\alpha(0)=0$ and $\alpha(\ell) > 0$ for $\ell > 0$ and non trivial interactions, we see immediately that the concavity of E_ℓ does not allow $\alpha(\ell)$ to be constant after some $\ell_o > 0$.

Proof of Theorem 8. With the definition

$$\beta(\ell) = E_\ell + \ell\alpha_\infty$$

the concavity of $\beta(\ell)$ follows from the concavity of E_ℓ. On the other hand the bound

$$E_\ell - E_{\ell'} \leq (\ell' - \ell)\, \alpha_\infty$$

for $\ell' \geq \ell$, established in the proof of Theorem 6, is equivalent to

$$\beta(\ell) \leq \beta(\ell') \quad \text{for } \ell \leq \ell'.$$

On the other hand we know from the proof of Theorem 4 that

$$\eta_2 \geq \eta(\ell) \geq t(\alpha(\ell) - \alpha(t))$$

and taking $\ell \to \infty$ we establish

$$\beta(t) \leq \eta_2$$

Therefore with the definition

$$\beta_\infty = \sup_\ell \beta(\ell) \qquad \text{we have}$$

$$\lim_{\ell\to\infty} \beta(\ell) = \beta_\infty .$$

In conclusion the author would like to thank E.R. Caianiello for the kind hospitality extended to him in Capri.

REFERENCES

1 A.S. WIGHTMAN: An introduction to some aspects of the relativistic dynamics of quantized fields. In: 1964 Cargese Summer School Lectures, ed. by M. Levy Gordon and Breach, New York 1967.

2 E. NELSON: A quantic interaction in two dimensions. In: Mathematical Theory of Elementary Particles", R. Goodman and I. Segal, Eds., Cambrige: M.I.T. Press 1966.

3 J. GLIMM and A. JAFFE: Quantum field models.In:"Statistical mechanics and quantum field theory", ed. by C.de Witt and R. Stora, Gordon and Breach, New York, 1971.

4 J. GLIMM and A. JAFFE: Boson quantum field models. In: "Mathematics of contemporany physics", ed. by R. Streater,Academic Press, New York, 1972.

5 L. ROSEN: A $\lambda\phi^{2n}$ field theory without cutoffs.Commun. math. Phys. 16, 157 (1970).

6 I. SEGAL: Construction of nonlinear local quantum processes I, Ann. Math. 92, 462 (1970).

7 B. SIMON and R. HOEGH-KROHN: Hypercontractive semigroups and two dimensional self-coupled Bose fields. J. Funct.Anal. 9, 121 (1972).

8 A.S. WIGHTMAN: Constructive field theory: Introduction to the problem. In: Proc. 1972 Coral Gables Conference.

9 J. GLIMM, A. JAFFE and T. SPENCER: The Wightman axioms and particle structure in the $P(\phi)_2$ quantum field model. Ann. Math., to appear.

10 F. GUERRA, L. ROSEN and B. SIMON: The $P(\phi)_2$ Euclidean quantum field theory as classical statistical mechanics. Ann. Math., to appear.

11 B. SIMON: The $P(\phi)_2$ Euclidean (Quantum) Field Theory, Princeton University Press, to appear 1974.

12 E. NELSON: Quantum Fields and Markoff Fields. In: "Pro - ceedings of Summer Institute of Partial Differential Equation", Berkeley 1971, Amer. Math. Soc., Providence, 1973.

13 J.GLIMM and A. JAFFE: The $\lambda(\phi^4)_2$ quantum field theory without cut-offs III, The Physical vacuum, Acta Math. 125,203 (1970).

14 F. GUERRA: Uniqueness of the vacuum energy density and Van Hove phenomenon in the infinite volume limit for two dimensional self-coupled Bose fields, Phys. Rev. Letts. 28,1213 (1972).

15 F. GUERRA, L. ROSEN and B. SIMON: Nelson's symmetry and the infinite volume behavior of the vacuum in $P(\phi)_2$, Commun. Math.Phys. 27, 10 (1972).

16 F. GUERRA, L. ROSEN and B. SIMON: The vacuum energy for $P(\phi)_2$: infinite volume limit and coupling constant dependence. Commun.Math.Phys. 29, 233 (1973).

17 E. NELSON: Construction of quantum fields from Markoff fields, J. Funct. Anal. 12, 97 (1973).

18 E. NELSON: The free Markoff field. J. Funct. Anal. 12, 211 (1973).

19 F. GUERRA, L. ROSEN and B. SIMON: Boundary conditions for the Euclidean Field Theory. Preprint in preparation.

20 E. NELSON: to appear.

21 B. SIMON: Correlation inequalities and the mass gap in II, Uniqueness of the vacuum for a class of strongly coupled theories. Ann.Math. to appear.

22 A. ALBEVERIO and R. HOEGH-KROHN: The Wightman axioms and the mass gap for strong interactions of exponential type in two dimensional space time. Oslo preprint, 1973.

PATH INTEGRALS AND GAUGE THEORIES*

Benjamin W. Lee

Institute for Theoretical Physics
State University of New York
Stony Brook, New York 11790

1. INTRODUCTION

Quantization of nonabelian gauge fields has been carried[1] out successfully in the path integral formulation of quantum theory due to Dirac[2] and Feynman[3]. Unfortunately the discussion of path integral quantization is completely ignored in most standard textbooks on field theory, or is given inadequately. I would like to describe the basic idea involved, so that one can proceed to the quantization of gauge theories by this method.

2. FUNDAMENTALS

Let us begin by considering a quantum mechanical system with one degree of freedom. Let $Q_H(t)$ be the position operator at t in the Heisenberg picture:

$$Q_H(t) = e^{iHt} Q_S e^{-iHt}$$
$$Q_S = Q_H(0) . \tag{1}$$

Let $|q,t>_H$ be the eigenstate of $Q_H(t)$ with eigenvalue q:

$$Q_H(t) |q,t>_H = q|q,t>_H ,$$

or

$$Q_S e^{-iHt}|q,t>_H = q e^{-iHt}|q,t>_H .$$

*Research supported by NSF Grant No. GP-32998X.

Now we define $|q\rangle$ as the eigenstate of Q_S with eigenvalue q:

$$Q_S|q\rangle = q|q\rangle.$$

Then we have

$$|q,t\rangle_H = e^{iHt}|q\rangle . \tag{2}$$

The most interesting quantity in quantum theory is the transformation matrix ${}_H\langle q',t'|q,t\rangle_H$ which is the probability amplitude between the state prepared at t to have the coordinate value q to the state of the coordinate value q' at time t'. According to (2), this is given by

$${}_H\langle q',t'|q,t\rangle_H = \langle q'|e^{-iH(t'-t)}|q\rangle . \tag{3}$$

We can imagine partitioning the time interval t'-t into n segments, so that each segment goes to zero as $n \to \infty$. For convenience, we shall divide the interval into n equal segments of duration Δ:

$$\Delta = (t'-t)/n .$$

Upon writing

$$e^{-iH(t'-t)} = e^{-iH(t'-t_{n-1})} e^{-iH(t_{n-1}-t_{n-2})} \cdots e^{-iH(t_1-t)}$$

and inserting complete sets of eigenstates of Q_S between the factors, we obtain

$$\langle q'|e^{-iH(t'-t)}|q\rangle = \int dq_1 \cdots \int dq_{n-1} \; \langle q'|e^{-iH\Delta}|q_{n-1}\rangle \times$$
$$\langle q_{n-1}|e^{-iH\Delta}|q_{n-2}\rangle \cdots \langle q_1|e^{-iH\Delta}|q\rangle. \tag{4}$$

Now, for sufficiently small Δ, we may evaluate $\langle q'|e^{-iH\Delta}|q\rangle$ as follows:

$$\langle q'|e^{-iH\Delta}|q\rangle \simeq \langle q'|1 - i\Delta H(P,Q)|q\rangle$$

where H(P,Q) is the quantum mechanical Hamiltonian expressed as a function of the operators P, and Q. When the Hamiltonian is of the form

$$H(P,Q) = \tfrac{1}{2}P^2 + V(Q) , \tag{5}$$

we can write

$$<q'|H(P,Q)|q> = <q'|\frac{P^2}{2}|q> + V\left(\frac{q+q'}{2}\right)\delta(q-q')$$

$$= \int \frac{dp}{2\pi} e^{ip(q'-q)}\left[\frac{1}{2}p^2 + V\left(\frac{q+q'}{2}\right)\right] .$$

To first order in Δ, we have

$$<q'|e^{-i\Delta H}|q> \simeq \int \frac{dp}{2\pi} e^{ip(q'-q)-i\Delta H\left(p,\frac{q'+q}{2}\right)} \tag{6}$$

where on the right hand side of Eq. (6), H(p,q) refers to the classical expression for the Hamiltonian.

So finally, substituting Eq. (6) into (4), we obtain

$$<q'|e^{-iH(t'-t)}|q> = \lim_{n\to\infty}\int\frac{dp_1}{2\pi}\int\frac{dp_2}{2\pi}\cdots\int\frac{dp_n}{2\pi}\int dq_1\cdots\int dq_{n-1}$$

$$\times \exp i \sum_{i=1}^{n}\left\{p_i(q_i-q_{i-1}) - \Delta H\left(p_i,\frac{q_i+q_{i-1}}{2}\right)\right\}$$

$$\equiv \int\left[\frac{dp\, dq}{2\pi}\right] e^{i\int_t^{t'} dt\, [p\dot{q}-H(p,q)]} \tag{7}$$

where the last line is to be understood as a symbolic way of writing the second line. Here, the boundary condition is

$$q_n = q', \qquad q_o = q .$$

A few remarks are in order here. First, insofar as the Hamiltonian is of the form of Eq. (5), the path integral (7) is rather insensitive to how we partition the time interval. Secondly, Eq. (7) makes unambiguous the functional measure with which the path integral is to be performed - it is over the canonical variables p and q, which are unambiguously defined by classical mechanics. [Note however that Eq. (7) is not invariant under canonical transformations of p and q variables.] In Eq. (7), whenever there is an ambiguity about the order of integrations, the p-integrations are to be performed first. Thus, when the Hamiltonian is of the form of Eq. (5), we can perform the p-integrations analytically, and obtain

$$<q'|e^{-iH(t'-t)}|q> = \lim_{n\to\infty}\int dq_1 \cdots \int dq_{n-1} \exp i \sum_{i=1}^{n} \Delta\left\{\left(\frac{q_i - q_{i-1}}{\Delta}\right)^2 - V\left(\frac{q_i + q_{i-1}}{2}\right)\right\}$$

$$\equiv \int [dq] \exp i\int_t^{t'} dt'\, L(q,\dot{q}) \tag{8}$$

which is the form written down by Feynman first. $L(q,\dot{q})$ is the Lagrangian of the system (5): $L(q\,\dot{q}) = \frac{1}{2}\dot{q}^2 - V(q)$. Thirdly, for velocity-dependent potentials, i.e., when the Hamiltonian is given by, for example,

$$H(p,q) = \tfrac{1}{2} p^2 f(q) + V(q) \tag{9}$$

the "derivation" of Eq. (7) does not really apply, and in fact the path integral formulation becomes ambiguous. This is not surprising because, in the usual operator formulation, there is also the ambiguity in ordering the operators in the term $p^2 f(q)$. One may postulate, as we shall do, that even for such Hamiltonians as (9), the prescription of Eq. (7) with the p-integrations performed first be used. [The operator ordering of the Hamiltonian (8) corresponding to this prescription is discussed by de Witt,[4] and more recently by Cheng.[5]]

3. GREEN'S FUNCTIONS

Expectation values of time-ordered products of Heisenberg operators may be expressed as path integrals. Thus,

$$_H<q',t'|T\, Q_H(t_1)Q_H(t_2)|q,t>_H$$

$$= <q'|e^{-iH(t'-t_>)} Q_S\, e^{-iH(t_>-t_<)} Q_S\, e^{-iH(t_<-t)}|q>$$

$$= \left[\frac{dq\,dp}{2\pi}\right] q(t_1)q(t_2)\, e^{i\int_t^{t'} dt[p\dot{q}-H(p,q)]} \tag{10}$$

where $t_>(t_<)$ is the larger (lesser) of t_1 and t_2, and where we have used the same reasoning as we did to derive Eq. (7). The Green's function, that is, the ground state expectation value of a time-order product of operators can be obtained similarly. By inserting complete sets of eigenstates of the operators $Q_H(t)$, $Q_H(t')$ in the definition

$$G(t_1,t_2) = \langle \text{ground}|TQ_H(t_1)\, Q_H(t_2)|\text{ground}\rangle$$

where $t' > t_1, t_2 > t$, we obtain

$$G(t_1,t_2) = \int dq \int dq' \; \phi_o^*(q',t')\; \phi_o(q,t)\langle q't'|TQ_H(t_1)Q_H(t_2)|q,t\rangle \tag{11}$$

where $\phi_o(q,t)$ is the ground state wave function

$$\phi_o(q,t) = {}_H\langle q,t|\text{ground state}\rangle. \tag{12}$$

One can use, in principle, Eq. (11) to compute the Green's functions in conjunction with Eq. (10). But this would entail knowing the ground state wave function (12) a priori, which is not always possible in practice.

The method we shall now describe allows us to compute the Green's functions without explicitly knowing the ground state wave function. Let $O(t_1,t_2)$ be $TQ_H(t_1)\, Q_H(t_2)$. We can write

$$\begin{aligned} &{}_H\langle q',t'|O(t_1,t_2)|q,t\rangle_H \\ &= \int dQ' \int dQ \; {}_H\langle q',t'|Q',T'\rangle_H \, {}_H\langle Q',T'|O(t_1,t_2)|Q,T\rangle_H \\ &\qquad \times {}_H\langle Q,T|q,t\rangle_H \end{aligned} \tag{13}$$

where we choose T and T' such that $t' \geq T' > t_1,t_2 > T \geq t$. Now recall that

$$\begin{aligned} {}_H\langle q',t'|Q',T'\rangle_H &= \langle q'|e^{-iH(t'-t)}|q\rangle \\ &= \sum_n \phi_n(q')\phi_n(Q')e^{-iE_n(t'-T')} \end{aligned} \tag{14}$$

where the subscript n refers to the n-th excited state of the system. In Eq. (14), the dependence on t' is explicit, so that we can easily compute the limit in which t' approaches $-i\infty$. Since $E_n \geq E_o$ for all n, we obtain

$$\lim_{t'\to -i\infty} {}_H\langle q',t'|Q',T'\rangle_H = \phi_o(q')e^{-E_o|t'|}\phi_o(Q')e^{iE_oT'} \tag{15}$$

the other terms vanishing faster than the term kept on the right of Eq. (15). [It is obvious that Eq. (15) holds, if there is a gap between the ground and the first excited states. When there is no such gap, the validity of Eq. (15) depends on the density of states at the tip of the spectrum. If it is no denser than that of a free system, Eq. (15) is still valid.] Now substituting Eq. (15) and an analogous one for ${}_H\langle Q,T|q,t\rangle_H$ in Eq. (13), and recalling Eq. (11), we obtain

$$\lim_{\substack{t'\to -i\infty \\ t\to i\infty}} {}_H\langle q',t'|O(t_1,t_2)|q,t\rangle_H$$

$$= \phi_o(q')e^{-E_o|t'|}\phi_o(q)e^{-E_o|t|}G(t_1,t_2),$$

or

$$G(t_1,t_2) = \lim_{\substack{t'\to -i\infty \\ t\to i\infty}} \frac{{}_H\langle q',t'|TQ_H(t_1)Q_H(t_2)|q,t\rangle}{{}_H\langle q',t'|q,t\rangle_H} \tag{16}$$

Note that Eq. (16) is independent of q and q'.

Higher order Green's functions can be generated from the functional of the external source J:

$$W[J] = \int \left[\frac{dpdq}{2\pi}\right] e^{i\int_t^{t'} dt[p\dot{q}-H(p,q)+J(t)q(t)]}$$

$$\sim \int [dq]\, e^{i\int_t^{t'} dt[L(q,\dot{q})+J(t)q(t)]} \tag{17}$$

where the second line follows if the Hamiltonian is of the form of Eq. (5). In Eq. (17), the limit

$$t' \to -i\infty\ , \qquad t \to i\infty \tag{18}$$

is understood, and the external source $J(t)$ is assumed to be "turned on" for a finite time interval:

$$J(t) \neq 0, \quad \text{only for } T' > t > T.$$

The "connected" Green's functions $G(t_1,t_2,\cdots,t_n)$ are given by

$$G(t_1,t_2,\cdots,t_n) = \frac{1}{i^n}\frac{\delta^n}{\delta J(t_1)\delta J(t_2)\cdots\delta J(t_n)} \ln W[J]\Bigg|_{J=0} . \tag{19}$$

Note that the absolute normalization of $W[J]$ does not matter in Eq. (19).

The boundary condition of Eq. (18) suggests that we consider the Schwinger-Symanzik functions[6] which are analytic continuations of the Green's functions defined in Eq. (19) to the imaginary time (or to the Euclidean space-time):

$$S(\tau_1,\tau_2,\cdots\tau_n) = (i)^n\, G(-i\tau_1,-i\tau_2,\cdots,-i\tau_n). \tag{20}$$

The generating functional W_E of the S functions:

$$S(\tau_1,\tau_2,\cdots,\tau_n) = \frac{\delta^n}{\delta J(\tau_1)\delta J(\tau_2)\cdots\delta J(\tau_n)} \ln W_E[J] \tag{21}$$

may be written as

$$W_E[J] = \int [dq]\, e^{\int_{-\infty}^{\infty} d\tau \left[-\frac{1}{2}\left(\frac{dq}{d\tau}\right)^2 - V(q) + J(\tau)q(\tau)\right]} . \tag{22}$$

Observe that the integrals over dq are now well-defined, being Gaussian integrals as $V(q)$ is bounded from below. Equation (22) is really the basis of path integral formalism. For mathematical rigor and incorporation of correct boundary conditions, Eqs. (17) and (19) should really be understood as a short hand notation for defining Green's functions as analytic continuations of the Schwinger-Symanzik functions generated by Eqs. (21) and (22).

4. FIELD THEORY

There is no difficulty in extending the formalism developed so far to systems of many degrees of freedom. Thus, for systems with n-degrees of freedom, we have

$$W[J_1, J_2, \cdots J_n] = \int \left[\prod_{i=1}^{n} \prod_t dq_i(t) \right] \exp i \int dt \mathcal{L}(q_1, q_2 \cdots q_n,$$

$$\dot{q}_1, \dot{q}_2 \cdots \dot{q}_n) \tag{23}$$

in place of Eq. (17). Equation (23) can be extended formally to systems with infinite degrees of freedom; the mathematical difficulties associated with infinite degrees of freedom are neither increased nor decreased by the use of path integral formalism, and should be handled as in other formalisms of field theory. In the following we will develop the machinery of the $\lambda\phi^4$ field theory, ignoring these subtleties.

Let us approximate the space-time continuum by a four dimensional lattice. Let ϕ_j be the value of the field strength at the lattice site j. In the continuum limit, $\phi_j \to \phi(x)$ and we shall write

$$\prod_j d\phi_j \to [d\phi(\underset{\sim}{x}, t)] = [d\phi(x)].$$

The generating functional of Green's functions is

$$W[J(\underset{\sim}{x}, t)] = \int [d\phi(x)] \exp i \int d^4\chi \, [\mathcal{L}(x) + J(x)\phi(x)]$$

$$= \int [d\phi(x)] \exp i \int d^4x \, [\tfrac{1}{2}(\partial_\mu \phi(x))^2 - \frac{\mu^2}{2} \phi^2(x)$$

$$- \frac{\lambda}{4} \phi^4(x) + J(x)\phi(x)]. \tag{24}$$

The corresponding expression in the Euclidean, or imaginary-time, formulation is

$$W_E[J(\underset{\sim}{x}, \tau)] = \int [d\phi(\underset{\sim}{x}, \tau)] \exp \left\{ \int d^4\chi \right.$$

$$\times \left[-\frac{1}{2}\left(\frac{\partial\phi}{\partial\tau}\right)^2 - \frac{1}{2}(\underset{\sim}{\nabla}\phi)^2 - \frac{\mu^2}{2}\phi^2 - \frac{\lambda}{4}\phi^4 \right.$$

$$\left. \left. + J(x, \tau)\phi(x, \tau) \right] \right\}. \tag{25}$$

Perturbation theory is based on the formula

$$W_E[J] = \left\{ \exp \int d^4\chi \mathcal{L}_I \left[\frac{1}{i}\frac{\delta}{\delta J}\right]\right\} W_{EO}[J] \tag{26}$$

where $\mathcal{L}_I[\phi]$ is the interaction Lagrangian

$$\mathcal{L}_I[\phi] = -\frac{\lambda}{4}\phi^4 \tag{27}$$

and $W_{EO}[J]$ is the free field generating functional:

$$W_{EO}[J] = \int [d\phi] \exp\left\{-\int d^4x \int d^4y \; \tfrac{1}{2}\,\phi(x)K(x,y)\phi(y) + \int d^4x \; J(x)\phi(x)\right\} \tag{28}$$

where

$$K(x,y) = \left(-\frac{\partial^2}{\partial\tau^2} - \nabla^2 + \mu^2\right)\delta^4(x-y).$$

The functional integral in Eq. (28) may be performed:

$$\int \prod_j dx_j \exp\left\{-\tfrac{1}{2}\, x_i\, K_{ij}\, x_j + b_i x_i\right\} = \frac{1}{\sqrt{\det A}} \exp \tfrac{1}{2}\, b^T K^{-1} b.$$

Thus, ignoring an overall multiplicative factor independent of J, we have

$$W_{EO}[J] \sim \exp \tfrac{1}{2}\int d^4x \int d^4y \; J(x)\, \Delta_E(x-y\,;\mu^2)\; J(y) \tag{29}$$

where the Euclidean propagator Δ_E is given by

$$\Delta_E(x;\mu^2) = \int \frac{d^4k}{(2\pi)^4}\, \frac{e^{ik\cdot x}}{k_o^2+k^2+\mu^2}\,. \tag{30}$$

Note that the analytic continuation Eq. (30) by the prescription of Eq. (20) gives the Feynman propagator with the correct boundary condition:

$$\Delta_E(x;\mu^2) \to i\Delta_F(x;\mu^2) = i\int \frac{d^4k}{(2\pi)^4}\,\frac{e^{ikx}}{k^2-\mu^2+i\varepsilon} \quad ,$$

$$k^2 = k_o^2 - \underset{\sim}{k}^2 \ . \tag{31}$$

To develop the perturbation series for $W_E[J]$, we merely expand the first factor on the RHS of Eq. (26) in powers of λ:

$$\exp\int d^4x\,\mathcal{L}_I\left[\frac{1}{i}\frac{\delta}{\delta J}\right] = \sum_{n=0}^{\infty}\frac{1}{n!}\left\{\int d^4x\,\mathcal{L}_I\left[\frac{1}{i}\frac{\delta}{\delta J}\right]\right\}^n \ . \tag{32}$$

What corresponds to Wick's theorem[8] in the Dyson-Wick expansion[7] in the expansion based on Eq. (26) in conjunction with (32) is simply the rule of functional differentiation:

$$\frac{\delta}{\delta J(x)}\,J(y) = \delta^4(x-y) \ . \tag{33}$$

5. GAUGE THEORIES

Before we can apply the path integral formalism to systems with gauge invariance, a preliminary remark[9] is in order.

Consider a system with a constraint, so that the Lagrangian is of the form

$$L = p\dot{q} - h(p,q,y) \tag{34}$$

where y is a variable which does not possess its canonical momentum, by virtue of the fact that $\partial L/\partial\dot{y} = 0$. In such a case, y is called a dependent variable, and y can be expressed in terms of p and q when we invert the equation of motion for y:

$$0 = \frac{\partial L}{\partial y} = \frac{\partial h}{\partial y} \equiv F(p,q,y) \Rightarrow y = y(p,q) \ . \tag{35}$$

[We shall assume that $F(p,q,y) = 0$ can be uniquely solved for y.] Then the Lagrangian which must be subjected to quantization is

$$L = p\dot{q} - h(p,q,y(p,q)) \equiv p\dot{q} - H(p,q)$$

where

$$H(p,q) = h(p,q,y(p,q)) \; . \tag{36}$$

[What is done here is familiar in quantum electrodynamics.[10] In Coulomb gauge, for example, A_o is a dependent variable. One expresses A_o in terms of A^{tr} and E^{tr}, and write down the Hamiltonian in terms of these independent variables. That this is a correct prescription is seen by noting that the equations of motion for p and q are the same for the system defined by the Hamiltonian (36) as for the system defined by Eq. (34)].

Now suppose that h(p,q,y) is at most quadratic in y:

$$h(p,q,y) = H_o(p,q) + B^T y + \tfrac{1}{2} y^T A y \tag{37}$$

where A and B are functions of p and q in general. We have written Eq. (37) in a form that can be generalized immediately to cases where there are more than one dependent variable. Equation (35) gives

$$y = -A^{-1}B$$

and so

$$H(p,q) = H_o(p,q) - \tfrac{1}{2} B^T A^{-1} B \; . \tag{38}$$

The path integral (7) for this system is

$$\int \left[\frac{dp\,dq}{2\pi}\right] \exp i \int dt\,[p\dot{q} - H_o(p,q) - \tfrac{1}{2} B^T A^{-1} B]$$
$$\sim \int [dy\,dp\,dq]\,\sqrt{\det A}\,\exp i \int dt\,[p\dot{q} - h(p,q,y)] \tag{39}$$

as one can verify immediately by recalling the structure (37) of h(p.q,y) and performing the y-integrations explicitly.

Equation (39) will be used in quantizing gauge theories. If the factor $\sqrt{\det A}$ is constant, quantization becomes simple. We shall illustrate the use of Eq. (39) in a gauge in which this factor is indeed a constant. The gauge in point is the axial gauge, first studied by Arnowitt.[11]

To be concrete, let us deal with an O(3) invariant gauge theory, whose Lagrangian is

$$L = -\tfrac{1}{2}\,\mathbf{F}_{\mu\nu}\cdot(\partial^\mu \mathbf{A}^\nu - \partial^\nu \mathbf{A}^\mu + g\mathbf{A}^\mu \times \mathbf{A}^\nu) + \tfrac{1}{4}\,\mathbf{F}^{\mu\nu}\cdot\mathbf{F}_{\mu\nu} \tag{40}$$

where the bold face characters refer to isovector quantities. Since Eq. (40) possesses a gauge invariance of the second kind, we must choose a gauge condition, which in our case is

$$\mathbf{A}_3 = 0\,. \tag{41}$$

The Euler-Lagrange equations are

$$D^\mu \mathbf{F}_{\mu\nu} = 0, \qquad D^\mu = \partial^\mu + g\mathbf{A}^\mu \times$$

$$\mathbf{F}_{\mu\nu} = \partial_\mu \mathbf{A}_\nu - \partial_\nu \mathbf{A}_\mu + g\mathbf{A}_\mu \times \mathbf{A}_\nu\,.$$

From these we find that $\mathbf{A}_i$, $\mathbf{F}_{oi}$, $i = 1,2$, are q- and p-variables, and

$$\mathbf{A}_o,\ \mathbf{F}_{i3},\ \mathbf{F}_{3o},\ \mathbf{F}_{12} = -\mathbf{F}_{21}$$

and y-variables.

It is not difficult to verify that the Lagrangian (40) has the structure of Eq. (34) with h given by Eq. (37). In fact, if we denote by y the column vector

$$y^T = (\mathbf{F}_{3i},\ \mathbf{F}_{03},\ \mathbf{A}_o,\ \mathbf{F}_{12}),$$

then A has the form

$$A = \left[\begin{array}{c|c|c|c} \delta_{ij} & & & \\ \hline & -1 & \partial_3 & \\ \hline & -\partial_3 & 0 & \\ \hline & & & 1 \end{array}\right] \delta^4(x-y)$$

so that A is independent of p- and q- variables.

The path integral (39) is of the form

$$\int [dF_{\mu\nu}][dA_\lambda] \prod_{x} \delta(\underset{\sim}{A}_3(x))$$

$$\exp i \int d^4x \, [- \tfrac{1}{2} \underset{\sim}{F}_{\mu\nu} \cdot (\partial^\mu \underset{\sim}{A}^\nu - \partial^\nu \underset{\sim}{A}^\mu + g\underset{\sim}{A}^\mu \times \underset{\sim}{A}^\nu) + \tfrac{1}{4} \underset{\sim}{F}_{\mu\nu} \cdot \underset{\sim}{F}^{\mu\nu}]$$

or, after the $F_{\mu\nu}$ integrations

$$\sim \int [dA_\lambda] \prod_{x} \delta(\underset{\sim}{A}_3(x))$$

$$\exp i \int d^4x \, [- \tfrac{1}{4} (\partial_\mu \underset{\sim}{A}_\nu - \partial_\nu \underset{\sim}{A}_\mu + g\underset{\sim}{A}_\mu \times \underset{\sim}{A}_\nu)^2] \, .$$

Finally, I will leave it as an exercise problem for the reader to verify that in Coulomb and Landau gauges, i.e., $\nabla_i \underset{\sim}{A}_i = 0$, $\partial^\mu \underset{\sim}{A}_\mu = 0$, respectively, $\sqrt{\det A}$ is not constant, but in fact equals the Jacobian factor given by Faddeev and Popov[1] for the respective gauges.

REFERENCES

1. R. P. Feynman, Acta Phys. Polonica 26, 697 (1963).
   B. S. deWitt, Phys. Rev. 162, 1195, 1239 (1967).
   L. D. Faddeev and V. N. Popov, Phys. Letters 25B, 29 (1967).

2. P. A. M. Dirac, Physicalische Zeitschrift der Sowjetunion 3, 64 (1933).

3. R. P. Feynman, Revs. Mod. Phys. 20, 267 (1948); Phys. Rev. 80, 440 (1950);
   R. P. Feynman and A. R. Hibbs, Quantum Mechanics and Path Integrals, (McGraw-Hill, New York, 1965).

4. B. S. deWitt, Revs. Mod. Phys. 29, 377 (1957).

5. K. S. Cheng, J. of Math. Phys. 13, 1723 (1972).

6. J. Schwinger, Proc. Nat. Acad. Sci. 44, 956 (1958).
   K. Symanzik in Proceedings of the International School of Physics "Enrico Fermi", Corse XLV, R. Jost (ed). (Academic Press, N.Y., 1969).

7. See, for example, Bjorken and Drell, Relativistic Quantum Fields, (McGraw-Hill, N.Y., 1965), Chapter 17, especially Eq. (17.22).

8. See, for example, Bjorken and Drell, loc. cit., Section 17.4.

9. See L. D. Faddeev, Theor. and Math. Phys. 1, 3 (1969); English translation 1, 1 (1969), Consultants Bureau, N.Y.
I am grateful to Sidney Coleman for teaching me the method described in this section.

10. See, for example, Bjorken and Drell, loc. cit., Secs. 14.2 and 15.2.

11. R. Arnowitt and S. I. Fickler, Phys. Rev. 127, 1821 (1962).

RENORMALIZABLE MODELS WITH BROKEN SYMMETRIES

J.H. Lowenstein, New York University and Max-Planck-Institut für Physik, Munich

A. Rouet, CNRS Marseille

R. Stora, CERN Geneva and CNRS Marseille

W. Zimmermann, New York University and Max-Planck-Institut für Physik, Munich

TABLE OF CONTENTS

I. GENERAL REVIEW OF PERTURBATIVE RENORMALIZATION

There exist at present several versions [1] of renormalized perturbation theory, which develop in various directions some of the ideas contained in the work of N.N. Bogoliubov [2] and coworkers. The general set up, as recently described by H. Epstein and V. Glaser [1, 3], is as follows:

$\mathcal{F}$: Fock space of a family of free fields $\underline{\Phi}_0(x)$

$U(a,A)$: the corresponding Fock representation of the covering of the Poincaré group

Ω : the vacuum state

$\underline{\mathcal{L}}_0(x)$: the "interaction" Lagrangian, a family of Wick monomials in $\underline{\Phi}_0$ and its derivatives which, together with a given monomial, contains all its submonomials.

$\underline{g}(x)$ a corresponding family of smooth space time dependent coupling constants, with fast decrease at infinity.

One can construct by recursion an operator valued formal power series in $\underline{g}$:

$$S(\underline{g}) = \mathbb{1} + \sum_{n=1}^{\infty} \frac{(i)^n}{n!} \int dx_1 \ldots dx_n \, T(\underline{x}_1, \ldots \underline{x}_n) \, \underline{g}(x_1) \ldots \underline{g}(x_n)$$

whose coefficients $T(\underline{x}_1, \ldots \underline{x}_n) \overset{\mathrm{Def.}}{=} T(\underline{\mathcal{L}}_0(x_1) \ldots \underline{\mathcal{L}}_0(x_n))$ are properly defined "time ordered" products of the Poincaré covariant interaction Lagrangians, densely defined in $\mathcal{F}$ as operator valued distributions. The antitime ordered products are defined as coefficients of the formal power series for $S^{-1}(\underline{g})$:

$$S^{-1}(\underline{g}) = \mathbb{1} + \sum_{n=1}^{\infty} \frac{(-i)^n}{n!} \int dx_1 \ldots dx_n \, \overline{T}(\underline{x}_1, \ldots, \underline{x}_n) \, \underline{g}(x_1) \ldots \underline{g}(x_n)$$

The recursive construction can be so carried out that the following causal factorization property holds:

$$S(\underline{g}_1 + \underline{g}_2) = S(\underline{g}_1) \, S(\underline{g}_2)$$

$$\text{if } \operatorname{supp} \underline{g}_1 \gtrsim \operatorname{supp} \underline{g}_2$$

$$\text{i.e. } (\operatorname{supp} \underline{g}_1 + \overline{V}_+) \cap \operatorname{supp} \underline{g}_2 = \emptyset$$

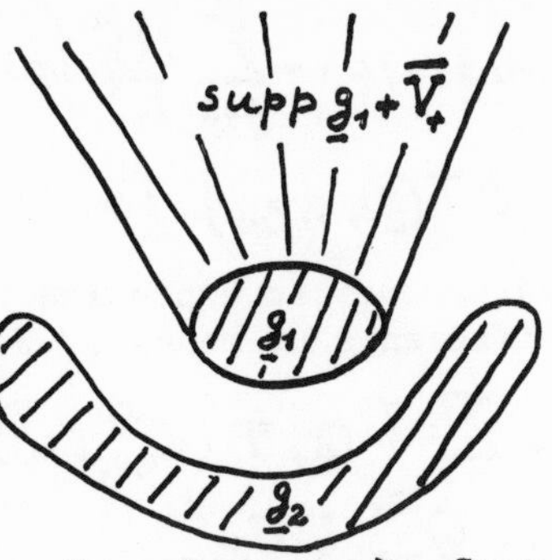

Fig. I Supp. $\underline{g}_1 \supsetneq$ Supp. $\underline{g}_2$

At each step of the recursive construction, there is an ambiguity whose minimality can be discussed via a precise theory of power counting. This ambiguity arises solely from the fact that causality and spectrum force the fields, and therefore their correlation functions, to be distributions. The problem to be solved concerns a certain "discontinuity" C_n which, in view of the causality assumption, has the union of two opposite closed cones $\Gamma_n^{\pm}$ as x-space support. One has to express C_n as the difference of a retarded and advanced distribution with respective supports Γ_n^+, Γ_n^-. This can be done, but not by mere multiplication by some Heaviside step function, and the solution is in general non-unique. At the nth step there arises an ambiguity in the definition of the operator valued distribution $T(x_1, \ldots, x_n)$ which is of the type

$$\Delta T(x_1, \ldots x_n) = \sum_i P_i(\partial)\, \delta(x_1 - x_2) \ldots \delta(x_{n-1} - x_n) : Q_{oi} :$$

where Q_{oi} is a Wick monomial in Φ_o and its derivatives, $P_i(\partial)$ a monomial of derivatives, and the sum ranges over all possible terms compatible with Wick's theorem such that

$$\deg. P_i + \omega(Q_{oi}) \geqslant \omega_n = \sum_{i=1}^{n} \left(\omega(\mathcal{L}_i) - 4\right) + 4 .$$

The index ω assigned to a Wick monomial adds up the number of derivatives and the number of fields each of which weighted by its naive dimension ($\frac{2j+2}{2}$ for a field carrying spin j if no better estimate holds due to the presence of an indefinite metric in $\mathcal{F}$).

Given two solutions, S and $\hat{S}$, characterized by coefficients $T(x_1, \ldots x_n)$, $\hat{T}(x_1 \ldots x_n) = T(x_1 \ldots x_n) + \Delta T(x_1 \ldots x_n)$, $\hat{S}$ is obtained by applying the prescription "T" to the interaction term

$$\int \hat{\mathcal{L}}_0(\underline{x},\underline{g})\,dx = \int \underline{\mathcal{L}}_0(x)\,\underline{g}(x)\,dx + \sum_{n=2}^{\infty} \frac{(i)^n}{n!} \int dx_1 \ldots dx_n \;\ldots$$

$$\ldots\; \Delta T(\underline{x}_1, \ldots \underline{x}_n)\; \underline{g}(x_1) \ldots \underline{g}(x_n)$$

Each operator time ordered product is expressed in terms of c-number distribution coefficients by means of Wick's theorem:

$$T\left(\mathcal{L}_{0,a_1}(x_1) \ldots \mathcal{L}_{0,a_n}(x_n)\right) = \sum \left(\Omega, T\left(\mathcal{L}^{r_1}_{0,a_1}(x_1), \ldots \mathcal{L}^{r_n}_{0,a_n}(x_n)\right)\Omega\right) \ldots$$

$$\ldots : Q^{r_1}_{0,a_1}(x_1) \ldots Q^{r_n}_{0,a_n} : C_{r_1, \ldots r_n} \;.$$

where $:\mathcal{L}^{r_i}_{0,a_i}\, Q^{r_i}_{0,a_i}: = \mathcal{L}_{0,a_i}$ and $C_{r_1 \ldots r_n}$ are some combinatorial factors. Thus, in p-space, the vacuum expectation values of time ordered products are ambiguous up to polynomials of degrees determined by power counting. If all components of $\underline{\phi}_0$ carry a non-vanishing mass, a "central" solution can be defined, which vanishes at zero momentum together with its ω first derivatives, ω being the power counting index. Non minimal solutions can be defined by assigning to some components of $\underline{\mathcal{L}}_0$ power counting indices larger than the naive index previously defined.

Physics usually requires to define an adiabatic limit which we now describe. Let us separate the components of $\underline{g}$ into two groups

$$\underline{\mathcal{L}}_0 \cdot \underline{g} = \underline{\mathcal{O}}_0 \cdot \underline{J} + \underline{\mathcal{L}}_0^{int} \cdot \underline{\lambda}$$

and let us inquire in what sense we can let the coupling constants $\underline{\lambda}$, whereas the space time varying source functions $\underline{J}$ will be used to generate fields and composite operators, such as currents, etc. The first term of $\underline{\mathcal{O}}$ as power series in $\underline{\lambda}$ is given by $\underline{\mathcal{O}}_0$. One can see that as $\underline{\lambda}(x) \to \underline{\lambda}$ the expression

$$\left(\Omega, S(J,\underline{\lambda})\Omega\right) / \left(\Omega, S(0,\underline{\lambda})\Omega\right)$$

has a limit whose coefficients of powers of $\underline{J}$ are the Green's functions of the interacting operators. It is, however, only when suitable normalization conditions are satisfied - whereby the vacuum to vacuum transition is so normalized that $(\Omega, S(0,\underline{\lambda})\Omega) \to 0$ as $\underline{\lambda}(x) \to \underline{\lambda}$, and the two point function is so defined that the one particle pole is fixed at its Fock space value[3] - that $S(J,\underline{\lambda})$ has a limit in the operator sense as $\underline{\lambda}(x) \to \underline{\lambda}$. In many cases one does not want to do this, but rather constructs a theory of Green's functions which, after the adiabatic limit is taken, leads to an interpretation in a Fock space $\mathcal{F}_{phys.}$ different from the one given initially. This point of view is often to be taken in the description of systems with a broken symmetry as we shall see later.

The connection with conventional methods[2] which use regularization procedures is the following: Let ϕ_0^{reg} be a regularized free field defined in a larger Fock space $\mathcal{F}^{reg.} \supset \mathcal{F}$ and $\mathcal{L}_0^{reg.}(\phi_0^{reg})$ a regularized form of the interaction Lagrangian. Under certain conditions (Pauli Villars regularization : $\mathcal{L}_0^{reg}(\phi_0^{reg}) = \mathcal{L}_0(\phi_0^{reg})$), it is possible to show that the central solution of the regularized theory goes over to the central solution of the non regularized theory in the sense of distributions, as the regularization is removed. It would be worthwhile to complete such a proof in more general cases, including that of the n-dimension regularization[4].

A particular solution of the regularized theory can easily be constructed in terms of Feynman graphs by mere multiplication of regularized Feynman propagators, and the corresponding central solution is uniquely determined by direct calculation of the regulator dependent Taylor expansions around p = 0 which can be gathered together into a redefinition of the interaction Lagrangian modulo regulator dependent ("infinite") counter terms.

Of particular interest are renormalizable theories for which $\omega(\mathcal{L}_0^{int}) \leq 4$ in which case possible counter terms occur in finite number, with coefficients formal power series in the coupling constants $\underline{\lambda}$. In all the following, we shall limit ourselves to the study of renormalizable models. In particular, the next chapter will be devoted to a detailed description and comparison of the various, possibly non central as well as non minimal, solutions, in the adiabatic limit. The combinatorial identities which will be established will be the main tools in the study of the models treated in the later chapters.

CHAPTER I: REFERENCES AND FOOTNOTES

This list of references is by no means exhaustive, but contains reviews where extensive reference to originals can be found.

[1] K. Hepp, Theorie de la Renormalisation. Lecture Notes in Physics Vol. 2. Springer Verlag New York 1969.
K. Hepp, in "Statistical Mechanics and Quantum Field Theory", Les Houches 1970, Gordon Breach New York (1971).
H. Epstein, V. Glaser, same volume.
W. Zimmermann, in "Lectures on Elementary Particles and Quantum Field Theory", Brandeis (1970) Vol. I. MIT Press, Cambridge Mass, (1970)
E.R. Speer: Feynman Amplitudes, Princeton University Press (1965).

[2] N.N. Bogoliubov, D.V. Shikov, "Introduction to the theory of quantized Fields", Interscience Pub. New York (1960).

[3] H. Epstein, V. Glaser: The role of locality in Perturbation Theory, CERN TH 1400, 16 September 1971, to appear in Ann. Institut Poincaré.
H. Epstein, V. Glaser, "Adiabatic Limit in Perturbation Theory" CERN TH 1344, 10 June 1971, in Meeting on Renormalization Theory, CNRS Marseille (1971).

[4] For a description of the n-dimensional regularization see, for instance, the forthcoming CERN report by G.'t Hooft, M. Veltman, also E.R. Speer, to appear.

II. DYNAMICS OF RENORMALIZABLE MODELS IN PERTURBATION THEORY

Within the framework of perturbation theory we will discuss the dynamical properties of renormalizable models based on the classical Lagrangian

$$\mathcal{L}_{cl.} = \mathcal{L}_{cl.o} + \mathcal{L}_{cl.}^{int} \quad . \tag{II.1}$$

$\mathcal{L}_{cl.o}$ denotes the free Lagrangian. The interaction part $\mathcal{L}_{cl.}^{int}$ is written as

$$\mathcal{L}_{cl.}^{int} = \sum_{j=1}^{J} \lambda_j \mathcal{L}_{cl.j} \tag{II.2}$$

where the coupling terms $\mathcal{L}_{cl.j}$ are monomials in the field components and their first derivatives. The Lagrangian is renormalizable if the naive dimension ω_j of each coupling term $\mathcal{L}_{cl.j}$ satisfies

$$\omega_j \leq 4$$

The components of all fields will be collected by a single field vector

$$\underline{\phi} = (\phi_1, \cdots, \phi_A)$$

The construction of field operators and Green's functions by the method of Epstein and Glaser (I.1) was described in the previous Chapter. The advantage of this method is that with mathematical rigour the field operators are constructed recursively in perturbation theory such as to satisfy the fundamental principles of quantum field theory. For practical reasons, however, other methods of renormalization are more convenient which work in the adiabatic limit right from the beginning. In the work that follows we will use such an alternative method which is based on a renormalized version of the Gell-Mann Low formula[1]. First, we assign a degree δ_j satisfying

$$\omega_j \leq \delta_j \leq 4$$

to each coupling term $\mathcal{L}_{cl.j}$. This assignment will determine the number of subtractions to be used for separating the finite part of a Feynman integral. The time ordered Green's functions are constructed by

$$\begin{aligned} &\langle T \phi_{\alpha_1}(x_1), \dots \phi_{\alpha_m}(x_m) \rangle = \\ &\langle T \, e^{i \int \mathcal{L}_{eff.o}^{int}(z) dz} \phi_{\alpha_1,o}(x_1) \cdots \phi_{\alpha_m,o}(x_m) \rangle^{norm.} \end{aligned} \tag{II.3}$$

The superscript "norm" indicates that vacuum diagrams (disconnected closed loops) should be omitted. $\mathcal{L}^{int}_{eff,o}$ denotes the effective interaction part of the Lagrangian and is given by

$$\mathcal{L}^{int}_{eff,o} = \sum \lambda_j \, N_{\delta_j} [\mathcal{L}_{cl.jo}] \tag{II.4}$$

The subscript $^{(o)}$ indicates that the free field propagators pertaining to $\mathcal{L}_{cl.o}$ are used to perform contractions. Time ordered functions involving the symbol N are defined as follows. The expansion of (II. 3) with respect to powers of λ_j leads to expressions of the form

$$\begin{aligned} &\langle T \, \phi_{\alpha_1,o}(x_1) \ldots \phi_{\alpha_m,o}(x_m) \\ &\qquad N_{\rho_1}[M_{1,o}(u_1)] \ldots N_{\rho_n}[M_{n,o}(u_n)] \rangle \end{aligned} \tag{II.5}$$

where the M_k may be any of the coupling terms $\mathcal{L}_{clj}$. We may as well consider the more general case of arbitrary non-linear monomials $M_{j,o}$ in the free field components and their derivatives. The integers ρ_j are restricted by

$$\rho_j \geq d_j$$

where d_j is the naive dimension of the monomial $M_{j,o}$. ρ_j is called the degree assigned to the monomial $M_{j,o}$. (II.5) represents a time ordered Green's function of the Wick products

$$: M_{k,o} : (u_k)$$

While the Wightman functions of Wick products (II. 5) are unique, the time ordered functions are only defined up to contact terms. The symbols $N_{\rho_1} \ldots N_{\rho_n}$ serve to specify a unique choice of time ordered functions, yet to be defined. As has been emphasized by Bogoliubov the renormalization of the Gell-Mann Low formula rests upon a proper definition of the time ordered functions of free fields and their Wick products.

Without going into details we give a rough sketch of the definition which is rather involved for arbitrary diagrams due to the phenomenon of overlapping divergencies. One first expands the formal expression

$$\langle T \, \phi_{\alpha_1,o}(x_1) \ldots \phi_{\alpha_m,o}(x_m) \, M_{1,o}(u_1) \ldots M_{n,o}(u_n) \rangle \tag{II.6}$$

with respect to Feynman diagrams using Wick's rules. Each diagram has m external lines labeled by $x_1, \ldots, x_m$ and n vertices at $u_1, \ldots, u_n$. To each diagram Γ belongs a Feynman integral which in general diverges. We amputate the external lines and denote the unrenormalized Feynman integral in momentum space by

$$\lim_{\varepsilon \to +0} \int dk_1 \ldots dk_t \; I_\Gamma \,(k_1, \ldots k_t; p_1, \ldots p_m; q_1, \ldots q_n) \tag{II.7}$$

The external momenta p_a, q_b correspond to the coordinates x_a, u_b. The momenta $k_1, \ldots, k_t$ form a set of independent integration variables. The unrenormalized integrand I_Γ is given by the usual Feynman rules pertaining to the Lagrangian (II. 1). The time ordered function (II.6) is defined by the finite part of (II.7) summed over all possible diagrams. The finite part will be formed by taking subtractions of the integrand I_Γ. To this end we introduce the degree

$$\delta(\Gamma) = 4 - B - \frac{3}{2} F + \sum_{j=1}^{n} (\delta_j - 4) \tag{II.8}$$

of proper diagrams Γ. B is the number of external boson lines, F the number of external fermion lines. We recall that δ_j is the degree assigned to the vertex at u_j. If the degrees δ_j are chosen to be the naive dimensions d_j

$$\delta_j = d_j \tag{II.9}$$

$\delta(\Gamma)$ is just the superficial divergence d (Γ) of the diagram

$$d(\Gamma) = 4 - B - \frac{3}{2} F + \sum_{j=1}^{m} (d_j - 4) \tag{II.10}$$

One has always

$$\delta(\Gamma) \geq d(\Gamma)$$

A proper diagram is called primitive if all proper subdiagrams have negative degree. For a primitive diagram the finite part is given simply by

$$\lim_{\varepsilon \to +0} \int dk_1 \ldots dk_t \left(1 - t_{p,q}^{\delta(\Gamma)}\right) \cdot I_\Gamma \,(k_1, \ldots k_t; p_1, \ldots p_m; q_1, \ldots q_n) \tag{II.11}$$

Here $t_{p,q}^{\delta(\Gamma)}$ denotes the Taylor operator in p, q up to order $\delta(\Gamma)$. If $\delta(\Gamma) < 0$, we set $t_{p,q}^{\delta(\Gamma)} = 0$. If the degrees equal the dimensions (see (II.9-10)) we precisely have Dyson's prescription for separating the finite part. In the general case oversubtractions are made which would not be required to render the integral finite. For arbitrary proper diagrams overlapping divergencies may occur which can be removed by applying Bogoliubov's combinatorial method (I.2). The finite part then is of the form

$$\text{(II.12)} \quad \lim_{\varepsilon \to +0} \int dk_1 \dots dk_t \left(1 - t_{p,q}^{\delta(\Gamma)}\right)(\bullet\bullet\bullet) \cdot \cdot I_\Gamma(k_1, \dots k_t; p_1, \dots p_m; q_1, \dots q_n) \cdot$$

where (...) indicated subtractions for subdiagrams. For details we refer to ref. [1] . This method of separating the finite part from a Feynman integral is thus an extension of the original work of Dyson and Salam. It can further be shown that the finite parts (II.12) equal those derived by Bogoliubov and Hepp with a somewhat different method[3] . Their results in turn are contained in the general class of solutions which was obtained by Epstein and Glaser in the adiabatic limit [I.3] .

As physical parameters of the theory we distinguish mass constants and coupling constants. Mass constants occur as coefficients of the free Lagrangian and are related to the discrete spectrum of the mass operator $P_\mu P^\mu$. The coupling constants may suitably be defined through various Green's functions. Let m_1, ..., m_A and g_1, ... , g_B be independent sets of mass or coupling constants which completely characterize the physics of the model. A corresponding set of renormalization conditions is then imposed on the Green's functions.

The renormalization conditions recursively determine $\lambda_1, \dots, \lambda_J$ as power series in g_1, ... g_B with finite, mass dependent coefficients. If all parameters λ_j vanish in zero order the expansion of (II.3) with respect to Feynman diagrams provides the power series expansion in g_j. However, if some λ_j do not vanish in zero order a series has to be summed for obtaining a finite order of perturbation theory. This difficulty is characteristic for some models with broken symmetries. A generalization of the present treatment will be proposed below in this Chapter which allows to avoid the summation problem.

We next turn to the definition of composite operators[1]. Let M be a non-linear monomial in the field components and their derivatives. We will construct a sequence

$$\text{(II.13)} \quad N_a[M(y)] \,, \quad a = d, d+1, d+2, \dots$$

of composite operators which are associated with this monomial. d is the naive dimension of M. Green's functions involving composite operators are defined by the following extension of (II.3)

(II.14) $$\langle T\, Z \rangle = \langle T\, e^{i\int \mathcal{L}^{int}_{eff,o}(z)\,dz}\, Z_o \rangle^{norm.}$$

with the abbreviation

(II.15) $$Z = \phi_{a_1}(x_1)\cdots\phi_{a_m}(x_m)\, N_{a_1}[M_1(y_1)]\cdots N_{a_n}[M_n(y_n)]$$

We now formulate a generalization of the formalism which is useful for models with broken symmetries[4]. In particular, it provides a simple remedy for the summation problem of such models. This will be demonstrated in Chapter IV on the Goldstone and the Higgs models. In this generalization we let the classical Lagrangian be a polynomial in a new parameter s

(II.16) $$\mathcal{L}_{cl}(s) = \mathcal{J}_4 + s\,\mathcal{J}_3 + s^2\,\mathcal{J}_2 + s^3\,\mathcal{J}_1$$

The naive dimensions d of the monomials occurring in $\mathcal{J}_\alpha$ are required to be less than or equal to

$$d \leq \alpha$$

Accordingly $\mathcal{J}_1$ is linear in the fields. The parameter s is allowed to vary within

(II.7) $$0 \leq s \leq 1$$

While s = 1 is the physically relevant case the dependence on s will play an important part in the subtraction procedure.

Before writing down the analogue of (II.14) we have to clarify the definition of the free Lagrangian which becomes less trivial in this generalization. The physical mass and coupling constants of the theory refer to the case s = 1. Accordingly, renormalization conditions are also imposed at s = 1. As usual, the free Lagrangian $\mathcal{L}_{cl,o}$ at s = 1 is taken to be the Lagrangian of the incoming and outgoing fields, apart from possible normalization factors. For general s we define the free Lagrangian by the sum of all bilinear terms of $\mathcal{L}_{cl}(s)$ with the coefficients replaced by their zero order values. The dependence of these coefficients on the coupling constants will be discussed shortly. For the moment we only note that the coefficients of $\mathcal{L}_{cl,o}(s)$ are quadratic functions of s, due to (II.16).

The Green's functions of the theory are constructed by

$$\langle T Z \rangle = \langle T_s \, e^{i\int N_4[\mathcal{L}^{int}_{d,o}(s,z)]dz} \, Z_o \rangle^{norm.} \tag{II.18}$$

with

$$\mathcal{L}^{int}_{d}(s) = \mathcal{L}_{d}(s) - \mathcal{L}_{d,o}(s) = \sum \lambda_j \mathcal{L}_{d,j}(s) \tag{II.19}$$

and

$$Z = \phi_{\alpha_1}(x_1) \cdots \phi_{\alpha_m}(x_m) \, N_{a_1}[O_1(y_1)] \cdots N_{a_n}[O_n(y_n)] \tag{II.20}$$

The formula defines the time ordered Green's functions of the fields, as well as Green's functions of composite operators. The symbol T_s indicates a special time ordering which we are going to define now. The expansion of (II.18) leads to expressions of the form

$$\langle T_s \, \phi_{\alpha_1}(x_1) \cdots \phi_{\alpha_m}(x_m) \, N_{\rho_1}[M_1(u_1)] \cdots N_{\rho_r}[M_r(u_r)] \rangle \tag{II.21}$$

In defining (II.21) we proceed as before but replace (II.12) by

$$\lim_{\varepsilon \to +0} \int dk_1 \ldots dk_t \, (1 - t^{\delta(\Gamma)}_{p,q,s})(\cdots) \cdot \tag{II.22}$$

$$\cdot \, I_\Gamma(s; k_1, \ldots k_t; p_1, \ldots p_m; q_1, \ldots q_r)$$

The new feature is that now subtractions are made in p_k, q_l and $\underline{s}$ simultaneously. In other words, $\underline{s}$ is treated like an external momentum, as far as subtractions indicated by (...) are modified similarly[5]. Finally, the renormalization conditions imposed at s = 1 determine the parameters λ_j as power series in g_j with mass dependent coefficients. Let $m_1, \ldots, m_A, g_1, \ldots, g_B$ be complete sets of independent mass and coupling constants at s = 1 [6]. Through the renormalization conditions the parameters $\lambda_1, \ldots \lambda_J$, become power series in the g_j with finite coefficients which depend on the mass constants[7]. Without proof we will now state some dynamical properties of Green's functions and field operators which hold to any order in perturbation theory. The dynamics is governed by the effective Lagrangian

$$\mathcal{L}_{eff}(s) = N_4(\mathcal{J}_4) + s\, N_3[\mathcal{J}_3] + s^2 N_2(\mathcal{J}_2) + s^3 \mathcal{J}_1 \,. \tag{II.23}$$

The principal dynamical laws can be stated as follows.

I. Action principle [2, 8]

$$\frac{\partial}{\partial \xi} \langle T \prod_j \phi_{\nu_j}(x_j) \rangle = i \langle T \int \frac{\partial \mathcal{L}_{eff}(s,z)}{\partial \xi} dz \prod_j \phi_{\nu_j}(x_j) \rangle \tag{II.24}$$

ξ is an arbitrary parameter. In $\partial \mathcal{L}_{eff} / \partial \xi$ the differential operator acts only on the coefficients of the field monomials.

II. Space-time differentiation of normal products[2].

$$\partial_\mu N_\delta [M] = N_{\delta+1} [\partial_\mu M] \tag{II.25}$$

$$\partial_\mu^z \langle T N_\delta [M(z)] \prod_j \phi_{\nu_j}(x_j) \rangle = \langle T N_{\delta+1} [\partial_\mu^z M(z)] \prod_j \phi_{\nu_j}(x_j) \rangle \tag{II.26}$$

III. Linear relations among normal products[1, 2].

If $\varphi < \delta$, a given normal product of degree φ and its Green's functions may be written as linear combinations

$$N_\varphi [M] = \sum a_i N_\delta [M_i] \tag{II.27}$$

$$\langle T N_\varphi [M(z)] \prod_j \phi_{\nu_j}(x_j) \rangle = \sum_i a_i \langle T N_\delta [M_i(z) \prod_j \phi_{\nu_j}(x_j) \rangle \tag{II.28}$$

where the sum extends over all monomials M_i of dimensions $d \leq \delta$ with appropriate quantum numbers and invariance properties.

IV. Equations of motion

$$\frac{\delta \mathcal{L}_{eff}}{\delta \phi_k} - \partial^\mu \frac{\delta \mathcal{L}_{eff}}{\delta \partial^\mu \phi_k} = 0 \tag{II.29}$$

(II.30)
$$\langle T\left[\frac{\delta\mathcal{L}_{eff}}{\delta\phi_k}(z)-\partial^\mu\frac{\delta\mathcal{L}_{eff}}{\delta\partial^\mu\phi_k}(z)\right]\prod_j\phi_{v_j}(x_j)\rangle = i\sum_j\delta_{k,j}\,\delta(z-x_j)\langle T\prod_{j\neq k}\phi_{v_j}(x_j)\rangle$$

In $\delta\mathcal{L}_{eff}/\delta\phi_k$, $\delta\mathcal{L}_{eff}/\delta\partial^\mu\phi_k$ the action of the differential operators on normal products is defined by

(II.31)
$$\frac{\delta}{\delta\phi_k}N_a[M] = N_{a-d_k}\left[\frac{\delta M}{\delta\phi_k}\right]$$
$$\frac{\delta}{\delta\partial^\mu\phi_k}N_a[M] = N_{a-d_k-1}\left[\frac{\delta M}{\delta\partial^\mu\phi_k}\right]$$

V. Equations of motion multiplied by a field component[2]

(II.32)
$$\phi_\ell\left[\frac{\delta\mathcal{L}_{eff}}{\delta\phi_k}-\partial^\mu\frac{\delta\mathcal{L}_{eff}}{\delta\partial^\mu\phi_k}\right]=0$$

(II.33)
$$\langle T\,\phi_\ell\left[\frac{\delta\mathcal{L}_{eff}}{\delta\phi_k}-\partial^\mu\frac{\delta\mathcal{L}_{eff}}{\delta\partial^\mu\phi_k}\right](z),\prod_j\phi_{v_j}(x_j)\rangle = i\sum_j\delta_{kj}\,\delta(z-x_j)\langle T\,\phi_\ell(x_j)\prod_{j\neq k}\phi_{v_j}(x_j)\rangle$$

Here multiplication of a normal product by a field component at the same point is defined by

(II.34)
$$\phi_\ell(z)\,N_a[M(z)] = N_{a+d_\ell}[\phi_\ell(z)M(z)]$$

VI. Equations of motion multiplied by a field monomial[9]

Let Q(x) be a monomial of dimension d in the fields and their derivatives. Then

(II.35)
$$\{Q\}\cdot\left[\frac{\delta\mathcal{L}_{eff}}{\delta\phi_k}-\partial^\mu\frac{\delta\mathcal{L}_{eff}}{\delta\partial^\mu\phi_k}\right]=0$$

(II.36)
$$\langle T\{Q\}\cdot\left[\frac{\delta \mathcal{L}_{eff}}{\delta \phi_k} - \partial^\mu \frac{\delta \mathcal{L}_{eff}}{\delta \partial^\mu \phi_k}\right](z) \prod_j \phi_{v_j}(x_j)\rangle =$$
$$i \sum_j \delta_{kj}\, \delta(z - x_j) \langle T\, N_d [Q(x_j) \prod_{l \neq j} \phi_{v_l}(x_l)\rangle$$

$\{Q\}$ acting on a normal product of a monomial M denotes the anisotropic normal prodct

(II.37)
$$\{Q\}\cdot N_a[M] = N_{a+d}[\{Q\}M]$$

which is defined in the following way. Let dim M denote the dimension of M. The Green's functions of (II.37) are again constructed by (II.22) with the subtraction degree $\delta(\gamma)$ assigned to each proper subgraph γ given by (II.10). However, the degree δ assigned to the normal product vertex is defined in an anisotropic manner. δ is equal to its minimal value, d + d(M), if γ has no internal lines arising from contractions with fields in M, and is equal to d + a otherwise. It should be noted that $N_{a+d}[\{Q\}M]$ can always be expanded in terms of isotropic normal products of degree a + d using relations similar to those of III.

Finally we discuss the connection between formulations involving the parameter $\underline{s}$ and the conventional treatment given in the beginning of this chapter. We start with the simplest case that the free Lagrangian of the $\underline{s}$-formulation is independent of $\underline{s}$. Then the rules for constructing Feynman integrals involve $\underline{s}$ only in form of powers assigned to some vertices of the diagram. $\underline{s}$ never occurs in the masses of the Feynman denominators. Analyzing the subtraction rules one finds that the model based on the s-dependent Lagrangian

(II.38)
$$\mathcal{L}_{cl}(s) = \mathcal{L}_{cl,o}(s) + \mathcal{L}_{cl}^{int}(s)$$
$$\mathcal{L}_{cl,o}(s) \text{ independent of } s$$
$$\mathcal{L}_{cl.}^{int}(s) = \mathcal{G}_4 + s\,\mathcal{G}_3 + s^2 \mathcal{G}_2 + s^3 \mathcal{G}_1$$

is equivalent to the model based on the conventional interaction Lagrangian

(II.39)
$$\mathcal{L}_{eff}^{int} = N_4[\mathcal{G}_4] + N_3[\mathcal{G}_3] + N_2[\mathcal{G}_2] + \mathcal{G}_1$$

Stated more precisely, the Green's functions evaluated from (II.18) with N_4 applied to the interaction part (II.38) are at s = 1 identical to the Green's functions defined through (II.14) with interaction part (II.39). Thus the generalization is not expected to give any new information for such cases.

If the unperturbed mass parameters depend on s the situation is quite different. Let $\mathcal{L}_{cl.,o}(s)$ and $\mathcal{L}_{cl}^{int}(s)$ be of the form

$$\mathcal{L}_{cl.,o}(s) = \mathcal{F}_4 + s\mathcal{F}_3 + s^2\mathcal{F}_2 \qquad \text{(II.40)}$$

$$\mathcal{L}_{cl.}^{int}(s) = \mathcal{G}_4 + s\mathcal{G}_3 + s^2\mathcal{G}_2 + s^3\mathcal{G}_1 \qquad \text{(II.41)}$$

with the coefficients of the monomials in $\mathcal{F}_\alpha$ independent of s and the coupling constants g_j. For the coefficients of the $\mathcal{G}_\alpha$ it is assumed that they vanish in zero order of g_j. Again there is a formal equivalence between the s-dependent Lagrangian

$$\begin{aligned} \mathcal{L}_{cl}(s) &= \mathcal{L}_{cl.,o}(s) + \mathcal{L}_{cl.}^{int}(s) \\ &= \mathcal{J}_4 + s\mathcal{J}_3 + s^2\mathcal{J}_2 + s^3\mathcal{J}_1 \end{aligned} \qquad \text{(II.42)}$$

and the conventional Lagrangian

$$\mathcal{L}_{eff} = N_4[\mathcal{J}_4] + N_3[\mathcal{J}_3] + N_2[\mathcal{J}_2] + \mathcal{J}_1 \qquad \text{(II.43)}$$

But it turns out that the appropriate form of the free Lagrangian is

$$\mathcal{L}_{eff,o} = N_4\left[\mathcal{F}_4 + \mathcal{F}_3 + \mathcal{F}_2\right] \qquad \text{(II.44)}$$

so that the interaction part becomes

$$\begin{aligned} \mathcal{L}_{eff.}^{int} = {} & N_4[\mathcal{G}_4] + N_3[\mathcal{G}_3] + N_2[\mathcal{G}_2] + \mathcal{G}_1 \\ & + N_3[\mathcal{F}_3] - N_4[\mathcal{F}_3] \\ & + N_2[\mathcal{F}_2] - N_4[\mathcal{F}_2] \end{aligned} \qquad \text{(II.45)}$$

The coefficients of $\mathcal{F}_2$ and $\mathcal{F}_3$ are independent of the coupling constants g_j. As was discussed above this leads to a summation problem since an infinite number of Feynman integrals contributes to a finite order in g_j. Since $\mathcal{F}_3$ and $\mathcal{F}_4$ are bilinear in the fields one deals with geometric series in momentum space which diverge in parts of the integration domains. Thus the formalism based on the Lagrangian (II.44 - 45) is not satisfactory. With the original s-formulation, however, each order in the g_j is represented by a finite number of Feynman integrals.

Despite of this it is always possible to construct a Lagrangian of the conventional type which is equivalent to (II.42), but in a non-trivial way. The disadvantage of this Lagrangian is that symmetries which the s-dependent Lagrangian may display could be completely distorted for the equivalent conventional Lagrangian. This happens to be the case for the Goldstone and the Higgs model.

We briefly state and prove this equivalence theorem [10].

EQUIVALENCE THEOREM

Let

$$\begin{aligned} \mathcal{L}'_{cl}(s) &= \mathcal{L}'_{cl,0}(s) + \mathcal{L}'^{\,int}_{cl}(s) \\ &= \mathcal{J}'_4 + (s-1)\,\mathcal{J}'_3 + (s^2-1)\,\mathcal{J}'_2 + (s^3-1)\,\mathcal{J}'_1 \end{aligned} \tag{II.46}$$

$\mathcal{J}'_\alpha$ are polynomials of their fields and their first derivatives with s-independent coefficients, the naive dimensions of the monomials in $\mathcal{J}'_4$ be less than or equal to four, the dimensions of monomials in $\mathcal{J}'_{1,2,3}$ be less than or equal to **1,2,3**. Let mass and coupling constants be defined through renormalization conditions on the time ordered Green's functions of fields. Then there exists a Lagrangian

$$\begin{aligned} \mathcal{L}''_{cl} &= \mathcal{L}''_{cl.0} + \mathcal{L}''^{\,int}_{cl.} \\ \mathcal{L}''^{\,int}_{eff} &= N_4\left[\mathcal{L}''^{\,int}_{cl.}\right] \end{aligned} \tag{II.47}$$

with the same renormalization conditions such that the models based on (II.46) and (II.47) are identical. More precisely, the time ordered functions of the fields are identical whether constructed by (II.18) with the Lagrangian (II.46) or by (II.3) with (II.47). If the coefficients of the interaction part of (II.46) vanish in zero order of the coupling constants the same is true for (II.47). Hence no summation problem arises with (II.47).

Remark: Green's functions involving normal products may change. This does not impair the equivalence of the models but only indicates that similarly constructed normal products must not be identified.

Proof: In order to eliminate the s-dependent terms we will construct a family of Lagrangians

$$\mathcal{L}_{cl}(s) = \mathcal{J}_4 + (s-1)\mathcal{J}_3 + (s^2-1)\mathcal{J}_2 + (s^3-1)\mathcal{J}_1 \tag{II.48}$$

which includes the Lagrangian (II. 46) with

$$\mathcal{J}_4 = \mathcal{J}_4' = \mathcal{L}_{cl}(1) \quad , \quad \mathcal{J}_\alpha = \mathcal{J}_\alpha' \quad \alpha = 1,2,3 .$$

and a Lagrangian

$$\mathcal{L}_{cl}''(s) = \mathcal{J}_4'' \quad , \quad \mathcal{J}_1'' = \mathcal{J}_2'' = \mathcal{J}_3'' = 0$$

with no s-dependent terms. For this family it will be shown that the Green's functions of fields do not change.

Let

$$M_{d1}, \ldots M_{d\,A(d)} \tag{II.49}$$

be all monomials of naive dimension d which can be formed out of the field components and their derivatives. We express $\mathcal{J}_4', \mathcal{J}_\alpha'$ as linear combinations of these monomials

$$\mathcal{J}_4' = \sum_{d=1}^{4} \sum_{j=1}^{A(d)} c_{dj}' M_{dj}$$

$$\mathcal{J}_\alpha' = \sum_{d=1}^{\alpha} \sum_{j=1}^{A(d)} t_{\alpha dj}' M_{dj} \quad \alpha = 1,2,3$$

With this notation the original Lagrangian becomes

$$\mathcal{L}_{cl}'(s) = \sum_{d=1}^{4} \sum_{j=1}^{A(d)} c_{dj}' M_{dj} + \sum_{\alpha=1}^{3} (s^{4-\alpha}-1) \sum_{d=1}^{\alpha} \sum_{j=1}^{A(d)} t_{\alpha dj}' M_{dj}$$

We now study the class

$$\mathcal{L}_{cl.}(s) = \sum_{d=1}^{4} \sum_{j=1}^{A(d)} c_{dj} M_{dj} + \sum_{\alpha=1}^{3} (s^{4-\alpha}-1) \sum_{d=1}^{\alpha} \sum_{j=1}^{A(d)} t_{\alpha dj} M_{dj} \tag{II.50}$$

of Lagrangians where the c_{dj} are functions of arbitrary parameters $t_{\alpha dj}$. Our intention is to place suitable restrictions on the functions c_{dj} such that at s = 1 the Green's functions remain independent of $t_{\alpha dj}$. To this end we form the derivative

$$\frac{\partial \langle T X \rangle}{\partial t_{\beta e k}} = i \sum_{d=1}^{a} \sum_{j=1}^{A(d)} \frac{\partial c_{dj}}{\partial t_{\beta e k}} \int dz \langle T N_4 [M_{dj}(z) X \rangle$$
$$+ i \int dz \langle T(N_\beta [M_{ek}(z)] - N_4 [M_{ek}(z)]) X \rangle$$

(II.28) then implies that $\langle T N_\beta [M_{ek}] X \rangle$ is a linear combination of Green's functions involving N_4-products

$$\langle T (N_\beta [M_{ek}(z)] - N_4 [M_{ek}(z)]) X \rangle =$$
$$= - \sum r_{\beta e k d j} \langle T N_4 [M_{dj}(z)] X \rangle$$

Thus

$$\frac{\partial \langle T X \rangle}{\partial t_{\beta e k}} = 0$$

holds provided

$$\frac{\partial c_{dj}}{\partial t_{\beta e k}} = r_{\beta e k d j}$$

The $r_{\beta e k d j}$ depend on $t_{\beta d k}$ directly and through the functionals c_{dj}. A more detailed discussion shows that a solution exists with the initial values

$$c_{dj} = c'_{dj} \quad \text{at} \quad t_{\beta d j} = t'_{\beta d j}$$

With these solutions as coefficients c_{dj} the family (II.50) of Lagrangians has all desired properties. We finally set $t_{\beta dj} = 0$ and find

$$\mathcal{L}''_{cl} = \sum_{j=1}^{A(d)} c''_{dj} M_{dj}$$

where c''_{dj} denotes the values at $t_{\beta dj} = 0$. Since the Green's functions of the basic fields do not change the same normalization conditions are satisfied throughout. In zero order of the coupling constants the coefficients $r_{\beta e k d j}$ vanish. Hence the zero order values of the c_{dj} do not change. Accordingly, the free part of the Lagrangian (II.50) remains the same at s = 1 and the coefficients of the interaction part $\mathcal{L}''_{eff}$ vanish in zero order. This completes the proof.

CHAPTER II: REFERENCES AND FOOTNOTES

[1] W. Zimmermann, Commun. math. Phys. 15, 208 (1969, and in "Lectures on Elementary Particles and Quantum Field Theory", Brandeis University, Vol. I, MIT Press, Cambridge Mass. (1970), Annals of Physics, to be published.

[2] J.H. Lowenstein, Phys. Rev. D4, 2281 (1971), and in "Seminars on Renormalization Theory", Vol. II, Maryland, Technical Report 73-068 (1972).

[3] K. Hepp, Commun, Math. Phys. 6, 161 (1967).

[4] J.H Lowenstein, M. Weinstein and W. Zimmermann, to be publ.

[5] The convergence of the finite part (II.22) will be proved in a paper by M. Gomes, J.H Lowenstein and W. Zimmermann, in preparation.

[6] It is assumed there that the number of mass parameters is not larger than the number of free parameters of the Lagrangian. For a discussion of this point see Chapter III, page 26

[7] In the models studied in Chapter IV one of the parameters λ_j is not determined by renormalization conditions, but directly given as constant or power series in g_j. It can be shown for the cases considered that the Green's functions do not depend on the value of this parameter .

[8] In unrenormalized form the action principle of quantum field theory is due to J. Schwinger, Phys.Rev. 91, 713 (1953). Using Caianiello's renormalization method related formulae were derived in E. Caianiello, M. Marinaro, Nuovo Cimento 27, 1185 (1963) and F. Guerra, M. Marinaro, Nuovo Cimento 42A, 306 (1966). The form (IV.24) is proved in ref. [2], it is also valid in the presence of anomalies.

[9] Y. Lam, Phys.Rev. D
M. Gomes and J.H Lowenstein, to be published.

[10] A. Rouet, to be published
J.H Lowenstein, M. Weinstein and W. Zimmermann, to be publ.

III. EXACT AND BROKEN SYMMETRIES: RENORMALIZATION EFFECTS

In the following, we shall exclusively consider renormalizable effective Lagrangians whose dimension four kinematical parts involve non vanishing mass terms. Superrenormalizable couplings may otherwise be present and will be weighted by a power of the previously introduced parameter $\underline{s}$ which measures their degree of superrenormalizability. The presence of superrenormalizable interactions is, as we shall see, essential to the definition of broken symmetries [1]. In all cases, the symmetry or lack of symmetry is best described by the effective Noether theorem, which follows from the various effective equations of motion described in chapter II and is summarized by a set of Ward identities. We shall first, for the purpose of orientation, look at the simple case of space time symmetries, and then will go into the subject of internal symmetries.

1. Space time symmetries. The energy momentum tensor.

We recall the effective equations of motion:

$$\langle T\left(\frac{\delta\mathcal{L}_{eff}}{\delta\Phi} - \partial_\mu \frac{\delta\mathcal{L}_{eff}}{\delta\partial_\mu\Phi}\right)(x)\, X\rangle = i\sum_{k\in X}\delta(x-x_k)\langle T X_{\hat{k}}\rangle$$

where, for each term in $\mathcal{L}_{eff}$, the corresponding derivative with respect to φ has to be counted with a power counting degree diminished by dim φ. The Noether current associated with space time translations:

$$\varphi(x) \rightarrow \varphi(x+a) \simeq \varphi(x) + a^\mu \partial_\mu \varphi(x)$$

is

$$j_\mu^{Noeth.\,\nu} = \pi_\mu \partial^\nu \varphi$$

with

$$\pi_\mu = \frac{\delta\mathcal{L}_{eff}}{\delta\partial_\mu\varphi}$$

According to the preceding rules $j_\mu^{Noeth.\,\nu}$ has to be counted with dimension four. Using the effective equations of motion under the quadratic combination $\pi_\mu \partial_\nu\varphi$, we have:

$$\partial^\mu\langle T\, \pi_\mu \partial^\nu\varphi X\rangle = \langle T\, \partial^\mu\pi_\mu \partial^\nu\varphi X\rangle + \langle T\pi_\mu\partial^\mu\partial^\nu\varphi X\rangle$$

$$= \langle T \frac{\delta \mathcal{L}_{eff}}{\delta \varphi} \partial^\nu \varphi X \rangle + \langle T \frac{\delta \mathcal{L}_{eff}}{\delta \partial_\mu \varphi} \partial^\nu \partial_\mu X \rangle$$

$$-i \sum_{k \in X} \delta(x - x_k) \partial^\nu_{(k)} \langle T X \rangle$$

$$= \partial^\nu \langle T \mathcal{L}_{eff}(x) X \rangle + \sum_{k \in X} \delta(x - x_k) \frac{1}{i} \partial^\nu_{(k)} \langle T X \rangle$$

Hence, the canonical energy momentum tensor [2]

$$\Theta_\mu^{\nu,c} = \pi_\mu \partial^\nu \varphi - g_\mu^\nu \mathcal{L}_{eff}$$

fulfills the Ward identity

$$\partial^\mu \langle T \Theta_\mu^{\nu,c}(x) X \rangle = \sum_{k \in X} \delta(x - x_k) \frac{1}{i} \partial^\nu_{(k)} \langle T X \rangle$$

The Ward identity associated with Lorentz invariance usually requires the use of the Belinfante energy momentum tensor, of the form

$$\Theta_\mu^{\nu,B} = \Theta_\mu^{\nu,c} + \partial^\lambda f_{\mu\lambda}^\nu$$

where $f_{\mu\lambda}^\nu$ is given by the usual formula [3], is formally antisymmetric in μ, λ, so that it does not spoil the Ward identity and has to be assigned dimension 3 . The symmetry of $\Theta_{\mu\nu}^B$ in μ and ν being a consequence of the equations of motion, results into the asymmetry identity

$$\langle T \Theta_{\mu\nu}^B(x) X \rangle - \langle T \Theta_{\nu\mu}^B(x) X \rangle = \sum_{k \in X} \delta(x - x_k) S_{\mu\nu}^{(k)} \langle T X \rangle$$

where the $S_{\mu\nu}^{(k)}$ are the relevant spin matrices.

The asymmetry identity, and the Ward identity put together, result into the Ward identity for the Lorentz current:

$$M_{\lambda,\mu\nu} = x_\mu \Theta_{\lambda\nu}^B - x_\nu \Theta_{\lambda\mu}^B$$

$$\partial^\lambda \langle T M_{\lambda,\mu\nu}(x) X \rangle = \sum_{k \in X} \delta(x - x_k) \left[\frac{1}{i} \left(x_\mu^{(k)} \partial_\nu^{(k)} - x_\nu^{(k)} \partial_\mu^{(k)} \right) + S_{\mu\nu}^k \right] \cdot$$

$$\cdot \langle T X \rangle$$

Finally, the dilation current which generates broken scale invariance can be conveniently written as

$$D_\lambda = x^\mu \Theta_{\lambda\mu}^{C.C.J.}$$

where the "improved" energy momentum tensor [4] of Callan, Coleman, Jackiw is given by

$$\Theta_{\lambda\mu}^{C.C.J.} = \Theta_{\lambda\mu}^{B} + \sum_s \frac{1+b_s}{6} \left(\Box g_{\mu\nu} - \partial_\mu \partial_\nu \right) \Phi_s^2$$

where the sum ranges over scalar fields and the factors $(1+b_s)$ are the finite wave function renormalization factors occurring in $\mathcal{L}_{eff}$ in front of the quadratic terms $\frac{1}{2}\partial_\mu \phi_s \partial^\mu \phi_s$

The corresponding Ward identity

$$\partial^\lambda \langle T\, D_\lambda(x)\, X \rangle = 2 \langle T\, M^2(x)\, X \rangle + \sum_{k\in X} \delta(x-x_{(k)}) \frac{1}{i} \cdot \left(x_{(k)}^k \partial^{(k)} + d^{(k)} \right) \langle T X \rangle$$

where $M^2(x)$ denotes the set of mass terms occurring in $\mathcal{L}_{eff}$, and thus contains dimension four terms, and $d^{(k)}$ is the canonical dimension of field $\varphi^{(k)}$, is a consequence of the Ward identity together with the trace identity:

$$\langle T\, \Theta_\mu^{\mu\; C.C.J.}(x)\, X \rangle = 2 \langle T\, M^2(x)\, X \rangle + \sum_{k\in X} \delta(x-x_k) \frac{1}{i} d^{(k)} \langle T X \rangle$$

The improvement thus does not yield a "soft" trace, and one can show that no other choice can be made so that the trace becomes "soft", at the expense of changing the canonical dimensions into abnormal dimensions [2],[5]. By integrating the Ward identity for the dilation current or combining the trace identity with the Ward identity for the energy momentum tensor at zero momentum, one obtains the so called Callan Symanzik equation [2] after full use of the reduction formulae of chapter II.

2. Internal Symmetries.

Let $\mathcal{L}_{eff}$ be of the form

$$\mathcal{L}_{eff} = \mathcal{L}_{sym.} + \mathcal{L}_{break.}$$

where the terms in $\mathcal{L}_{break.}$ have dimensions strictly less than four and all mass terms of dimension four in $\mathcal{L}_{sym.}$ are non vanishing. Symmetry and symmetry breaking are understood with respect to a transformation of the type

$$\underline{\varphi} \to {}^g\underline{\varphi} = \mathcal{D}(g)\,\underline{\varphi} \;,\; g \in G \;.$$

where G denotes a compact Lie group. $\mathcal{D}$ a finite dimensional unitary representation of G. The infinitesimal transformation law is of the type

$$\underline{\varphi} \to \underline{\varphi} + i\vec{\theta}.\vec{\delta\omega}\,\underline{\varphi}$$

where $\vec{\theta}$ represents the Lie algebra $\mathcal{G}$ of G in $\mathcal{D}$. When $\mathcal{D}$ is the adjoint representation, we shall denote by $\vec{t}$ the representatives of $\mathcal{G}$. Defining

$$\vec{j}_\mu^{\,Noeth.} = iN_3\ \pi_\mu \vec{\theta}\,\varphi$$

and using the effective equations of motion, we obtain the Ward identity

$$\partial^\mu \langle T\, \vec{j}_\mu^{\,Noeth.}(x)\ X\rangle = \langle T \frac{\delta \mathcal{L}_{break}}{\vec{\delta\omega}}(x) X\rangle + \sum_{k\in X} \delta(x-x_k)\vec{\theta}_k \langle TX\rangle$$

which has to be understood, as the whole theory in the sense of multiple formal series in all the parameters of $\mathcal{L}_{break}$ and some of the parameters of $\mathcal{L}_{sym}$. One could similarly establish current algebra Ward identities[6] which assume a normal form provided a suitable time ordered product involving two currents is defined, which differs from the time ordered products we have used so far by allowed ambiguities.

It is clear that the treatment given so far calls for resummation procedures which allow a decent interpretation within the physical Fock space alluded to in chapter I. This can be done as follows. First of all, in general $\mathcal{L}_{break}$ will induce nonvanishing field vacuum expectation values $\langle\underline{\varphi}\rangle = \underline{F}$. One can first perform the change of variable

$$\underline{\varphi}' = \underline{\varphi} - \underline{F} \qquad \langle\varphi'\rangle = 0$$

One can show[7] that the Green functions for $\underline{\varphi}'$ can be computed from the Lagrangian

$$\mathcal{L}'(\underline{\varphi}') = \mathcal{L}(\underline{\varphi}' + \underline{F})$$

with the dimension assignment given by

$$N_\delta\ (\underline{\varphi}' + \underline{F})^p = \sum_q \binom{p}{q}\ \underline{F}^q\ N_{\delta-|q|}\,\underline{\varphi}'^{\,p-q}$$

and the Ward identity reads

$$\partial^\mu \langle T\, \vec{j}'_\mu(x)\ X'\rangle = \langle T\, \vec{D}'(x)\ X'\rangle + \sum_{k\in X} \delta(x-x_k)\ \vec{\theta}_k \langle TX'\rangle + \sum_k \delta(x-x_k)\,(\vec{\theta}F)_k \langle T X'_{\hat{k}}\rangle$$

where $\vec{J}'_\mu$ and $\vec{D}'$ are respectively deduced from $\vec{J}_\mu^{\text{Noeth}}$, $\frac{\delta \mathcal{L}^{\text{break}}}{\delta \vec{\omega}}$ by the substitution rule $\varphi \to \varphi' + \underline{F}$ according to the above formula. The introduction of the $\underline{s}$ parameter and its use as indicated in Chapter II allow to perform the necessary resummations. On the other hand, the equivalence theorem described in chapter II then asserts the existence of a dimension four effective Lagrangian (without a linear term) which yields the Green's functions $\langle T X' \rangle$. One should however be careful that composite operators $\vec{J}_\mu$, $\vec{D}$ such that

$$\langle T \vec{J}_\mu(x), X' \rangle_{\mathcal{L}_4} = \langle T \vec{J}'_\mu(x)\, X' \rangle_{(\mathcal{L}_{sym} + \mathcal{L}_{break})(\varphi' + F)}$$

$$\langle T \vec{D}(x)\, X' \rangle_{\mathcal{L}_4} = \langle T \vec{D}'(x)\, X' \rangle_{(\mathcal{L}_{sym} + \mathcal{L}_{break})(\varphi' + F)}$$

are only known to have the same maximum dimension as $\vec{J}'_\mu$, $\vec{D}'$ respectively[7].

The determination of $\mathcal{L}_4$ by the requirement that Ward identities be fulfilled is thus in general incomplete at this stage, since only the dimension of $\vec{D}$ goes unchanged through the change of effective Lagrangian, except in the case of a linear symmetry breaking. In the case of a non linear symmetry breaking, the structure of $\vec{D}$ is not arbitrary. Together with a term of given dimension and covariance under G, it must contain all terms corresponding to the ambiguities compatible with covariance and power counting[7]. The question then arises to recover in the $\mathcal{L}_4$ version the information which characterizes the covariance of the breaking.

It is conjectured that this information can be recovered by studying the high momentum behaviour of two point and three point Green's functions via e.g. the Callan Symanzik equations.

Alternatively, let $\mathcal{L}_i$ be an irreducible tensorial term in $\mathcal{L}_{break}$, $\frac{\delta \mathcal{L}_i}{\delta \omega_\alpha} = \mathcal{L}_i^\alpha = \tau_{ij}^\alpha \mathcal{L}_j$ for some representation $\{\tau\}$ of $\mathcal{G}$.

Then, the following Ward identity holds:

$$\partial^\mu \langle T J_\mu^{\text{Noeth.}\beta}(x)\, \mathcal{L}_i^\alpha(y) X \rangle = \langle T \frac{\delta \mathcal{L}_{break}}{\delta \omega_\beta}(x)\, \mathcal{L}_i^\alpha(y) X \rangle$$

$$+ \delta(x-y)\, t_\gamma^{\beta\alpha} \langle T \mathcal{L}_i^\gamma(y) X \rangle + \delta(x-y)\, \tau_{ij}^\beta \langle T \mathcal{L}_j^\alpha(y) X \rangle$$

$$+ \sum_{k \in X} \delta(x - x_k)\, \Theta_{(k)}^\beta \langle T \mathcal{L}_i^\alpha(y) X \rangle$$

One may thus look for D^α in the form $D^\alpha = \sum D_i^\alpha$ if $\mathcal{L}_{break}$ was of the form $\sum \mathcal{L}_i$

and determine the covariant operators D_i^α characterized by their maximum dimension and the fulfillment of Ward identities of the type[8]

$$\partial^\mu \langle T\, j_\mu^\beta(x)\, D_i^\alpha(y)\, X' \rangle = \langle T\, D^\beta(x)\, D_i^\alpha(y)\, X' \rangle$$

$$+ \; \delta(x-y)\, t_\gamma^{\beta\alpha} \langle T\, D_i^\gamma(y)\, X' \rangle + \; \delta(x-y)\, \tau_{ij}^\beta \langle T D_j^\alpha X' \rangle$$

$$+ \; \sum_{k\in X} \delta(x-x_k)\, \theta_k^\beta \langle T\, D_i^\alpha(y)\, X' \rangle$$

$$+ \; \sum_{k\in X} \delta(x-x_k)\, (\theta^\beta F)_k \langle T\, D_i^\alpha(y)\, X'_{\hat{k}} \rangle$$

where the time ordered products involving pairs of composite operators may eventually not be the conventional ones.

Once the characteristics of the initial Lagrangian have been recovered, through highly non linear, recursively soluble relations between the coefficients of $\mathcal{L}_4$ (values of vertex functions at zero momentum) two kinds of situations may occur. If enough parameters are left undetermined so that all physical masses of stable particles can be chosen as free parameters, one has a theory in a physical Fock space $\mathcal{F}_{phys}$ the perturbation parameter being $\hbar$, which counts the number of loops in Feynman diagrams. If not, no perturbative treatment known at present can describe the situation in terms of the correct physical Fock space.

The algebraic complexity of the general situation looks at the moment forbidding, and the more symmetric treatment indicated in Chapter IV is by far more attractive since there all operators retain all the symmetric aspects which are completely hidden in the $\mathcal{L}_4$ formalism. The only reason, other than computational, which calls for a study of the $\mathcal{L}_4$ formalism is the present lack of treatment in the present framework of systems for which, because of the group structure, some symmetric mass parameters vanish. This is admittedly a weakness of the formalism and calls for a more complete study of the interrelations between the various solutions to the decomposition problems posed by the causality requirement. In these more difficult cases, one may either first suitably approximate the theory by one in which no vanishing mass parameter is involved and study some zero mass limit. This is the method examplified in Chapter IV. Alternatively one may modify the starting point as will be shown in Chapter V and directly define the desired theory.

CHAPTER III: REFERENCES AND FOOTNOTES

[1] This is best explained in
K. Symanzik, "Lectures on Symmetry Breaking" in Cargese (1970). Gordon and Breah, New York (1972).

[2] J.H. Lowenstein, P.R. D4, 2281 (1971).
A. Rouet, R. Stora, Nuovo Cim. Lett. 4, 136, 139 (1972).

[3] R. Jost, The General Theory of Quantized Fields. A.M.S. Providence (1965).

[4] The formal "improved" energy momentum tensor was defined in:
CG. Callan, S. Coleman, R. Jackiw, Ann. Phys. 59, 42 (1970).

The present finite version is due to M. Bergere (private communication). The asymmetry identity was first found by A. Rouet, unpublished, and A. Rouet, R. Stora[2].
The general form of $\theta_{\mu\nu}$ compatible with Ward identities was first given by K. Symanzik and K. Wilson, private communications (1970) and is now best described in terms of normal products, as done here.

[5] That the trace of the energy momentum tensor becomes soft at the GellMann Low value of the coupling constant, is shown in:
B. Schroer, Lett. Nuov.Cim. 2, 867 (1971).

[6] K. Symanzik, unpublished and J. Lowenstein, Seminars on Renormalization Theory, Vol. II, Maryland Technical Report 73 - 068 (1972).

[7] Y.P. Lam, B. Schroer, to be published.
A. Rouet, Equivalence theorems for Effective Lagrangians, Marseille CNRS preprint 73/P.528. March 1973.

[8] Arguments of this type may be found in:
A. Becchi: "Absence of strong interactions to the axial anomaly in the σ model", CERN TH 1611. January (1973). Comm. Math. Phys. to appear.

IV. Gauge Invariant Quantization of the Goldstone and Higgs Model

The general formulation of the previous sections will now be applied to the examples of the Goldstone and the Higgs model. We begin with the discussion of the Goldstone model. B. Lee and K. Symanzik developed two alternative methods of quantizing the Goldstone model[1,2] . In B. Lee's work the model is regularized and quantized in a gauge invariant manner. It is then shown that the regularization can be removed for the renormalized Feynman amplitudes. In the treatment that follows we use B. Lee's method of gauge invariant quantization but without introducing a regularization. Instead we will deal directly with the unregularized, but properly renormalized, Feynman amplitudes. It will be shown that the desired properties of the model follow easily as an application of the general theorems given in Section II and III. The connection to Symanzik's method will be discussed later.

As classical Lagrangian we propose

(IV.1)

$$\mathcal{L}_{cl.}(s) = \partial_\mu \varphi^* \partial^\mu \varphi - (\eta_0^2 - s^2 w^2)\varphi^*\varphi - h_0(\varphi^*\varphi)^2 - \frac{\delta_0(s)}{\sqrt{2}}(\varphi + \varphi^*)$$

with

(IV.2)

$$\varphi = \frac{1}{\sqrt{2}}\left(\tau_1 + s v_0 + i\tau_2\right)$$

$$\tau_1 = Z_2^{1/2}\psi\ ,\quad \tau_2 = Z_2^{1/2}\chi\ ,\quad v_0 = Z_2^{1/2} v$$

ψ and χ denote the properly normalized fields. Their vacuum expectation values are required to vanish

(IV.3) $$\langle\psi\rangle = \langle\chi\rangle = 0$$

while it is assumed that

(IV.4) $$v_0 \neq 0$$

Expressed in terms of the fields ψ and χ (IV.1) is to be interpreted as an s-dependent Lagrangian in the sense of Chapter II. We recall that the parameter s only serves to specify the subtraction procedure and is set equal to one finally.

$\delta_0(s)$ is a polynomial in s with

$$\delta_0(0) = 0$$

of which only the value

$$\delta_0(1) = \delta_0$$

at s = 1 is relevant.

The parameter w is restricted to a permissable range which will be given later. It can be shown that the Green's functions of the theory do not depend on the value of w [4].

For $\delta_o = 0$ (IV.1) is the Lagrangian of the Goldstone model. Formally it is gauge invariant, but the gauge symmetry is spontaneously broken due to (IV.4). The model describes two particles σ and π which are associated with the fields ψ or χ resp. The π-particle has zero mass and represents the Goldstone particle. The σ-particle is unstable since it can decay into π-particles by the interaction term $\psi\chi^2$ present in the Lagrangian (IV.1).

For $\delta_o \neq 0$ the gauge symmetry is explicitly broken by the term proportional to $\varphi + \varphi^*$. In this case the model is called the explicitly broken Goldstone model. It describes two massive particles σ and π of mass M and μ associated with the fields ψ or χ. We always assume $\mu < M$. The σ-particle is stable only for $M < 2\mu$. In the unstable case M denotes an appropriate mass parameter related to the complex pole of the ψ-propagator [5]. In the Goldstone limit $\delta_o \to 0$ the π-particle becomes massless, i.e. $\mu \to 0$.

Unfortunately, the Lagrangian (IV.1) is not meaningful for $\delta_o = 0$ since some of its coefficients are infrared divergent. In order to bypass infrared problems the case $\delta_o \neq 0$, or $\mu \neq 0$, is considered first. Then the perturbation expansion is well-defined with finite coefficients of the Lagrangian (IV.1). Eventually the Goldstone limit $\mu \to 0$ is applied yielding Green's functions and S-matrix of the Goldstone model.

Apart from $\underline{s}$ and w the independent parameters of the theory will be the σ-mass M (or an appropriate mass parameter M in the unstable case), the π-mass μ and a suitably defined coupling constant g. Five renormalization conditions are required to hold at s = 1 [6]. These conditions uniquely determine the parameters η_o, h_o, δ_o, v_o, z_2 as power series in g with coefficients depending on M, μ and w, but independent of $\underline{s}$.

We now determine the free part of the Lagrangian (IV.1) following the instructions of Chapter II. Let $\mathcal{L}_{eff\,Q}$ denote the sum of all terms in $\mathcal{L}_{eff}$ which are quadratic in ψ and χ or their derivatives.

$$\begin{aligned} \mathcal{L}_{eff\,Q} = {} & \frac{1}{2} z_2 (\partial_\mu \psi \partial^\mu \psi) + \frac{1}{2} z_2 (\partial_\mu \chi \partial^\mu \chi) \\ & - \frac{1}{2} \left(\eta_o^2 - s^2 w^2 + 3 h_o s^2 v_o^2 \right) \psi^2 \\ & - \frac{1}{2} \left(\eta_o^2 - s^2 w^2 + h_o s^2 v_o^2 \right) \chi^2 \end{aligned} \tag{IV.5}$$

The free part of the effective Lagrangian is defined by (IV.5) with the coefficients replaced by their zero order values:

$$\mathcal{L}_{\text{eff}\,0} = \frac{1}{2}\,\partial_\mu\psi\,\partial^\mu\psi + \frac{1}{2}\,\partial_\mu\chi\,\partial^\mu\chi - \frac{1}{2}\,M^2(s)\,\psi^2 - \frac{1}{2}\,\mu^2(s)\,\chi^2\,, \tag{IV.6}$$

$$\lim_{g\to 0} Z_2 = 1,$$

$$M^2(s) = \lim_{g\to 0}\left(\eta_0^2 - s^2 w^2 + 3\,h_0\,s^2\,v_0^2\right), \tag{IV.7}$$

$$\mu^2(s) = \lim_{g\to 0}\left(\eta_0^2 - s^2 w^2 + h_0\,s^2\,v_0^2\right).$$

The mass parameters M^2, μ^2 equal the values of (IV.7) at s = 1

$$M^2 = M^2(1)\quad,\quad \mu^2 = \mu^2(1)$$

Since η_0, w, h_0, v_0 are independent of s we obtain

$$\lim_{g\to 0} h_0\,v_0^2 = \frac{M^2 - \mu^2}{2} \tag{IV.8}$$

$$M^2(s) = \frac{3}{2}\,\mu^2 - \frac{1}{2}\,M^2 + w_0^2 + s^2\left(\frac{3}{2}\,(M^2 - \mu^2) - w_0^2\right)$$

$$\mu^2(s) = \frac{3}{2}\mu^2 - \frac{1}{2}\,M^2 + w_0^2 + s^2\left(\frac{1}{2}\,(M^2 - \mu^2) - w_0^2\right) \tag{IV.9}$$

w_0 denotes the zero order value of w.

Since we need the theory in the range $0 \leq s \leq 1$ we impose the consistency condition

$$M^2(s) \geq 0 \qquad \mu^2(s) \geq 0$$

As long as $M > 0$, $\mu > 0$ the model is not expected to suffer from infrared problems. One should therefore avoid vanishing mass values in the range $0 \leq s \leq 1$ by imposing the stronger condition

$$M^2(s) > 0\,,\ \mu^2(s) > 0 \quad \text{if}\quad M^2 > 0,\ \mu^2 > 0. \tag{IV.10}$$

This restricts the permissable values of w by

$$w_0^2 > \frac{1}{2}\,M^2 - \frac{3}{2}\,\mu^2 \tag{IV.11}$$

Particular convenient is a choice of w for which

$$w_0^2 = \lim_{g\to 0} h_0 v_0^2 = \frac{1}{2}(M^2 - \mu^2) > \frac{1}{2}M^2 - \frac{3}{2}\mu^2 \qquad \text{(IV.12)}$$

With this the s-dependence (IV.9) of the masses becomes

$$\begin{aligned} M^2(s) &= \mu^2 + s^2(M^2 - \mu^2) \\ \mu^2(s) &= \mu^2 \end{aligned} \qquad \text{(IV.13)}$$

Hence the π -mass is independent of s while the σ -mass has the simple form (IV.13).

With the free Lagrangian (IV.6) involving the s-dependent masses (IV.7) the perturbation expansion (II.14) of the Green's functions is completely determined.

The main advantage of this gauge invariant formulation is that partial current conservation follows quite naturally, essentially by following the classical derivation. The current operator is defined by taking the minimal normal product of the classical Noether current

$$j_\mu = i\,N\left[\varphi\,\partial_\mu \varphi^* - \varphi^*\,\partial_\mu \varphi\right] \qquad \text{(IV.14)}$$

More precisely, the symbol N means that the minimal normal product should be applied to each monomial of the current expressed in terms of ψ and χ. According to Chapter III the Green's functions of this current operator satisfy the Ward identities

$$\begin{aligned} \partial_\mu \langle T\, j^\mu(x)\, \varphi(u_1)\ldots\varphi^*(v_1)\ldots\rangle &= i\sum_k \delta(x-u_k)\langle T\varphi(u_1)\ldots\varphi^*(v_1)\ldots\rangle \\ &\quad - i\sum_k \delta(x-v_k)\langle T\varphi(u_1)\ldots\varphi^*(v_1)\ldots\rangle \\ &\quad + i\,\frac{\delta_0(s)}{\sqrt{2}}\langle T\left(\varphi(x)-\varphi^*(x)\right)\varphi(u_1)\ldots\varphi^*(v_1)\ldots\rangle \end{aligned}$$

$$\begin{aligned} \partial_\mu \langle T\, j^\mu(x)\, \varphi(u_1)\ldots\varphi^*(v_1)\ldots\rangle &= \\ &\sum_k \delta(x-u_k)\langle T\varphi(u_1)\ldots\varphi^*(v_1)\ldots\rangle \\ &- \sum_k \delta(x-v_k)\langle T\varphi(u_1)\ldots\varphi^*(v_1)\ldots\rangle \\ &+ i\,\frac{\delta_0(s)}{\sqrt{2}}\langle T\left(\varphi(x)-\varphi^*(x)\right)\varphi(u_1)\ldots\varphi^*(v_1)\ldots\rangle \end{aligned} \qquad \text{(IV.15)}$$

In operator form the law of partial current conservation becomes

$$\partial_\mu j^\mu(x) = i\,\frac{\delta_0(s)}{\sqrt{2}}\left(\varphi(x)-\varphi^*(x)\right) \qquad \text{(IV.16)}$$

(IV.16) follows from (IV.15) by applying the reduction formulae to the fields ψ and χ of arguments u_j and v_j.

It can be shown that in the Goldstone limit $\mu \to 0$ the time-ordered Green's function and the S-matrix exist [7]. The current of the Goldstone model is conserved.

It is characteristic for the gauge invariant approach that - apart from the linear term - the fields ψ and χ only appear in the gauge invariant combinations of the Lagrangian (IV.1). Without destroying the gauge invariant form of the non-linear part the Lagrangian (IV.1) may be replaced by an s-independent Lagrangian of type (II.43-45). As was discussed in Chapter II such Lagrangians suffer from a summation problem in finite order of perturbation theory. In the present case the Lagrangian contains terms of the form

$$N_4\,\psi^2 - N_2\,\psi^2\,, \quad N_4\,\chi^2 - N_2(\chi^2) \tag{IV.17}$$

where the coefficients do not vanish in zero order [8]. Thus an infinite number of Feynman diagrams appears in any given order of perturbation theory.

On the other hand the equivalence theorem (equ. (II.46-47)) allows to construct an equivalent Lagrangian which contains only N_4-products and does not involve a summation problem in finite order. The Lagrangian is of the form

$$\mathcal{L}_{eff} = \sum_{j=1}^{9} C_j\, N_4(M_j) \tag{IV.18}$$

where the M_j denote the monomials

$$\psi^2,\ \psi^3,\ \psi^4,\ \chi^2,\ \chi^4,\ \psi\chi^2,\ \psi^2\chi^2,\ \partial_\mu\psi\,\partial^\mu\psi\,,\ \partial_\mu\chi\,\partial^\mu\chi$$

The perturbation expansion based on (IV.18) represents Symanzik's method of renormalizing the explicitly broken Goldstone model [3]. While the Lagrangian is not manifestly gauge invariant, the coefficients of the coupling terms are correlated in such a way that the Ward identities hold in the desired form.

The renormalization of the Higgs model was first developed by B. Lee by applying Symanzik's method to a regularized version [9]. B. Lee and Zinn-Justin extended the method of gauge invariant quantization to Higgs-Kibble models including the non-Abelian case [10]. In the remainder of this chapter we use the approach of gauge invariant quantization to renormalize the Abelian Higgs model by applying the general methods of Chapter II and III. As s-dependent Lagrangian we propose

$$\begin{aligned}\mathcal{L}_{cl} = {} & (D_\mu\varphi)^*(D^\mu\varphi) - (\eta_0^2 - s^2 w^2)\varphi^*\varphi - h_0(\varphi^*\varphi)^2 \\ & - \frac{1}{4} B_{\mu\nu} B^{\mu\nu} + \frac{1}{2} m_0^2 B_\mu B^\mu - \frac{1}{2\alpha_0}(\partial_\mu B^\mu)^2 \\ & + \frac{\delta_0(s)}{\sqrt{2}}(\varphi + \varphi^*)\end{aligned} \tag{IV.19}$$

with

$$D_\mu = \partial_\mu - i e_o B_\mu \;;\quad B_{\mu\nu} = \partial_\mu B_\nu - \partial_\nu B_\mu$$

(IV.20) $$B_\mu = Z_3^{1/2} A_\mu\,,\; e_o = Z_3^{-\frac{1}{2}} e\,,\; \alpha_o = Z_3 \alpha\,,\; m_o = Z_3^{-\frac{1}{2}} m$$

$$\varphi = \frac{1}{\sqrt{2}}\left(\tau_1 + s v_o + i \tau_2\right)$$

$$\tau_1 = Z_2^{1/2}\psi\,,\; \tau_2 = Z_2^{1/2}\chi\,,\; v_o = Z_2^{1/2} v$$

ψ, χ and A_μ denote the properly normalized fields. Their vacuum expectation values are required to vanish while it is assumed that

(IV.21) $$v_o \neq 0$$

The gauge class used in (IV.19) is the analogue of the Gupta-Bleuler or Stueckelberg gauge in (massive) electrodynamics. The treatment of the Higgs model in the 't Hooft gauge will be discussed in Chapter V.

For $\delta_o = 0$ and m = 0 (IV.19) is the Lagrangian of the Higgs model. The model described by (IV.19) for $\delta_o = 0$, but non-vanishing original photon mass $m \neq 0$ will be called the pre-Higgs model. In both cases the Lagrangian (IV.19) is formally gauge invariant, but the gauge symmetry is spontaneously broken due to (IV.21).

For $\delta_o \neq 0$ the gauge symmetry is explicitly broken by the term proportional to $\varphi + \varphi^*$. In this case the model is called the explicitly broken pre-Higgs model. No infrared problems occur for $\delta_o \neq 0$ and the Lagrangian (IV.19) can be used for setting up renormalized perturbation theory without difficulties. For $\delta_o = 0$ infrared divergencies occur for some coefficients of (IV.19), but the Goldstone limit $\delta_o \to 0$ and the subsequent Higgs limit $m \to 0$ (in the Landau gauge) may be applied yielding the Green's functions of the pre-Higgs and the Higgs model.

The free Lagrangian of the explicitly broken Higgs model is

(IV.22) $$\begin{aligned}\mathcal{L}_{eff\,o} = &-\frac{1}{4} F_{\mu\nu} F^{\mu\nu} + \frac{1}{2} m_t^2(s) A_\mu A^\mu - \frac{1}{2\alpha}\left(\partial_\mu A^\mu\right)^2 \\ &+ \frac{1}{2}\left(\partial_\mu\psi\,\partial^\mu\psi + \partial_\mu\chi\,\partial^\mu\chi\right) + w(s)\, A_\mu \partial^\mu \chi \\ &- \frac{1}{2} M^2(s)\,\psi^2 - \frac{1}{2}\mu^2(s)\,\chi^2 .\end{aligned}$$

$$F_{\mu\nu} = \partial_\mu A_\nu - \partial_\nu A_\mu$$

The s-dependence of the coefficients becomes (IV.9) and

$$m_t^2(s) = m^2 + s^2 w^2$$
$$w(s) = s w \tag{IV.23}$$

with

$$w_{(0)} = \lim_{e \to 0} w , \quad W = \lim_{e \to 0} e v$$
$$\lim_{e \to 0} h_0 v^2 = \frac{M^2 - \mu^2}{2} \tag{IV.24}$$

For M, m, μ, w > 0 the free theory can be shown to be consistent with positive masses in the range $0 \leq s \leq 1$ if μ is sufficiently small and w chosen to satisfy (IV.11).

The particles of the theory are determined from the free Lagrangian (IV.22) at $\underline{s} = 1$. Following is a table of the particles, their masses and associated fields.

Particle	Mass [11]	Field
Vector meson	m_t	A_μ
σ	M	ψ
π	κ	$\partial_\mu A^\mu, \chi$
Ghost particle	λ	

with

$$m_t^2 = m^2 + w^2$$
$$\kappa^2 = \frac{1}{2}(\alpha m^2 + \mu^2) - \frac{1}{2}\sqrt{(\alpha m_t^2 + \mu^2) - 4\alpha \mu^2 m_t^2} \tag{IV.25}$$
$$\lambda^2 = \frac{1}{2}(\alpha m^2 + \mu^2) + \frac{1}{2}\sqrt{(\alpha m_t^2 + \mu^2) - 4\alpha \mu^2 m_t^2}$$

In the Goldstone limit $\delta_0 \to 0$, or equivalently $\mu \to 0$, the π-particle becomes massless ($\kappa \to 0$) and represents the Goldstone particle of the massive Higgs model.

As usual an indefinite metric formulation is employed in order to quantize the Lagrangian (IV.19). In general, the S-matrix will not be unitary since the ghost particles of negative probabilities participate in the interaction. No physical interpretation of the model is possible then. In the Goldstone limit, however, the ghost particles are expected to decouple from the rest of the system. The argument proceeds as in electrodynamics. The ghost particles are described by the divergence $\partial_\mu A^\mu$ of the vector potential. The field equation (II.29) of the vector potential reads

$$-\partial_\nu F^{\mu\nu} + \frac{1}{\alpha}\partial^\mu \partial_\nu A^\nu + m_t^2 A^\mu = j^\mu \qquad \text{(IV.26)}$$

with the current operator

$$j^\mu = ie\, N\left[\varphi (D^\mu \varphi)^* - \varphi^* D^\mu \varphi\right] + (Z_3 - 1)\,\partial_\nu F^{\mu\nu} \qquad \text{(IV.27)}$$

The current operator is partially conserved

$$\partial_\mu j^\mu = \frac{i}{\sqrt{2}}\,\delta e\,(\varphi - \varphi^*) = \delta e\, Z_2^{1/2} \chi \qquad \text{(IV.28)}$$

as follows from Ward identities similar to (IV.15). (IV.26) and (IV.28) yield

$$(\Box + \alpha m^2)\,\partial_\mu A^\mu = i\alpha\,\delta e\, Z_2^{1/2}\chi \qquad \text{(IV.29)}$$

as field equation of $\partial_\mu A^\mu$. In the Goldstone limit $\delta_0 \to 0$ the divergence $\partial_\mu A^\mu$ becomes a free field

$$(\Box + \alpha m^2)\,\partial_\mu A^\mu = 0 \qquad \text{(IV. 30)}$$

and the ghost particles decouple. Accordingly the S-matrix of the massive Higgs model is unitary. By using differential equations of the Callan-Symanzik type it can also be shown that the Green's functions are well defined in the Goldstone limit $\mu \to 0$ [12].

The only stable particles of the massive Higgs model are the π-particle and the free ghost particle.

In the limit $m \to 0$, the pre-Higgs model passes over into the Higgs model. The zero-mass π-particles decouple, with the massive spin-one unstable particles of the pre-Higgs model becoming stable in the limit. The massive spin-zero particles also become stable, provided $M < 2w$. The Green's functions of the A_μ, ψ and χ fields can be shown to exist in the Higgs limit, but only in Landau gauge ($\alpha = 0$) [12].

The equivalence theorem (II.46 - 47) can be applied to construct an equivalent Lagrangian consisting of N_4-products only. In contradistinction to (IV.19) the non-linear part of the N_4-Lagrangian is not manifestly gauge invariant. This N_4-version of the Higgs model represents B. Lee's original approach in the language of the normal product formalism.

Chapter IV: References and Footnotes

[1] The material of this section will be published in a series of papers by J.H Lowenstein, M. Weinstein, W. Zimmermann (part I and II), and J.H. Lowenstein, B. Schroer (part III).

[2] B. Lee, Nucl. Phys. B9, 649 (1969).

[3] K. Symanzik, Lett. Nuovo Cimento 2, 10 (1969) and Commun. Math. Phys. 16, 48 (1970).

[4] For the proof see part II of ref. [1].

[5] For problems concerning unstable particles in perturbation theory we refer to M. Veltman, Physica 29, 122 (1969) and part III of ref. [1].

[6] For the formulation of the renormalization conditions see ref. [1].

[7] K. Symanzik, Lett. Nuovo Cimento 2, 10 (1969) and Commun. Math. Phys. 16, 48 (1970).
F. Jegerlehner and B. Schroer, to be published.

[8] By appropriate choice of $\mathfrak{w}$ one of the coefficients can be made to vanish in zero order, but not both.

[9] B. Lee, Phys. Rev. D5, 823 (1972).

[10] B. Lee and J. Zinn-Justin, to be published.

[11] The values given are the masses in zero order. Only for stable particles may the zero order values be identified with the exact masses by suitable normalization conditions.

[12] See ref. [1], part III.

V. Models with vanishing Symmetric Mass Parameters

As stressed in chapter III, the combinatorics of intermediate renormalization comes into conflict with the possible occurrence of vanishing mass parameters, and, in particular, another description of symmetry breaking has to be found if the group structure implies the vanishing of some mass parameters. Typical examples of such symmetries are chiral symmetries, when spin 1/2 fields are involved, and gauge symmetries, although no difficulty should in principle arise if the physical masses are to be non vanishing. In such cases, one may first consider the classical theory as a limit of a theory where no vanishing mass parameter occurs, quantize the latter and let the spurious mass parameters go to zero. This is the road chosen in Chapter IV. The only alternative strategy which is known at present [1] is to investigate the structure of the classical Lagrangian which describes the tree approximation of the theory to be constructed, characterize the symmetry via the Ward identities which express the classical Noether theorem in presence of external sources, and look for an L_4 Lagrangian for which Ward identities of the type found in the tree approximation hold in finite renormalized form, in the sense that no composite operator different from those found in the tree approximation occur, with, however, possibly different coefficients. In case such a program cannot be completed, with composite operators occurring with their naive dimension, renormalized Ward identities are said to contain anomalies - e.g. of the type found in the trace identity for the energy momentum tensor cf. Ch. III. There exists at the moment, unfortunately, no general theorem which allows to predict from the group structure the presence of anomalies.

Since the present program is still at a very experimental stage, we shall content ourselves with the description of two examples: the σ model with nucleons and the Abelian Higgs Kibble model treated in 't Hooft's gauge.

1. The σ model involving nucleons [2]

Choosing for simplicity the chiral group to be $U(1) \times U(1)$, and denoting the meson field (π, σ), the nucleon field, ψ , the Lagrangian in the tree approximation is obtained from the formal Lagrangian

$$\mathcal{L}_{\text{formal}} = \bar{\psi} \gamma \frac{1}{i} \partial \psi + g\bar{\psi}(\sigma + i\pi\gamma_5\psi) + \frac{1}{2}\left(\partial_\mu \pi \partial^\mu \pi + \partial_\mu \sigma \partial^\mu \sigma\right)$$

$$- \frac{m^2}{2}(\pi^2 + \sigma^2) - \frac{\lambda}{4}(\pi^2+\sigma^2)^2 + C\sigma$$

by the field translation $\sigma = \sigma' + F$, under the constraint that no term linear in σ' remains.
We thus obtain

$$\mathcal{L}_{tree} = \bar\Psi \gamma \tfrac{1}{i}\partial \psi + gF\bar\Psi\psi + g\bar\Psi(\sigma' + i\pi\gamma_5)\psi$$
$$+ \tfrac{1}{2}\left(\partial_\mu\pi\partial^\mu\pi + \partial_\mu\sigma'\partial^\mu\sigma'\right) - \frac{m^2+\lambda F^2}{2}\pi^2 - \frac{m^2+3\lambda F^2}{2}\sigma'^2$$
$$- \frac{\lambda}{4}(\pi^2+\sigma'^2)^2 - \lambda F\sigma(\sigma^2+\pi^2)$$

with the constraint $C = \lambda F^3 + m^2 F$
This Lagrangian may be written as follows

$$\mathcal{L}_{tree} = \bar\Psi\left(\gamma\tfrac{1}{i}\partial + M\right)\psi + g\bar\Psi(\sigma' + i\pi\gamma_5)\psi$$
$$+ \tfrac{1}{2}\partial_\mu\pi\partial^\mu\pi - \frac{m_\pi^2}{2}\pi^2 + \tfrac{1}{2}\partial_\mu\sigma'\partial^\mu\sigma' - \frac{m_\sigma^2}{2}\sigma'^2$$
$$- g^2\frac{m_\sigma^2 - m_\pi^2}{8M^2}(\pi^2+\sigma'^2)^2 - g\frac{m_\sigma^2-m_\pi^2}{2M}\sigma'(\pi^2+\sigma'^2)$$

where the following change of parameters was made:

$$M = gF$$
$$m^2 + \lambda F^2 = m_\pi^2$$
$$m^2 + 3\lambda F^2 = m_\sigma^2$$

which implies $\lambda = g^2\dfrac{m_\sigma^2 - m_\pi^2}{2M^2}$

(Instead of M, m_π, m_σ, g we could have chosen as parameters M, m_π, m_σ, F).

The Ward identities are most easily obtained by applying Noether's theorem to $\mathcal{L}_{formal}$ + source term and performing the field translation afterwards. The result is

$$\partial^\mu \langle T j_\mu(x) X\rangle^{tree} = \sum_{k\in X}\delta(x-x_k)\Theta_k\langle T X\rangle$$
$$+ \sum_{\pi\in X}\delta(x-x_\pi)F\langle T X_{\hat\pi}\rangle$$
$$+ C\langle T\pi(x)X\rangle$$

where θ_k represents an infinitesimal chiral transformation on field (k) ($\pi \to \sigma$, $\sigma \to -\pi$, $\psi \to i\gamma_5\psi$ $\bar{\psi} \to -\bar{\psi} i\gamma_5$)
and

One now looks for an effective Lagrangian

$$\begin{aligned}\mathcal{L}_4 = & (1+B)\bar{\psi}\gamma\frac{1}{i}\partial\psi + (M+A)\bar{\psi}\psi + g_\sigma(1+h_\sigma)\bar{\psi}\sigma\psi \\ & + g(1+h_\pi)\bar{\psi}\, i\pi\gamma_5\psi + \frac{1}{2}(1+b_\pi)\partial_\mu\pi\,\partial^\mu\pi - \frac{1}{2}(m_\pi^2+a_\pi)\pi^2 \\ & + \frac{1}{2}(1+b_\sigma)\partial_\mu\sigma\,\partial^\mu\sigma - \frac{1}{2}(m_\sigma^2+a_\sigma)\sigma^2 \\ & - g^2\frac{m_\sigma^2-m_\pi^2}{8M^2}\left[(1+l_\pi)\pi^4 + 2(1+l_{\pi\sigma})\pi^2\sigma^2 + (1+l_\sigma)\sigma^4\right] \\ & - g\frac{m_\sigma^2-m_\pi^2}{2M}\left[(1+k_\sigma)\sigma^3 + (1+k_\pi)\sigma\pi^2\right]\end{aligned}$$

which depends on 13 parameters, and for a current

$$\begin{aligned}J_\mu = & (1+\alpha_\pi)\pi\partial_\mu\sigma - (1+\alpha_\sigma)\sigma\partial_\mu\pi + F(1+\beta_\pi)\partial_\mu\pi \\ & + (1+\gamma)\bar{\psi}\gamma_\mu\gamma_5\psi\end{aligned}$$

such that a Ward identity identical with that obtained in the tree approximation holds. This is possible and leaves freedom, without varying F, to fix the masses of π, σ, ψ at their zeroth order values m_π, m_σ, M and to set equal to unity the residues of the propagators of the π and the ψ fields at their respective poles. The number of free parameters is just the number of parameters occurring in the most general formal Lagrangian invariant under chiral transformations, except for a linear breaking term.

2. The Abelian Higgs Kibble model in the 't Hooft gauge.[3]

The interest of this model within the general class of spontaneously broken gauge models is that it exhibits most features characteristic of these models in so far as ultraviolet behaviour is concerned. Infrared difficulties, on the other hand, are avoided.

One starts from the formal Lagrangian

$$\mathcal{L}_{formal} = -\frac{1}{4} G_{\mu\nu} G^{\mu\nu} + (D_\mu \varphi)^* (D^\mu \varphi) + \mu^2 \varphi^* \varphi$$
$$- g(\varphi^* \varphi)^2 - \frac{1}{2\alpha} (\partial_\mu A^\mu + \rho \varphi_2)^2 + \frac{1}{\alpha} \bar{c} m c$$

where, for the time being, the Faddeev-Popov term, $\bar{c}mc$, whose convenience will be seen later, is ignored. The mode corresponding to broken symmetry is obtained by making the substitution

$$\varphi = \frac{\varphi_1 + v + i\varphi_2}{\sqrt{2}}$$

v being determined by the condition that no term linear in φ_1 survives:

$$\mu^2 = g v^2$$

A translation induced photon mass term yields a transverse photon mass

$$m^2 = e^2 v^2$$

The mass corresponding to the φ_1 field is

$$M^2 = -\mu^2 + 3 g v^2 = 2 g v^2$$

The mass matrix corresponding to the coupled $\varphi_2, \partial_\mu A^\mu$ quadratic form yields a degenerate eigenvalue at

$$\lambda^2 = \rho m$$

Before the introduction of the Faddeev Popov part of the Lagrangian the only non gauge invariant part of the Lagrangian is given by the gauge term

$$-\frac{\mathcal{G}^2}{2\alpha} = -\frac{(\partial_\mu A^\mu + \rho \varphi_2)^2}{2\alpha}$$

In terms of the new variables, the Lagrangian reads:

$$\mathcal{L}_{tree} = \frac{1}{2} \partial_\mu \varphi_1 \partial^\mu \varphi_1 - \frac{M^2}{2} \varphi_1^2 + \frac{1}{2} \partial_\mu \varphi_2 \partial^\mu \varphi_2 - \frac{\rho^2}{2\alpha} \varphi_2^2$$
$$+ \left(m - \frac{\rho}{\alpha}\right) \varphi_2 \partial_\mu A^\mu - \frac{1}{2\alpha} (\partial_\mu A^\mu)^2 - \frac{1}{4} G_{\mu\nu} G^{\mu\nu} + \frac{m^2}{2} A_\mu A^\mu$$
$$- e A_\mu (\varphi_1 \partial^\mu \varphi_2 - \varphi_2 \partial^\mu \varphi_1) + \frac{e^2}{2} A_\mu A^\mu (\varphi_1^2 + \varphi_2^2)$$
$$+ e m A_\mu A^\mu \varphi_1 - \frac{e^2 M^2}{2 m^2} (\varphi_1^2 + \varphi_2^2)^2 - \frac{2 e M^2}{m} \varphi_1 (\varphi_1^2 + \varphi_2^2)$$

where g was eliminated against e through the relations

$$2 g v^2 = M^2, \quad e^2 v^2 = m^2, \quad \Rightarrow \quad \frac{2g}{e^2} = \frac{M^2}{m^2}$$

The restricted 't Hooft gauge is obtained for $\rho = \alpha m$ whereby the $\varphi_2 \partial_\mu A^\mu$ cross term vanishes. The gauge invariance is best expressed by applying the integrated Noether theorem corresponding to the gauge group to $\mathcal{L}_{formal}$+ source terms, and performing the field translation on the corresponding Ward identity, which assumes the form

$$0 = -\int \frac{1}{\alpha} \frac{\delta \mathcal{G}_x}{\delta \Lambda_y} \mathcal{G}_x + \text{source terms} = \frac{1}{\alpha} \int \mathcal{M}_{yx} \mathcal{G}_x + \text{source terms}$$

On the other hand, the question of the gauge invariance of the physical scattering amplitudes first proceeds through the study of the variation of the connected Green's functional under the change of the gauge parameters γ (here, α, ρ):

$$\frac{\partial G^c(J)}{\partial \gamma} = -i \int \frac{\partial}{\partial \gamma} \left(\frac{\mathcal{G}^2}{2\alpha} \right) = -i \int \frac{\mathcal{G}}{\sqrt{\alpha}} \frac{\partial}{\partial \gamma} \left(\frac{\mathcal{G}}{\sqrt{\alpha}} \right)$$

where use has been made of the fact that the Green's functional is the Legendre transform of the Lagrangian.[1],[2] Gauge invariance is achieved if

$$\langle \mathcal{G} \rangle_{phys} = 0$$

for a suitable definition of physical states. In any event, it is desirable to convert the Ward identity into an identity of the Slavnov type:[3]

$$\mathcal{G}_x = \cdots$$

which requires the inversion of the - in general field dependent - differential operator $\mathcal{M}$. This is best achieved by introducing the Faddeev Popov fields with Lagrangian $\frac{1}{\alpha} \bar{C} \mathcal{M} C$ and source terms $\bar{\xi} C + \bar{C} \xi$. The Ward identity then reads

$$0 = \int -\frac{1}{\alpha} \mathcal{M}_{yx} \mathcal{G}_x + \frac{1}{\alpha} \bar{C}_z \frac{\delta \mathcal{M}_{zt}}{\delta \Lambda_y} C_t + \text{source terms}$$

and the Slavnov identity is obtained by integrating over y after multiplication by $\bar{C}_y$[5]. Using the equations of motion, the first term becomes

$$\int \bar{\xi}_x \mathcal{G}_x$$

whereas the second term vanishes if the Faddeev Popov ghost fields obey Fermi's statistics:

$$\int dy\, dz\, dt\ \bar{c}_y \bar{c}_z \frac{\delta G_t}{\delta\Lambda_y \delta\Lambda_z} c_t = 0$$

by virtue of the abelianness of the gauge group. In the non abelian case,[3] this term does not vanish, but assumes a particularly simple form, independent of the gauge function: $\bar{c}\vec{t}\bar{c}\,\vec{\xi}$ where $\vec{t}$ denotes the infinitesimal generator of the internal Lie algebra of the gauge group, only in the case where the Faddeev Popov ghost fields obey Fermi's statistics.

Written in full, in terms of the translated field variables, the Slavnov identities read in the tree approximations

$$\begin{aligned}
&\langle T\,(\partial_\mu A^\mu + \rho\varphi_2)(x)\ \varphi_1(x_1)\cdots\varphi_1(x_n)\ A_{\mu_1}(y_1)\cdots A_{\mu_m}(y_m)\ \varphi_2(z_1)\cdots\varphi_2(z_p) \\
&\qquad c(t_1)\cdots c(t_k)\ \bar{c}(u_1)\cdots\bar{c}(u_k)\rangle^c = \\
&- e\sum_{i=1}^{n} \langle T\, c(x)\ \varphi_1(x_1)\cdots(\varphi_2\bar{c})(x_i)\cdots\varphi_1(x_n)\cdots\bar{c}(u_k)\rangle^c \\
&+ \sum_{i=1}^{m} \langle T\, c(x)\ \varphi_1(x_1)\cdots\varphi_1(x_n)\ A_{\mu_1}(y_1)\cdots\partial_{\mu_i}\bar{c}(y_i)\cdots\bar{c}(u_k)\rangle^c \\
&+ m\sum_{i=1}^{p} \langle T c(x)\varphi_1(x_1)\cdots\varphi_1(x_n)A_{\mu_1}(y_1)\cdots A_{\mu_m}(y_m)\varphi_2(z_1)\cdots\bar{c}(z_i)\cdots\bar{c}(u_k)\rangle^c \\
&+ e\sum_{i=1}^{p} \langle T c(x)\varphi_1(x_1)\cdots\varphi_1(x_n)A_{\mu_1}(y_1)\cdots A_{\mu_m}(y_m)\varphi_2(z_1)\cdots(\varphi_2\bar{c})(z_i)\cdots\bar{c}(u_k)\rangle^c \\
&- \sum_{i=1}^{k} \langle T c(x)\cdots c(t_1)\cdots(\partial_\mu A^\mu + \rho\varphi_2)(t_i)\cdots c(t_k)\,\bar{c}(u_1)\cdots\bar{c}(u_k)\rangle^c
\end{aligned}$$

The reason for including Faddeev Popov fields within the Slavnov identities is that among other things, it is believed that they will turn out to be relevant to the unitarity problem of the physical S operator for the fully renormalized theory.

One can show that conversely, if such identities are to hold in the tree approximation the Lagrangian must be of the form initially postulated except for the possible occurrence of wave function renormalization factors Z_A, Z_ϕ in front of the terms $-\frac{1}{4} G_{\mu\nu} G^{\mu\nu}$, $(D_\mu \varphi)^* (D_\mu \varphi)$, respectively. Written in full, the Faddeev Popov contribution to $\mathcal{L}_{tree}$ is

$$\frac{1}{\alpha}\left(\partial_\mu \bar{c}\, \partial^\mu c + \rho m\, \bar{c} c + \rho e \varphi_1 \bar{c} c \right)$$

The Lagrangian then depends on the following parameters:

$$Z_A, m; Z_\phi, M, \rho, e, \alpha$$

which can be characterized as follows:

Z_A, m are related to the residue and pole of the transverse photon propagator

Z_ϕ, M are related to the residue and pole of the φ_1 field propagator

ρ is related to the common $[\partial_\mu A^\mu, \varphi_2, C]$ ghost propagator pole [4]

α is related to the residue of the Faddeev Popov ghost propagator

e is related to the $\varphi_1 \varphi_1 \varphi_1$ or $\varphi_1 \varphi_1 A^T$ scattering.

Alternatively, g is related to $\varphi_1 \varphi_1 \varphi_1 \varphi_1$ scattering. In all the following, we shall assume m, M and the ghost mass to be restricted by inequalities which insure the stability of all three types of particles. In order that the physical scattering amplitudes be gauge invariant, one sees that, by virtue of the Slavnov identities, physical states should only contain φ_1, and A^T quanta.

The question is now to construct the most general dimension four effective Lagrangian whose zeroth order approximation in terms of the loop counting parameter coincides with $\mathcal{L}_{tree}$ and inquire whether constraints can be put on the coefficients in such a way that Slavnov type identities may hold.

Let then

$$\begin{aligned}
\mathcal{L}^{(4)}_{eff} = & -\frac{1}{4}(1+b)G_{\mu\nu}G^{\mu\nu} + \frac{m^2+a}{2}A_\mu A^\mu - \frac{1}{2\alpha}(1+c)(\partial_\mu A^\mu)^2 \\
& + \frac{d}{4}(A_\mu A^\mu)^2 + \frac{1}{2}(1+B_1)\partial_\mu\varphi_1\partial^\mu\varphi_1 - \frac{M^2+A_1}{2}\varphi_1^2 \\
& + \frac{1}{2}(1+B_2)\partial_\mu\varphi_2\partial^\mu\varphi_2 - \frac{1}{2}\left(\frac{\rho^2}{\alpha}+A_2\right)\varphi_2^2 + \left(\frac{\rho}{\alpha}-m-\tilde{f}_2\right)A_\mu\partial^\mu\varphi_2 \\
& + e(1+f_1)A_\mu\varphi_2\partial^\mu\varphi_1 - e(1+f_2)A_\mu\varphi_1\partial^\mu\varphi_2 + \frac{e^2}{2}(1+g_1)A_\mu A^\mu\varphi_1^2 \\
& + \frac{e^2}{2}(1+g_2)A_\mu A^\mu\varphi_2^2 + em(1+\tilde{g}_1)A_\mu A^\mu\varphi_1 \\
& - \frac{e^2M^2}{2m^2}(1+h_1)\varphi_1^4 - \frac{e^2M^2}{2m^2}(1+h_2)\varphi_2^4 - \frac{e^2M^2}{m^2}(1+h_{12})\varphi_1^2\varphi_2^2 \\
& - \frac{2eM^2}{m}(1+k_1)\varphi_1^3 - \frac{2eM^2}{m}(1+k_2)\varphi_2^2\varphi_1 \\
& + \frac{1+H}{\alpha}\Big[\partial_\mu\bar{c}\,\partial^\mu c - (\rho m+A)\bar{c}c - \rho e(1+\ell)\bar{c}\varphi_1 c\Big] \\
& \qquad + r_1\bar{c}\varphi_1^2 c + r_2\bar{c}\varphi_2^2 c + s\,\bar{c}A_\mu A^\mu c \;\Big]
\end{aligned}$$

We now look for an identity of the Slavnov type, recalling the conservation of the number of Faddeev Popov ghost pairs, and the invariance under charge conjugation:

$$A_\mu \to -A_\mu\,;\quad \varphi_2 \to -\varphi_2\,;\quad \varphi_1\to\varphi_1\,;\quad c\to c\,;\quad \bar{c}\to\bar{c}\,.$$

One can use the effective equations of motion for $\bar{c}, \partial_\mu A^\mu, \varphi_1, \varphi_2$ under the following combinations: [3]

$$\begin{aligned}
\int N_5\,\bar{c}\,\Box\,\partial_\mu A^\mu &= \mathrm{cst}\int \delta_{\bar{c}}\,\partial_\mu A^\mu + \int N_5\,\delta_{\bar{c}}\,\partial_\mu A^\mu \\
&= \mathrm{cst}\int \bar{c}\,\partial_\mu\,\delta_{A_\mu} + \int \bar{c}\,\partial_\mu J^\mu
\end{aligned}$$

$$\int N_4 \bar{c} \Box \varphi_2 = cst \int \varphi_2 \delta_{\bar{c}} + \int N_4 j_{\bar{c}} \varphi_2$$
$$= cst \int \bar{c} \delta_{\varphi_2} + \int N_4 \bar{c} j_2$$

$$\int N_5 \bar{c} \Box (\varphi_1 \varphi_2) = cst \int \delta_{\bar{c}} N_2 (\varphi_1 \varphi_2) + N_5 j_{\bar{c}} \{\varphi_1 \varphi_2\}$$
$$= \int 2 \bar{c} \partial_\mu \varphi_1 \partial^\mu \varphi_2 + cst \int \bar{c} \varphi_2 \delta_{\varphi_1}$$
$$- \int N_5 j_1 \{\bar{c} \varphi_2\} + cst \int \bar{c} \varphi_1 \delta_{\varphi_2} + \int N_5 j_2 \{\bar{c} \varphi_1\}$$

where the brackets $\{\cdot\}$ indicate anisotropic normal products and $j_1, j_2, j_{\bar{c}}, J_\mu$ are simply related (within the addition of mass terms and the multiplication through wave function renormalization coefficients) to the corresponding field sources.

The reduction of anisotropic to isotropic normal product, as well as the anisotropic use of the equations of motion for φ_1, φ_2 in the evaluation of $\partial^\mu J_\mu$ allows to cast these three identities into the form

$$\int \partial_\mu A^\mu \delta_{\bar{c}} \quad = \text{wanted terms} + \text{unwanted terms}$$
$$\int \varphi_2 \delta_{\bar{c}} \quad = \text{wanted terms} + \text{unwanted terms}$$
$$\int N_2 \varphi_1 \varphi_2 \delta_{\bar{c}} \quad = \text{wanted terms} + \text{unwanted terms}$$

Wanted terms are of the form $\int \bar{c} \partial_\mu \delta_{A_\mu}$, $\int \bar{c} \delta_{\varphi_2}$, $\int N_2 \bar{c} \varphi_1 \delta_{\varphi_2}$, $\int N_2 \bar{c} \varphi_2 \delta_{\varphi_1}$

After repeated use of Zimmermann's identities unwanted terms can be cast into the form of itegrals of the following monomials, all counted with dimension 5:

$\bar{c} \varphi_2$,

$\bar{c} \partial_\mu A^\mu$, $\bar{c} \varphi_1 \varphi_2$

$$\bar{c}\varphi_2\varphi_1^2,\ \bar{c}\varphi_2 A_\mu A^\mu,\ \bar{c}\varphi_2^3,\ \bar{c}\varphi_1\partial_\mu A^\mu,\ \bar{c}A_\mu\partial^\mu\varphi_1,$$
$$\bar{c}\varphi_2\varphi_1^3,\ \bar{c}\varphi_2^3\varphi_1,\ \bar{c}A_\mu A^\mu\varphi_1\varphi_2,\ \bar{c}\partial_\mu A^\mu\varphi_1^2,\ \bar{c}A_\mu\partial^\mu\varphi_1\varphi_1,$$
$$\bar{c}\,\partial_\mu A^\mu\varphi_2^2,\ \bar{c}\varphi_2\partial_\mu\varphi_2 A^\mu,\ \bar{c}\,\partial_\mu A^\mu A_\lambda A^\lambda,\ \bar{c}A_\lambda\partial^\lambda(A_\mu A^\mu),$$
$$\bar{c}\,\partial_\mu\varphi_1\partial^\mu\varphi_2,\ \bar{c}\,\partial_\mu\bar{c}\,A^\mu c\,.$$

There are 19 unwanted terms whereas the Lagrangian depends on 25 coefficients. One of which is connected with the normalization of the Faddeev Popov ghost field. It turns out that one can express the unwanted term $N_5\,\bar{c}\varphi_1\varphi_2$ in terms of $\int\varphi_2\,\delta_{\bar{c}}$, wanted terms, and other unwanted terms, and, similarly $\int N_5\bar{c}\,\partial_\mu\varphi_1\partial^\mu\varphi_2$ in terms of $\int N_2\,\varphi_1\varphi_2\;\delta_{\bar{c}}$, wanted terms and other unwanted terms. Imposing then the vanishing of the remaining 17 unwanted terms, one gets Slavnov identities of the form:

$$\langle T\left[\partial_\mu A^\mu+(\rho+\delta\rho)\varphi_2+\delta\rho_{12}N_2(\varphi_1\varphi_2)\right]\varphi_1(x_1)\cdots\varphi_1(x_n)$$
$$A_{\mu_1}(y_1)\cdots A_{\mu_m}(y_m)\,\varphi_2(z_1)\cdots\varphi_2(z_p)\,c(t_1)\cdots c(t_k)\,\bar{c}(u_1)\cdots\bar{c}(u_k)\rangle^c$$
$$=-e\left(\frac{1+\beta_1}{1+H}\right)\sum_{i=1}^{n}\langle T\,c(x)\varphi_1(x_1)\cdots(\varphi_2\bar{c})(x_i)\cdots\varphi_1(x_n)\cdots\cdots\bar{c}(u_k)\rangle^c$$
$$+\left(\frac{1+\beta_A}{1+H}\right)\sum_{i=1}^{m}\langle T\,c(x)\cdots\cdots A_{\mu_1}(y_1)\cdots\partial_{\mu_i}\bar{c}(y_i)\cdots\cdots\bar{c}(u_k)\rangle^c$$
$$+\,m\left(\frac{1+\beta_{12}}{1+H}\right)\sum_{i=1}^{p}\langle T\,c(x)\cdots\cdots\varphi_2(z_1)\cdots\bar{c}(z_i)\cdots\varphi_2(z_p)\cdots\cdots\bar{c}(u_k)\rangle^c$$
$$+e\left(\frac{1+\beta_2}{1+H}\right)\sum_{i=1}^{p}\langle T\,c(x)\cdots\cdots\varphi_2(z_1)\cdots(\varphi_1\bar{c})(z_i)\cdots\varphi_2(z_p)\cdots\cdots\bar{c}(u_k)\rangle^c$$
$$-\sum_{i=1}^{k}\langle T\,c(x)\cdots\cdots c(t_1)\cdots\left[\partial_\mu A^\mu+(\rho+\delta\rho)\varphi_2+\delta\rho_{12}N_2(\varphi_1\varphi_2)\right](t_i)\cdots$$
$$c(t_k)\cdots\cdots\bar{c}(u_k)\rangle^c\,.$$

where the coefficients $\delta\rho$, $\delta\rho_{12}$, β_1, β_A, β_{12}, β_2 are finite formal power series in $\hbar$. Apart from the fact that α only occurs in the combination $\alpha/(1+H)$ it is not known at present whether simple relations automatically hold between these coefficients because of the complicated way in which they were obtained. In particular, there is not a priori reason why $\delta\rho_{12}$ should vanish. In other words, had we started from a gauge function of the form $B = \partial_\mu A^\mu + \rho\varphi_2 + \rho_{12}\varphi_1\varphi_2$ in the tree approximation- the most general expression of dimension 2 odd under charge conjugation - we would have obtained a Slavnov identity involving a renormalized gauge function of the form $B^{Ren.} = \partial_\mu A^\mu + (\rho+\delta\rho)\varphi_2 + (\rho_{12}+\delta\rho_{12}) N_2\, \varphi_1\varphi_2$ and even when ρ_{12} vanishes the induced $\delta\rho_{12} N_2(\varphi_1\varphi_2)$ term may survive. In fact, if we believe that the tree approximation property

$$\langle T\ B(x)\, B(y)\rangle \propto \delta(x-y)$$

survives through radiative corrections, one can see that the $N_2(\varphi_1\varphi_2)$ induced term has to be present.

Once the Slavnov identities have been obtained, there are, besides the combination $\alpha/(1+H)$, seven parameters left free which can be used to fulfill the following normalization conditions: position of the poles, residues being unity at these poles for the transverse photon and φ_1 propagator, mass shell value of the three or four φ_1 (or $\varphi_1\varphi_1 A^T$) vertices equal to the tree approximation value, and finally double vanishing of the inverse determinant of the coupled $\partial_\mu A^\mu$, φ_2 propagator matrix at $p^2 = \rho m^2$.[6] The last condition combined with the Slavnov identity implies that the Faddeev Popov ghost propagator has a simple pole at $p^2 = \rho m^2$, whose residue can be normalized to one by means of the parameter H, if one wishes to do so. The theory is then completely interpretable within a Fock space with indefinite metric whose structure will be reported elsewhere together with the appropriate asymptotic theory.

The crucial test for the correctness of this renormalization scheme of course relies on the check of unitarity and gauge invariance of the physical S operator which has not yet been attacked within the present framework, and, next the existence of local gauge invariant observables which leave the physical subspace invariant, in the same way as in quantum electrodynamics.[7]

Chapter V: References and Footnotes

[1] This is the strategy advocated by J. Schwinger in: J. Schwinger "Particles Sources and Fields", Addison Wesley Pub. Co., Reading, Mass. (1970).
J. Schwinger, "Particles and Sources", Gordon & Breach New York (1970) (Brandeis 1967). See also:
A. Rouet, R. Stora, Lectures given at the Universities of Geneva and Lausanne (1973).

[2] Ch. III, Refs. [1] and [8]. The determination of an L_4 Lagrangian such that Ward identities hold, performed in Ref. [7] of Ch. III requires one more constraint than is allowed by the number of parameters at disposal. That constraint, a relation between some r coefficients, can be shown to be automatically fulfilled by an argument concerning the high momentum behaviour of vertex functions. This argument is due to O. Piguet and similar to one used by him in the construction of an L_4 Lagrangian for the Higgs Kibble model with massive photons of Chapter IV).
(O. Piguet, private communication).

[3] More details can be found in A. Rouet, R. Stora, Ref. [1], and A. Rouet, article in preparation.

[4] We wish to thank B.W. Lee for a discussion on this point.

[5] This way of obtaining the Slavnov identities can be found in: L. Quaranta, A. Rouet, R. Stora, E. Tirapegui "Spontaneously broken gauge invariance: Ward identities, Slavnov identities, gauge invariance", in "Renormalization of Yang Mills fields and Applications to particle Physics", CNRS Marseille, June 19-23 (1972), where however the treatment of renormalization is formal due to the a priori possible occurrence of infrared difficulties which does not happen in the treatment given in this chapter.

[6] Such a normalization condition was used in a version of the renormalization of the Higgs Kibble model in Stueckelberg's gauge, by J.H. Löwenstein, M. Weinstein, W. Zimmermann.

[7] J.H. Löwenstein, B. Schroer, PR. D 6, 1553 (1972).

GROUP THEORETICAL APPROACH TO CONFORMAL INVARIANT QUANTUM FIELD THEORY

G. MACK

Institut für Theoretische Physik

der Universität Bern, Switzerland

1. - INTRODUCTION - In this set of lectures we will address ourselves to the question of the construction and properties of a non-trivial exactly conformal invariant quantum field theory (QFT). As has been explained in some detail in Symanzik's lectures (1) such a theory, if it exists, has a good chance of being relevant to the description of the real world in certain high energy limits - or possibly at intermediate energies as was suggested by Wilson (2). The basic hypothesis and some predictions of the theory can therefore in principle be tested by experiment. Besides, conformal QFT is also interesting as a laboratory, because it can be analysed to a remarkable extent by strictly non-perturbative methods, i.e. without recourse to iterative techniques. It therefore offers welcome insight into the structure of local quantum field theory. To one's surprise one finds structures much reminiscent of dual resonance models. We hope to illustrate this in what follows. Lastly, it offers an example of how the geometry of spacetime can affect and to some extent determine the structure of a theory of fundamental processes. This aspect is less trivial than might be thought, see the discussion below in Secs.3 (fixed points) and especially Sec.5. There is also another approach to conformal QFT which is based on iterative techniques; it has already been reviewed in the author's Kaiserslautern lectures (3).

Our starting point will be the fundamental integral equation which Green functions in Lagrangean field theory are known to obey (4). They are an infinite set of coupled nonlinear integral equations for all the n-point Greenfunctions, $n = 2...\infty$. Because of this they have up to now had a reputation of being intractable. However we shall show here that in a conformal invariant theory all these equations can be simultaneously diagonalized, i.e. reduced to algebraic equations. This is done with the help of a group theoretical method, the conformal partial wave expansion. We will be interested in theories for which the conformal partial waves are analytic functions in a complex χ-plane*. The fundamental integral equations then amount to requiring presence of simple poles with factorizable residues. In the special case of the 4-point function the residues must be positive numbers because of axiomatic positivity. We will also show how to define Green functions involving tensor fields of higher rank. Using these we will finally be able to derive operator product expansions à la Wilson-Zimmermann. The c-number coefficients therein will of course obey the constraints from conformal symmetry which have previously been obtained by Ferrara et al. [5]. We conclude with a brief discussion of the obstacles which are still in the way of actually constructing an acceptable theory. Some of them come from the fact that the Green functions must be crossing symmetric, i.e. symmetric in their arguments $x_1 \ldots x_n$, while the fundamental integral equations are not manifestly crossing symmetric.

Graphical notation : In these lectures we shall study quantum field theory by looking at its connected Green functions :

$$G(x_1 \ldots x_n) = \langle 0|T(\phi(x_1)\ldots\phi(x_n))|0\rangle_c = \tag{1.1}$$

A graphical notation will be used as indicated, in which an n-point Green function will be represented by a bubble with n long legs. This notation will occasionally be used also for n = 2 , but usually the dressed 2-point function will be represented by a line

$$G(x_1 x_2) = \text{———} , \tag{1.2}$$

and sawing off a leg from a bubble means amputation with the dressed inverse propagator $G^{-1}(x_1 x_2)$.
Because of the spectrum condition the Green functions can be continued to the Euclidean domain

$$x^4 = ix^o \text{ real}$$

Later we shall consider such Euclidean Green functions most of the time. This simplifies the group theoretical analysis.

---

*Validity of an analogous property in canonical perturbation theory is implicit in Weinberg's power counting theorem.

2. - FUNDAMENTAL INTEGRAL EQUATIONS - To build a theory one must first of all have a set of equations from which the Green functions should be determined. It is well known that the Green functions in Lagrangian QFT satisfy an infinite set of coupled non-linear integral equations, derivable e.g. from a cutoff Lagrangian with subsequent removal of the cutoff . For simplicity I will write down the equation for ϕ^3-theory (in $D \leq 6$ space time dimensions) as they were obtained by Symanzik long ago (4); they are given in figure 1 below. Similar equations exist for other theories.

Summation in the last term of equation (a) and (c) is over partitions of the lower legs into 2 groups containing each at least one leg.

Some words of explanation may be needed. The equations involve not only the Green functions themselves but also certain auxiliary amplitudes.

Firstly one defines amplitudes which are 1-particle irreducible in one channel, they are constructed out of Green functions proper according to

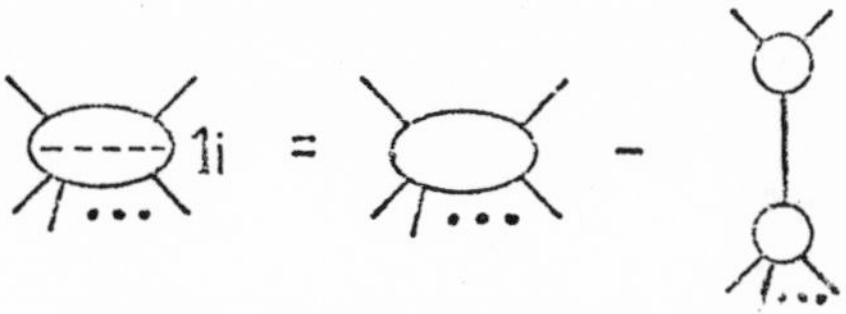

(2.1)

Here and below the dotted lines in the bubble may serve to remind one that the amplitude is not symmetric in all its arguments but only in those attached to the upper and lower legs separately.

The other auxiliary amplitudes are the Bethe-Salpeter kernel and the so-called "2-particle-irreducible kernels", they are distinguished by a symbol "B" resp. "2i". For our purposes they should be thought of as determined in terms of the Green functions by the Bethe-Salpeter equations, figures 1(i). In the case of the 2i-kernels this is a straight definition, while for the Bethe-Salpeter kernel one must solve an integral equation.

If the BS- and 2i-kernels are thought of as so expressed in terms of the Green functions, then the dynamical integral equations, figure 1(ii), are equations relating in general the n-point function to an n + 1 - point function and lower ones, so they are coupled. There are infinitely many, one for every n; they are integral equations since there is always a term which involves an integration over loop momentum. Two of them are in fact integro-differential equations, they are the renormalized Schwinger Dyson (SD)-equations, for vertex and propagator. The latter has been written here in a peculiar form (6) which differs from ref.4 but is equally correct (see below). A slight modification of the derivation in ref.4 would instead lead to the equation shown in Fig.2.

The primes ' stand for differentiation with respect to momentum

(i) Bethe Salpeter equations → define kernels

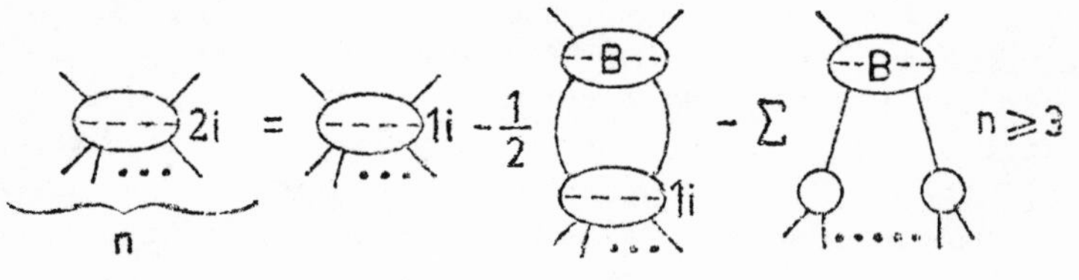

(a) define "2i-kernel"

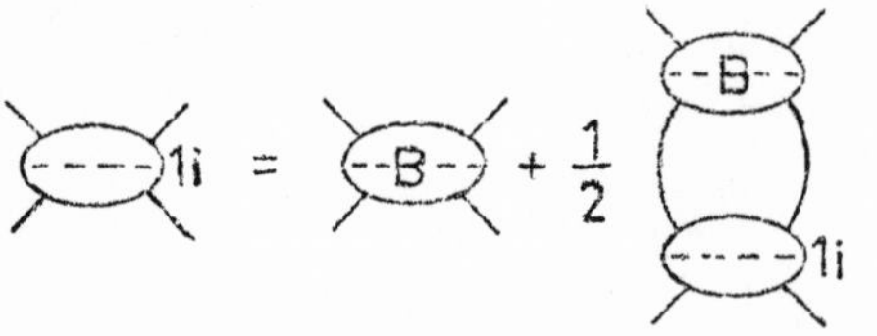

defines BS-kernel (b)

(ii) Dynamical equations (for ϕ^3-theory)

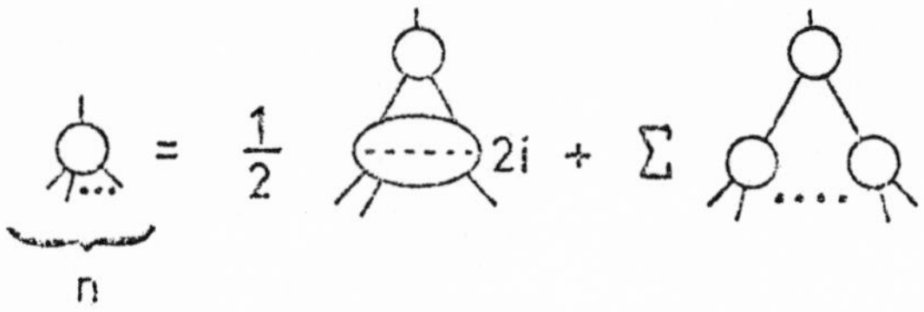

(c)

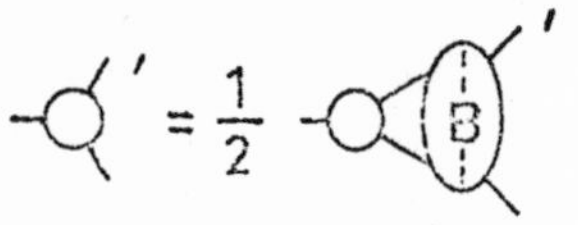

SD-Eq. for Vertex (d)

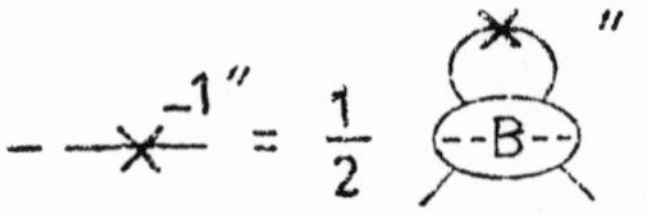

ren. SD-Eq. for Propagator (e)

Fig.1. Fundamental integral equations = renormalized integral equations for n-point functions.

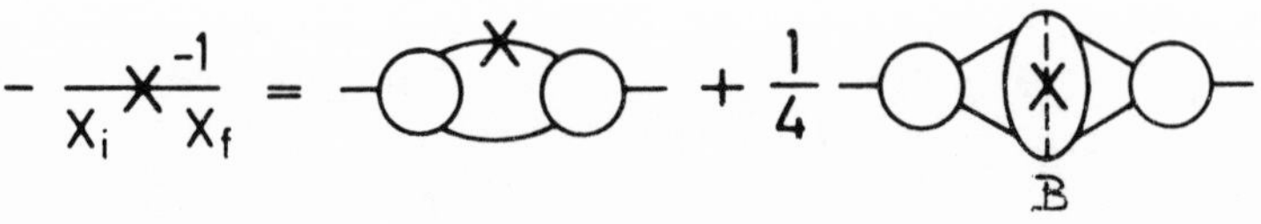

Fig.2. Alternative form of the ren. SD-Equation for the propagator. Let the (amputated) Bethe-Salpeter kernel be denoted by $B(x_1 x_2; x_3 x_4)$. Then the "crossed" BS-kernel is $\sum_{i=1}^{4} x_i^\mu \frac{\partial}{\partial x_i^\nu} B(x_1 x_2; x_3 x_4)$, $\mu \neq \nu$; it appears in the last term on the RHS. For other notation see text. $x_i \neq x_f$.

flowing between two external legs, or multiplication with a coordinate difference,

$$' = (x_i^\nu - x_j^\nu) \qquad (2.2)$$

Similarly the cross $\times = (x_1^\nu - x_2^\nu) \dfrac{\partial}{\partial x_1{}^\mu}$, $\mu \neq \nu$.

<u>Physical significance</u>: The fundamental integral equations, Fig.1, are meaningful and true in canonical perturbation theory. However, they are more general. First of all they can be derived from a Lagrangean, with no reference to iterative techniques* (4). Moreover, something can be said about their rôle in enforcing basic principles.

In a local theory there ought to exist a conserved stress energy tensor $\theta_{\mu\nu}(x)$, which provides us with a Hamiltonian $H = P^o$, and generators P^i of space translations. They are given by $P^\mu = \int d\underline{x}\theta^{\mu o}(x)$ and must satisfy correct commutation relations with the fields, viz. $[\phi(x), P^\mu] = i\partial^\mu\phi(x)$. These will hold if the Green functions involving the stress tensor

$$G_{\mu\nu}(x;y_1\ldots y_n) = \langle T\left(\theta_{\mu\nu}(x)\phi(y_1)\ldots\phi(y_n)\right)\rangle \qquad (2.3)$$

satisfy the correct Ward-Takahashi-identities (WTI) (7). It will be explicitly shown below how the Green functions (2.3) are constructed in a conformal invariant theory. Moreover it has been verified by the author (8) without recourse to iterative techniques, that validity of the WTI follows from the fundamental integral equations, Fig.1, plus one technical assumption: One has to assume that a certain system of homogeneous linear integral equations has no nontrivial (crossing symmetric and sufficiently nonsingular) solution, this guarantees that the solution of the corresponding inhomogeneous equations is unique.

<u>Boundary conditions</u>: Because some of the fundamental integral equations are in fact integro-differential equations one must in addition supply boundary conditions.

In canonical perturbation theory one solves all the fundamental integral equations by iteration. At each iteration step one uses as boundary condition the usual renormalization conditions. In momentum space:

---

*Strictly speaking one obtains in this way Fig.2 for propagator's integral equation, instead of Fig.1e (6) . The latter equation is however more fundamental, it is needed to construct the stress tensor's 3-point function. Conversely, both equations are equivalent modulo validity of the Ward-Takahashi-identities mentioned below.

$$G^{-1}(p,-p) = 0 \quad \text{at } p^2 = m^2, \qquad \frac{\partial}{\partial p^2} G^{-1}(p,-p) = -iz \quad \text{at } p^2 = \mu^2$$

$$V(\hat{p}_1 \hat{p}_2 \hat{p}_3) = -igz^{3/2} \quad \text{for} \quad \hat{p}_i \cdot \hat{p}_j = \frac{1}{2}\mu^2 (3\delta_{ij} - 1) \tag{2.3}$$

The parameter z fixes the normalization of the field and may be chosen according to convenience. Customarily one puts $z = 1$ and $\mu = m$. The vertex V is the amputated 3-point Green function.

In a conformal invariant theory, in contrast, the role of boundary condition is played by the requirement of conformal symmetry. One can in this case drop the primes ' in Figs.1d,e if in Fig.1e one restricts attention to noncoinciding external arguments. This suffices since a dilatation invariant 2-point function is determined uniquely (as a distribution) by its values for noncoinciding arguments.

3. - DILATATION AND CONFORMAL INVARIANCE - In the following we shall inquire into the possibility of attacking the fundamental integral equations of Fig.1 nonperturbatively, i.e. without recourse to iteration. Our starting point will be the observation that the equations (Fig.1) have a lot of symmetry.

The crucial point is this: the equations of figure 1 do not depend explicitely on mass m or coupling constant g nor any other external parameters, but only on the Green functions themselves. Mass and coupling constants enter only in the boundary conditions (2.3). Thus in particular there is no dimensional parameter in this whole system of equations, therefore one will naturally expect that they are exactly dilatation invariant. And indeed they are that and more.

Let us be more precise. Dilatations are coordinate transformations.

$$x^\mu \to \rho x^\mu \,, \quad \rho > 0 \quad \text{so that } ds^2 \to \rho^2 ds^2$$

for the line element. The field transformation law under infinitesimal dilatations can be found from the theory of induced representations [9], viz. $\delta\phi = -i\varepsilon\hat{D}\phi$ with

$$\hat{D}\phi(x) = i(d + x_\nu \partial^\nu)\phi(x) \tag{3.1}$$

The "dimension" of ϕ will be taken to be a real number (we thus exclude the possibility of a non-diagonalizable matrix), it is a new quantum number analogous to spin. Through (3.1) also the transformation law of the Green functions is fixed, and one can check that the whole system of equations (Fig.1) is indeed exactly invariant, for any d. Following Wilson we shall allow non-canonical, i.e. noninteger dimensions d.

Moreover the equations (Fig.1) allow for an even larger symmetry: they are also exactly invariant under conformal transformations

$$x^\mu \to \sigma(x)^{-1}(x^\mu - c^\mu x^2) \quad \text{where } \sigma(x) = 1 - 2c\cdot x + c^2x^2$$

where c^μ characterizes the transformation. Conformal transformations are a kind of space time dependent dilations [4], as is evident from the transformation law of the line element

$$ds^2 \to \sigma(x)^{-2}ds^2$$

The field transformation law is

$$\hat{K}_\mu\phi(x) = -i(2dx_\mu + 2x_\mu x_\nu\partial^\nu - x^2\partial_\mu - 2ix^\nu\Sigma_{\mu\nu})\phi(x) \tag{3.2}$$

where $\Sigma_{\mu\nu} = 0$ for scalar fields ϕ.

Summing up we emphasize once more the

FACT : The fundamental integral equations (Fig.1) are exactly dilatation- and conformal invariant.

Euclidean Green functions: It is convenient to look at the Green functions in the Euclidean domain, i.e. for imaginary times (cf. Sec.1). The conformal group in the Euclidean domain is SO(5,1) and the Lorentz group is compact SO(4), resp. SO(D+1,1) and SO(D) for D space time dimensions.

While in Minkowski space one can only consider infinitesimal conformal transformations, the Green functions in Euclidean space are globally invariant under the conformal group, according to the hypothesis of "weak conformal invariance" [10]. To exploit this it is however necessary to compactify the Euclidean space by adding points at infinity. This is best done by using the familiar manifestly conformal invariant formalism. One introduces a projective space of positive light-like 6-vectors (resp. D+2 - vectors) related to Euclidean coordinates $\{x\}$ by

$$\xi^\mu = \kappa x^\mu \quad (\mu = 1\ldots4) \; , \quad \kappa = \xi^6 - \xi^5 \; , \quad \xi^6 > 0$$

$$\xi^2 = g_{AB}\xi^A\xi^B = 0 \qquad \text{with} \qquad g_{AB} = \text{diag}(++++,+-)$$

Euclidean points at ∞ are mapped into points on the cone with $\kappa = 0$

The Green functions can be considered as functions of the ξ_i by virtue of the definition

$$G(\xi_1\ldots\xi_n) = \kappa_1^{-d}\ldots\kappa_n^{-d}\; G(x_1\ldots x_n)$$

with d the dimension of the field $\phi(x)$. As a consequence they are homogeneous in each variable separately

$$G(\rho\xi_1,\xi_2\ldots\xi_n) = \rho^{-d}G(\xi_1\ldots\xi_n) \quad \text{etc.} \quad \text{for } \rho > 0$$

Requirement of conformal invariance (to be imposed later) will then read

$$G(\Lambda\xi_1 \ldots \Lambda\xi_n) = G(\xi_1 \ldots \xi_n)$$

for any pseudo-orthogonal matrix $\Lambda \in SO(5,1)$. In the following we shall, for simplicity of presentation, write formulae in x-space and ignore subtleties associated with points at infinity.

The conformal invariant 2-point function is determined (up to physically irrelevant normalization) by conformal invariance, viz.

$$G(x,y) = (2\pi)^{-\frac{1}{2}D}\, \Gamma(d)\, \left(\tfrac{1}{2}[x-y]^2\right)^{-d} / \Gamma(\tfrac{1}{2}D-d) \tag{3.3a}$$

The inverse in the convolution sense is

$$G^{-1}(x,y) = (2\pi)^{-\frac{1}{2}D}\Gamma(D-d)\left(\tfrac{1}{2}(x-y)^2\right)^{-D+d} / \Gamma(d-\tfrac{1}{2}D) \tag{3.3b}$$

Fixed points and asymptotic limits These questions have been dealt with in some detail in Symanziks lectures at this school [1]. We will only add few remarks on how these considerations fit into a theory where one starts from the fundamental integral equations of figure 1.

We have emphasized the fact that these equations are exactly dilatation and conformal invariant. Suppose then that we are given some Poincaré-invariant solution of these equations (which may or may not possess a higher symmetry). Let us moreover assume that any such solution is uniquely determined by the values of parameters m^2, g, z defined through Eqs.(2.3). The fundamental integral equations being invariant, the transform of any solution must evidently be another solution. In general this new solution must be expected to satisfy different boundary conditions. Let us consider dilatations by a factor $\rho \equiv \lambda^{-1}$. In this case the transformed solution is again Poincaré-invariant and will satisfy boundary conditions of the same form (2.3) characterized by parameters m'^2, g', z' :

original solution		transformed solution
m^2, g, z	$\longrightarrow$ dilatations by λ^{-1}	m'^2, g', z'

These are exactly the facts expressed by the renormalization group [11] ($m^2 = 0$) resp. Callan-Symanzik equations [12] ($m^2 \neq 0$). It is easy to show that

$$m'^2 = \lambda^2 m^2 \quad \text{while} \quad g' = g'(\lambda, g)$$

is a function of λ and g which satisfies a certain composition law that follows from the group property of dilatations. As a consequence it is uniquely determined by $\beta(g) \equiv \frac{\partial}{\partial\lambda^2} g'(\lambda,g)_{\lambda=1}$.

With the equations being invariant it is natural to ask whether they also have some invariant solutions. A trivial one is found right away: For a massless free field all the equations are of the form 0 = 0 and the solution is invariant if the dimension d = 1 (resp. $\frac{1}{2}(D-2)$) in the transformation laws (3.1), (3.2). We shall make the HYPOTHESIS: There exists a nontrivial dilatation and conformal invariant solution, with the coupling constant $g_\infty \neq 0$.

Evidently this requires that $g'(\lambda,g_\infty) = g_\infty$, i.e. $\beta(g_\infty) = 0$. The Green functions of the conformal invariant solution will temporarily be denoted by $G_{GML}(x_1 \ldots x_n)$, their physical significance depends on the slope $\beta'(g_\infty)$. The well known result of the Callan Symanzik analysis, [5,6] is that in momentum space

$$G(\lambda p_1 \ldots \lambda p_n) = G_{GML}(\lambda p_1 \ldots \lambda p_n)$$

$$\text{as} \quad \begin{cases} \lambda \to \infty \text{ if } \beta'(g_\infty) < 0 \\ \lambda \to 0 \text{ if } \beta'(g_\infty) > 0 \text{ and } m^2 = 0 \end{cases} \tag{3.4}$$

except for overall normalization, g_∞ is then called an UV-stable resp. IR-stable fixed point. The general case of several coupling constants has been discussed by Wilson [2]. Assuming the coupling constants are sufficiently close to the eigenvalues there will be a range of λ for which approximate equality (3.4) holds. This range may or may not extend to ∞ ("intermediate energy scaling").

For all the known theories in D > 2 dimensions $\beta(g) \neq 0$ (this can be legitimately concluded from perturbation theory). Therefore one expects at most a discrete set of solutions of the eigenvalue condition $\beta(g_\infty) = 0$. As a consequence, a conformal invariant theory has no continuous free parameter if the above mentioned assumption holds that m^2, g, z suffice to determine a solution of Eqs. (Fig.1). In the following we shall however not assume this, and the question of uniqueness is then still open.

From now on we shall concentrate on the properties of exactly conformal invariant theories and their Green functions will be simply denoted by $G(x_1 \ldots x_n)$.

4. - CONFORMAL PARTIAL WAVE EXPANSIONS - The expansions are analogous to the familiar Jacob-Wick partial wave expansion with respect to the rotation group. The main difference is that we will expand Green functions in coordinate space, and that our group is noncompact. We shall first state the result, the group theoretical interpretation will be given in the next section.

Irreducible representations of SO(5,1) [resp. SO(D+1,1)] are characterized by $\chi = [\ell,\delta]$, where ℓ specifies a representation of the Lorentz group SO(4) [resp. SO(d)], and δ is a complex number. Here we will only need completely symmetric traceless tensor representations of the Lorentz-group, their rank will also be denoted by ℓ.

The partial waves of a Euclidean n-point Green function $G(x_1 \ldots x_n)$ will be denoted by

$G_\alpha^\chi\ (x|x_3 \ldots x_n) \equiv$ 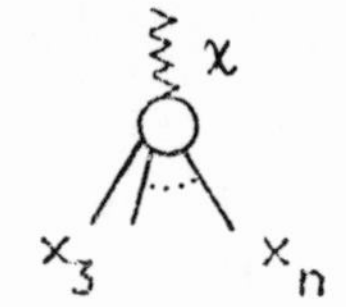 (4.1)

here $\alpha = (\alpha_1 \ldots \alpha_\ell)$ is a Lorentz multi-index, with the same ℓ as in $\chi = [\ell,\delta]$.

They are conformal invariant and transform with respect to $x_3 \ldots x_n$ in the same way as the Green functions, while they transform with respect to x according to equations (3.1), (3.2) with d replaced by δ, and $\Sigma_{\mu\nu}$ the completely symmetric traceless ℓ-th rank tensor representation of the Lorentz group, acting on indices $\alpha_1 \ldots \alpha_\ell$.

To write down the conformal partial wave expansion one needs in addition the representation functions, or rather Clebsch-Gordan kernels

$\Gamma_\alpha^{-\chi}(x_1 x_2|x) \equiv$ 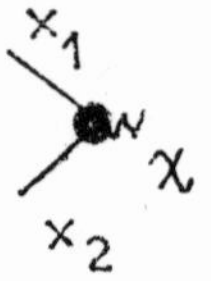 (4.2)

They are uniquely determined conformal invariant 3-point functions and will be given explicitly below; again $\alpha = (\alpha_1 \ldots \alpha_\ell)$, and $-\chi \equiv [\ell, D-\delta]$.

The expansion is then

 $= \displaystyle\oint\!\!\!\!\!\sum d\chi$ 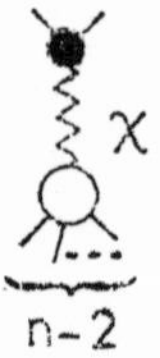 $= \displaystyle\oint\!\!\!\!\!\sum d\chi \int dx\, \Gamma_\alpha^{-\chi}(x_1 x_2|x) \cdot G_\alpha^\chi\ (x|x_3 \ldots x_n)$ (4.3)

In comparison with ordinary partial wave expansion χ plays the role of angular momentum ; x,α are "magnetic quantum numbers", and $\Gamma^{-\chi}$ replace the Jacobi-polynomials resp. Legendre functions. Integration is over the principal series of unitary irreducible representations* with $\ell = 0,2,4,\ldots$ (by Bose symmetry), viz.

$$\oint\!\!\!\!\!\sum d\chi = \frac{1}{2\pi i} \sum_\ell \int c(\chi)\, d\delta \qquad (4.4)$$

with integration running from $\frac{1}{2}D - i\infty$ to $\frac{1}{2}D + i\infty$, and the Plancherel

*except in D = 2 space time dimensions, if $d < \frac{1}{2}$. In this case there may occur in addition a discrete supplementary series contribution with $\chi = [0, D-2d]$, cp. ref. [13].

weight $c(\chi)$ is some polynomial in δ for even D. Explicitly, with $p \equiv \frac{1}{2}D$,

$$c(\chi) = \frac{1}{\ell!}\Gamma(p+\ell)\frac{1}{2}(2\pi)^{p}n(\chi)n(-\chi) \qquad (4.4b)$$

where

$$n(-\chi) = (2\pi)^{-p}\frac{\Gamma(2p-\delta+\ell)}{\Gamma(-p+\delta)}\{(\delta-1)\delta\ldots.(\delta-2+\ell)\}^{-1} \qquad (4.4c)$$

A special case of interest is the expansion of the 4-point function $G(x_1\ldots x_4)$. In this case the partial wave amplitude is a conformal invariant 3-point function and therefore its x-dependence is completely fixed

$G^{\chi}_{\alpha}(x|x_3x_4) \equiv$ 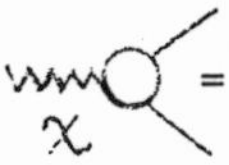 $= g(\chi)$ 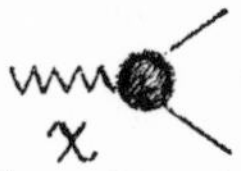 $= g(\chi)\Gamma^{\chi}_{\alpha}(x_3x_4|x)$ (4.5)

All the dynamical information is thus in the factor $g(\chi)$ (analog to conventional partial wave amplitudes a_ℓ.) Our graphical notation uses the group theoretical fact that Γ^{χ} can be obtained from $\Gamma^{-\chi}$ by amputation with a suitable propagator (see below).

The expansion for 4-point function and Bethe-Salpeter kernel can then be written as [14],

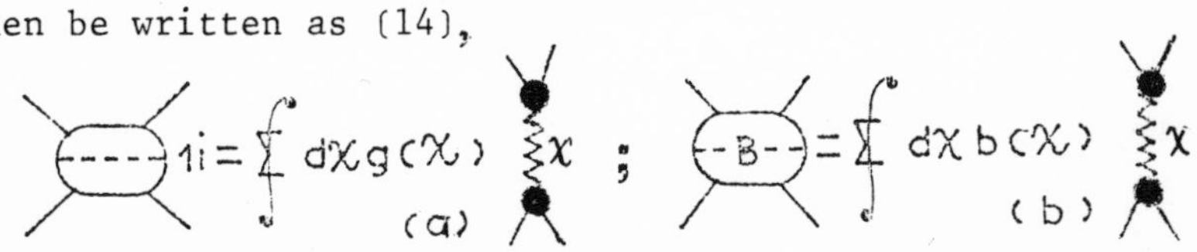

(4.6)

with partial wave amplitudes $g(\chi)$ resp. $b(\chi)$ depending only on χ.

There is a Plancherel-theorem which expresses completeness and orthogonality of the expansion functions. It gives [15]

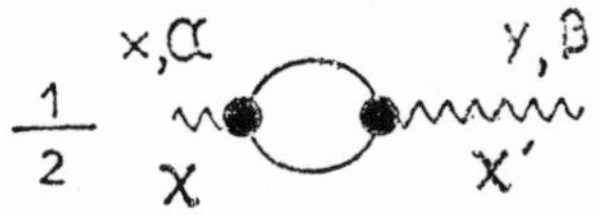

$$= \frac{1}{2}\delta_{\alpha\beta}\delta(\chi,\chi')\delta(x-y) + \qquad (4.7)$$

$$+ \frac{1}{2}\delta(\chi,-\chi')\ \Delta^{-\chi}_{\alpha\beta}(x-y) \quad ,$$

where $\delta(\chi,\chi')$ is essentially a Kronecker-δ in ℓ and a δ-function in dimensions δ. More precisely

$$\sum\!\!\!\!\!\!\int d\chi'\delta(\chi,\chi')f(\chi') = f(\chi)$$

Let us now give the explicit expression [15] for the Clebsch-Gordan kernels Γ^{χ}

$$\Gamma^{\chi}_{\alpha_1\ldots\alpha_\ell}(x_1x_2|x_3) = N(\chi)(\tfrac{1}{2}x^2_{12})^{-d+\frac{1}{2}\delta-\frac{1}{2}\ell} \cdot (\tfrac{1}{2}x^2_{13}\cdot\tfrac{1}{2}x^2_{23})^{-\frac{1}{2}(\delta-\ell)}$$

$$\cdot(\hat{x}_{\alpha_1}\ldots\hat{x}_{\alpha_\ell} - \text{traces}) \qquad (4.8)$$

where

$$x_{ij} = (x_i - x_j) \quad \text{and} \quad \hat{x}_\alpha = \frac{(x_{13})_\alpha}{\frac{1}{2}x_{13}^2} - \frac{(x_{23})_\alpha}{\frac{1}{2}x_{23}^2}$$

It is convenient to introduce also the intertwining kernel

$$\underset{\chi}{\sim\!\sim\!\sim\!\sim} \equiv \Delta^{\chi}_{\alpha_1\ldots\alpha_\ell,\beta_1\ldots\beta_\ell}(x)$$

$$= n(\chi)\left(\tfrac{1}{2}x^2\right)^{-\delta}\{g_{\alpha_1\beta_1}(x)\ldots g_{\alpha_\ell\beta_\ell}(x) - \text{traces}\} \qquad (4.9)$$

where $\quad g_{\alpha\beta}(x) = -g_{\alpha\beta} + 2x_\alpha x_\beta/x^2$

Normalization factors have been chosen such that Eq.(4.7) holds and

$$\Gamma^{-\chi}_{\alpha}(x_1x_2|z) = \int dz'\Gamma^{\chi}_{\beta}(x_1x_2|z')\,\Delta^{-\chi}_{\beta\alpha}(z'-z) \qquad (4.10)$$

Explicitly, $n(\chi)$ is given by (4.4c) and $N(\chi)$ may be chosen as (4.11)

$$N(\chi) = (2\pi)^{-p}\left\{\frac{\Gamma(-p+d_+ +\frac{1}{2}\delta+\frac{1}{2}\ell)\Gamma(d_+ -\frac{1}{2}\delta+\frac{1}{2}\ell)\Gamma(-d_- +\frac{1}{2}\delta+\frac{1}{2}\ell)\Gamma(d_- +\frac{1}{2}\delta+\frac{1}{2}\ell)}{\Gamma(2p-d_+ -\frac{1}{2}\delta+\frac{1}{2}\ell)\Gamma(p-d_+ +\frac{1}{2}\delta+\frac{1}{2}\ell)\Gamma(p+d_- -\frac{1}{2}\delta+\frac{1}{2}\ell)\Gamma(p-d_- -\frac{1}{2}\delta+\frac{1}{2}\ell)}\right\}^{\frac{1}{2}}$$

where $p \equiv \frac{1}{2}D$, and $d_+ = d$, $d_- = 0$ for our model of one scalar field of dimension d. The necessary computations to establish this have been done by Dobrev, Petkova, Petrova, Todorov and the author [16].

Using orthogonality relation (4.7) we can invert the expansion to obtain the partial waves from the Green-function, viz.

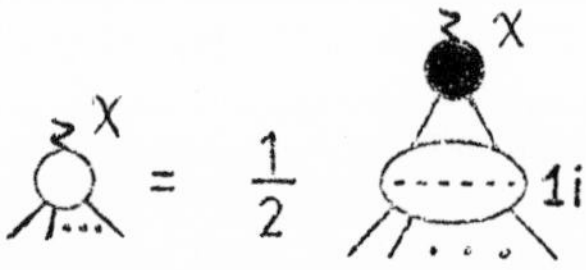
(4.12)

Using symmetry relation (4.10) of Clebsch-Gordon kernels we see that the partial waves satisfy an analogous symmetry property :

$$G^{-\chi}_{\alpha}(x|x_3\ldots x_n) = \int dx'\Delta^{-\chi}_{\alpha\beta}(x-x')G^{\chi}_{\beta}(x'|x_3\ldots x_n) \qquad (4.13a)$$

and therefore in particular

$$g(\chi) = g(-\chi) \tag{4.13b}$$

Finally the reader may wonder why we expand partly 1-particle irreducible Green functions rather than full ones. The reason will become clear in the next section. We remark that an expansion for the full connected Green functions also holds, it differs from expansion (4.3) only in the path of the χ-integration. This is however a nontrivial consequence of the dynamical integral equations as will be shown later. The result will be as follows:

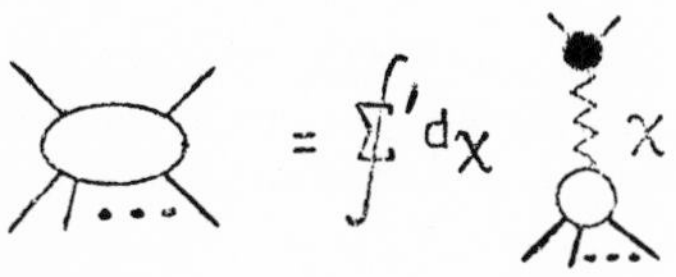

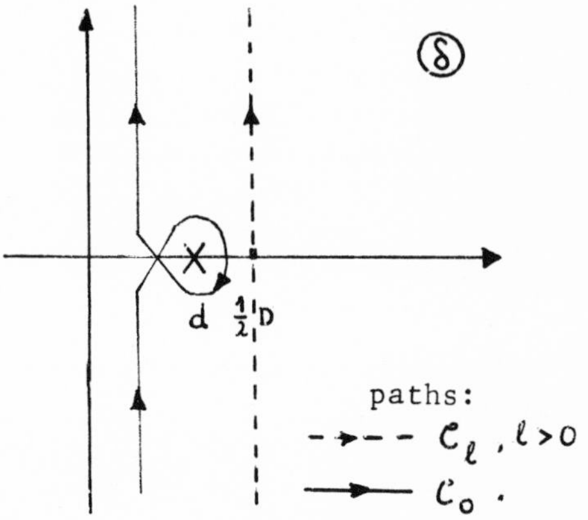

where (4.14)

$$\sum\!\!\!\!\!\!\int' d\chi = \frac{1}{2\pi i} \sum_{\ell} \int_{\mathcal{C}_\ell} c(\chi)\, d\delta$$

with the paths of integration running in the complex δ-plane as shown on the right. We emphasize that all these results are valid for Euclidean Green functions.

5. - GROUP THEORETICAL ORIGIN - First we will have to discuss briefly the unitary irreducible representations of the Euclidean conformal group SO(5,1) resp. (SO(D+1,1). It follows from a general theorem of Harish-Chandras that they can all be obtained as irreducible components of induced representations which are constructed as follows.

One starts from the fact that every element Λ of the Euclidean conformal group $G \equiv SO(D+1,1)$ can be uniquely decomposed as

$$\Lambda = kan \tag{5.1}$$

where n is a special conformal transformation, a is a dilatation, and k is an element of the maximal compact subgroup $K \equiv SO(D+1)$. Eq.(5.1) is known as an "Iwasawa decomposition" [18], naming the corresponding subgroups K,A,N, respectively, one also writes it as $G = KAN$.

The maximal compact subgroup K contains as a proper subgroup the Euclidean Lorentz group $U = SO(D)$. Its elements commute with dilatations a , and U is in fact the maximal subgroup of K with this property. N is abelian and isomorphic to flat $\mathbb{R}^D$.

Harish Chandra tells us that we need only consider induced representations on the privileged homogeneous space

$$\{x\} = G/UAN \tag{5.2}$$

and the inducing representation of the stability group UAN is to be chosen <u>finite dimensional</u>, and trivial when restricted to N .

But the homogeneous space (5.2) is nothing but (the conformal compactification of) our familiar Euclidean space.To see this we only need to recall that the stability subgroup of the point x = 0 in Euclidean space consists of dilatations, special conformal transformations and homogeneous Lorentz transformations. We see here emerging the geometrical origin of the finite component nature of local quantum field theory, this will become clearer later.

Explicitly, a space H^χ carrying such an induced representation consists of finite component wave functions on compactified Euclidean space, viz. $H^\chi = \{\varphi_\alpha(x)\}$ with transformation law

$$(T^\chi(\Lambda)\varphi)_\alpha(x) = \rho_\Lambda\, D^\chi_{\alpha\beta}(t_x^{-1}\Lambda t_{x'})\varphi_\beta(x') \quad \text{with} \quad x' = \Lambda^{-1}x \tag{5.3}$$

Herein t_x is a translation* taking 0 into x ; the argument $s = t_x^{-1}\Lambda t_{x'}$ of D is an element of the stability subgroup, viz. $s\hat{x} = \hat{x}$ for $\hat{x} = 0$, and may therefore be factorized as $s = uan$. The inducing finite dimensional representation of UAN is characterized by $\chi = [\ell,\delta]$ and given by

$$D^\chi_{\alpha\beta}\,(uan) = \sigma(a)\, D^\ell_{\alpha\beta}\,(u)$$

Here D^ℓ is a finite dimensional representation of the Lorentz group U , and α,β label a basis in the corresponding representation space. For 1-valued representations one can choose a tensor basis $\alpha = (\alpha_i)$ with indices α_i running from 1...4 . Finally, $\sigma(a)$ is a 1-dimensional representation of A specified by δ and given by

$$\sigma(a) = \exp\,\tau(\delta-\tfrac{1}{2}D) \quad \text{for} \quad a = \exp -i\tau\hat{D}$$

---

*We ignore subtleties associated with points at infinity, cp.Sec.3.

There appears in (5.3) also a multiplier ρ_Λ which corrects for the non-invariance of the measure d^4x. It could be absorbed into D^χ, viz. $\rho_\Lambda\sigma(a) = \exp \tau\delta$. Note that the subgroup N is represented trivially, viz. $D^\chi(n) = 1$. When written in infinitesimal form, transformation law (5.3) reduces to (3.1), (3.2) and the familiar Poincaré transformation law.

The induced representations $\chi = [\ell,\delta]$ and $-\chi = [\ell,D-\delta]$ are equivalent (except for a discrete set of values of χ , the socalled "integer points"); the corresponding representation spaces are mapped into each other with the help of the intertwining kernel (4.9).

There are several series of unitary irreducible representations [19] . For us the only important ones will be the principal series

$$\chi = [\ell,\delta] \quad , \quad \delta - \frac{D}{2} \text{ pure imaginary}$$

and the scalar representations of the supplementary series

$$\chi = [0,\delta] \quad , \quad 0 < \delta < D \, .$$

The induced representations are already irreducible for such χ (and more generally for all χ except integer points). The positive definite invariant scalar product in these two cases is given by

$$(\psi,\phi) = \int d^Dx\bar{\psi}_\alpha(x)\psi_\alpha(x) \quad \text{resp.} \quad (\psi,\phi) = \int d^Dxd^Dy\ \bar{\psi}(x)G^{-1}(x-y)\phi(y)$$

with propagator G^{-1} from (3.3).

Let us next consider functions $f(x_1,x_2)$ of two variables which transform according to a representation $\chi_o = [0,d]$ of the supplementary series in each variable. We consider the space H of functions f with finite norm (f,f) ,

$$(f,g) \equiv \frac{1}{2}\int d^Dx_1d^Dx_2d^Dy_1d^Dy_2\ \bar{f}(x_1x_2)G^{-1}(x_1-y_1)G^{-1}(x_2-y_2)g(y_1y_2)$$

H carries a unitary representation of the conformal group because the scalar product (f,g) is positive and invariant. The representation is known as Kronecker product of two representations χ_o of the supplementary series. Like every unitary representation, it can be decomposed into unitary irreducible representations : $H = \oint d\hat{\chi}\ H^\chi$.

Each subspace $H^{\hat{\chi}}$ carries a unitary irreducible representation $\hat{\chi}$. Therefore it may be identified with an irreducible subspace of some induced representation space $H^{\chi} = \{\varphi_\alpha(x)\}$ according to what was said before.

Since H and H^{χ} are function spaces, the projection from H onto $H^{\hat{\chi}} \subseteq H^{\chi}$ can be implemented by a Clebsch Gordon kernel $\Gamma_\alpha^{-\hat{\chi}}(x_1 x_2|x)$, and the decomposition of any function $f \in H$ may be written as

$$f(x_1 x_2) = \int\!\!\!\!\!\sum d\hat{\chi} \int dx \Gamma_\alpha^{-\hat{\chi}}(x_1 x_2|x)\varphi_\alpha^{\hat{\chi}}(x)$$

The last question is, which unitary representations appear in the decomposition? For $D = 2$ the answer was given by Naimark [13], but for general D a rigorous mathematical treatment appears not to have been given in the literature. The author has performed a heuristic analysis (see Appendix A), the result is that for $D > 2$ only the principal series is relevant*. The Clebsch-Gordon kernels are uniquely determined by conformal covariance to have the form (4.8), they are nonvanishing only for completely symmetric tensor representations ℓ of the Lorentz group U.

We now want to apply the expansion to the Green functions, viz. $f(x_1 x_2) = G(x_1 x_2 x_3 \ldots x_n)$ (smeared over $x_3 \ldots x_n$ with some test function). To do this one has to assume that the norm (f,f) defined by (5.4) is finite for this function. Unfortunately this assumption is violated by the Born term [2nd term on RHS of Eq.(2.1)] which must therefore be subtracted before we expand. We obtain then an expansion formula for the partly 1-particle irreducible Green functions as was written down in Sec.4. We remark paranthetically that adding the Born term amounts to including in addition an irreducible supplementary series contribution.

Summing up, the conformal quantum numbers are Lorentz spin ℓ and dimension δ, and the conformal partial wave expansion amounts to a decomposition into terms with exchange of definite $\chi = [\ell,\delta]$. Its distinctive feature is the finite component nature of the objects exchanged in (4.3), this traces back to the geometrical structure of the Euclidean conformal group.

---

*When one wants to extend the theory to fundamental fields with spin, representations of the discrete series may enter into the partial wave expansion (4.3) for odd space time dimension D.

6. - SOLUTION OF INTEGRAL EQUATIONS - If the partial wave expansion for all the $n \geq 4$ point Green functions and kernels is inserted into the integral equations (Fig.1), it turns out that they are thereby diagonalized due to the orthogonality relation (4.7), and thus reduce to algebraic (as opposed to integral) equations. Moreover, these algebraic relations amount to simple factorization properties.

We will consider functions of $\chi = [\ell, \delta]$ as analytic functions of the complex variable δ ; they will of course be uniquely determined by their values for χ in the principal series, viz. $\mathrm{Re}\delta = \frac{1}{2}D$. We define the residue at $\chi_a = [\ell_a, \delta_a]$

$$\operatorname*{res}_{\chi=\chi_a} f(\chi) \equiv - \operatorname*{res}_{\delta=\delta_a} c(\ell_a, \delta) f(\ell_a, \delta)$$

with c the Plancherel weight appearing in (4.4).

Let us now for a moment leave the SD-equation (Fig.1e) for the propagator out of consideration, and consider all the others, including Fig.2.
The result is then as follows:

The integrands of the expansions (4.3),

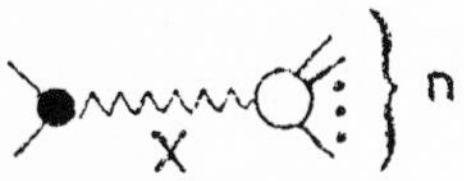
have poles in δ at $\chi = \chi_o = [0.d]$ (6.1)

for all $n \geq 2$, with d = dimension of the field ϕ.
Moreover the residue must factorize

$\operatorname*{residue}_{\chi=\chi_o} = \frac{1}{2}$ 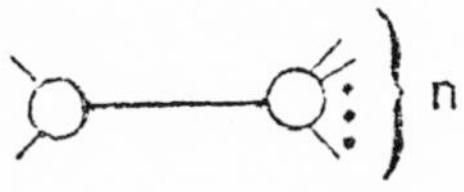 (6.2)

i.e. is expressible in terms of the ordinary Green functions. The factor $\frac{1}{2}$ comes because in (4.3) every inequivalent representation is counted twice.
Thus the integral equations amount to simple FACTORIZATION PROPERTIES of the partial waves.

Finally, the propagator's SD-equation, Fig.1e, requires that for $n = 2$ the LHS of (6.1) has a pole in δ also at

$$\chi = \chi_2 = [2, D] \tag{6.3}$$

for D space time dimensions. The Bethe-Salpeter equations indicate then that such a pole should be expected for all n, barring cancellations. As we shall see, the pole at $\chi = \chi_2$ is related to the existence of a stress tensor with dimension D, and the pole at $\chi = \chi_o$ to the existence of the fundamental scalar field $\phi(x)$ with dimension d.

Let us show for an example how these results obtain. Consider e.g. the SD-equation for the 3-point vertex. Because conformal invariant 3-point functions are unique

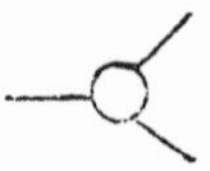
$= g$
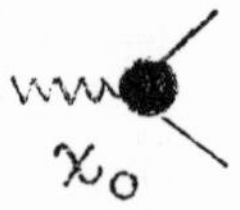

for $\chi_o = [0,d]$ (6.4a)

The proportionality factor g will be called the coupling constant. Inserting this and the partial wave expansion (4.6b) of the BS-kernel gives

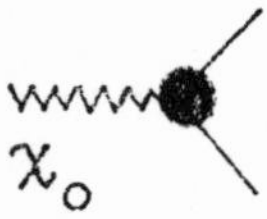

$= \frac{1}{2}\int d\chi\, b(\chi)$
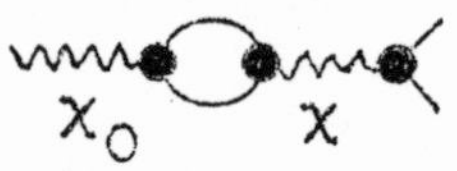

(6.4b)

The orthogonality relation (4.7) cannot be applied directly, since χ_o is not in the principal series, but it can be used after a process of analytic continuation. As a result equation (6.4b) simplifies to

$$b(\chi_o) = 1 \tag{6.5}$$

From the Bethe Salpeter equation defining the BS-kernel one finds in a similar way (15) that $g(\chi) = b(\chi)[1-b(\chi)]^{-1}$. Thus the partial wave amplitude for the 4-point function and therefore the integrand of expansion (4.6a) have a pole at $\chi = \chi_o$. Because of Eq.(6.4a) the residue will be $2/g^2 b_o$ times the RHS of Eq.(6.2), where

$$b_o^{-1} = \underset{\chi=\chi_o}{\text{res}}\ g(\chi)$$

In a similar way one derives factorization properties for the higher n-point functions from the integral equations (Fig.1c).

To do so, one must of course first write down the relation between the partial wave amplitudes of the Green functions and the 2i-kernels which follows from the BS-equations (Fig.1a). This is straightforward by Eq.(4.7).

Finally the propagator equation (Fig.2) can also be analyzed

with the help of a trick that is due to Parisi [20] (s.Sec.10 below).
It is found to require that

$$\tfrac{1}{2}g^2 b_0 = 1 \qquad \text{i.e.} \qquad \operatorname*{res}_{\chi=\chi_0} g(\chi) = \tfrac{1}{2}g^2 \qquad (6.6)$$

Lastly let us note that the factorization property (6.2) permits to write down an expansion for the full Green function in the
form (4.14), assuming some analyticity (s.below).

7. - TENSOR FIELDS - In general the partial wave amplitudes are
expected to have poles not only at $\chi = \chi_0$ and $\chi = \chi_2$ but at many
other points $\chi = \chi_a$. In fact the poles come in pairs at χ_a and
$-\chi_a$ due to symmetry property (4.13) of partial wave amplitudes.
The physical significance of these poles is as follows:

To every pole at some $\chi_a = [\ell_a, \delta_a]$ with $\delta_a > \tfrac{1}{2}D$, there corresponds a symmetric traceless tensor field $0^a_{\alpha_1 \dots \alpha_{\ell_a}}(x)$ of rank
ℓ_a and with dimension δ_a .
A single exception is made by the pole at $-\chi_0$ which does not correspond to a physical field (s.below). Let

$$G^a_{\alpha_1 \dots \alpha_\ell}(x; y_1 \dots y_m) = <T(0^a_{\alpha_1 \dots \alpha_\ell}(x)\phi(y_1)\dots\phi(y_m)>_c$$

analytically continued to Euclidean x, y_i . These Green functions
are given by

$$G^a_\alpha(x; x_3 \dots x_n) \equiv \quad \text{[diagram]} \quad = \quad \operatorname*{res}_{\chi_a} \text{[diagram]} \qquad (7.1)$$

This definition is justified by observing that it will satisfy the
standard integral equations (7)

$$\text{[diagram]} \quad = \quad \tfrac{1}{2} \text{[diagram]} 2i \div \Sigma \text{[diagram]} \quad , \qquad (7.2a)$$

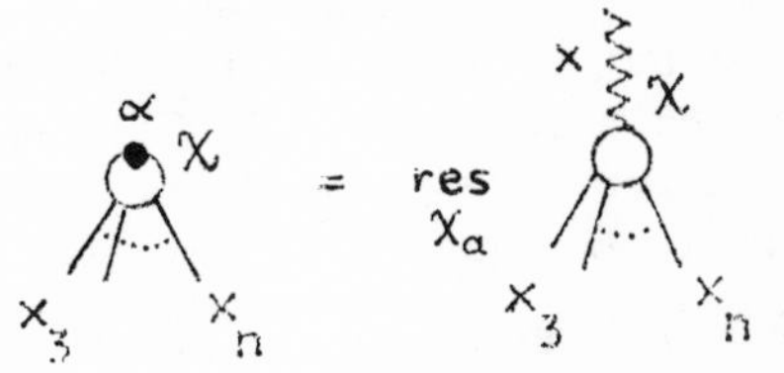

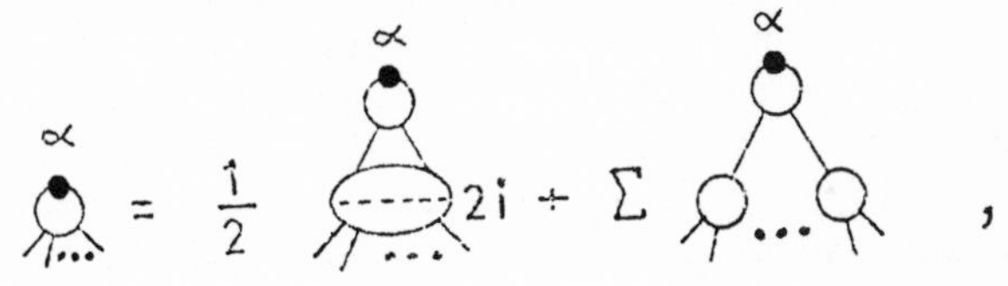

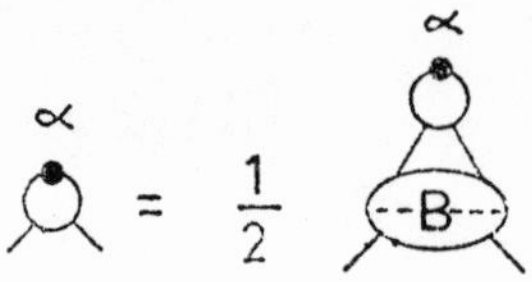

(7.2b)

as may be readily checked with the help of orthogonality relation (4.7) and the aforementioned relation between the partial wave amplitudes of the Green functions and the 2i-kernels.

A special case is the stress tensor, which is associated with the pole at χ_2. If we extend the theory to include internal symmetries (e.g. a ϕ^3-theory with isospin 2) there will also be symmetry currents associated with poles of the appropriate partial wave amplitudes at $\chi_1 = [1,D-1]$. In both cases we must include in Eq.(7.1) an n-independent overall factor in order that the currents are correctly normalized. The normalization is fixed by the requirement that the 3-point functions $G_\alpha(x,y_1y_2)$ satisfy the correct Ward Takahashi identities; this has been discussed elsewhere [6].

The physical significance of the poles of the partial waves suggest the natural POSTULATE:

> $g(\chi)$ is a meromorphic function of χ (i.e. in δ for all ℓ) and so are the partial wave amplitudes of all higher n-point functions.

This analyticity property cannot be derived from the integral equation alone but is a necessary requirement for the validity of operator produced expansions à la Wilson [21] and Ferrara et al. [5]. Similar analyticity properties hold in each order of perturbation theory as a consequence of Weinberg's power counting theorem.

There is one technical point however. From inversion formula (4.12) we see that partial wave amplitudes G^χ are proportional to normalization factor $N(\chi)$ because Γ^χ is, by (4.8). However, due to equivalence of representations χ and $-\chi$ only the product $N(\chi)N(-\chi)$ has a physical meaning, but the ratio $N(\chi)/N(-\chi)$ does not, and has only been fixed by convention in (4.11). Physical postulates should be independent of such conventions. We are thus led to demand analyticity for $N(\chi)^{-1}G^\chi(\ldots)$ in χ; this implies analyticity of $g(\chi)$ as a special case (n=4). Accordingly we will in the following use a modified Eq.(7.1) in which the RHS has been replaced by res $N(\chi)^{-1}G^\chi$.

8. - OPERATOR PRODUCT EXPANSIONS - From the preceding discussion it is now clear how one should proceed to derive operator product expansions. Essentially one will do this by shifting the path of the δ-integration in (4.3) to the right, picking up the contribution from any pole in χ in the partial wave amplitudes which the path crosses. Because of (7.1) their residues are related to matrix elements of tensor field operators.

Before any path shifting can be done, one must however go over to representation functions of the second kind which have good asymptotic properties as $x_1 \to x_2$. This is well known in Regge theory where one splits Legendre functions P_ℓ into Q_ℓ and $Q_{-\ell-1}$. Here an analogous split is possible. It is convenient to carry out a Fourier transform with respect to the last argument x in (4.2). One can write

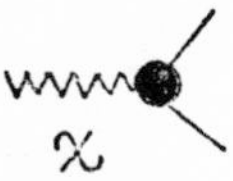 $\equiv \Gamma^{\chi}_{\alpha}(x_1x_2|p) = Q^{\chi}_{\alpha}(x_1x_2|p) + \ldots Q^{-\chi}_{\ldots}(x_1x_2|p)$ (8.1)

where we have dropped some factors.

An explicit expression for Q^χ will be given in Appendix B. It has the following properties.

$Q^{\chi}_{\alpha}(x_1x_2|p)$ has good asymptotic properties, i.e. the strength of the singularity at $x_1=x_2$ decreases when Reδ is decreased. Moreover it is a meromorphic function of χ and possesses a power series expansion in p .

We will use a graphical notation

 $= Q^{-\chi}(\ldots)$ (8.2)

Using symmetry of partial waves in $\chi \to -\chi$ one can rewrite the partial wave expansion as

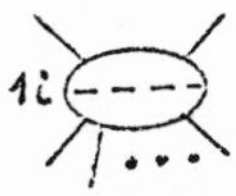 $= 2\sum\!\!\!\!\int d\chi$  (8.3)

Now we are ready to shift the path of the δ-integration to the right. This will produce an asymptotic expansion because the Q^χ have good asymptotic properties. Contributions come from all poles of the integrand. Using definition (7.1) of matrix elements of tensor fields, the result reproduces the conformal invariant operator product expansions of Ferrara et al. Rewritten in terms of operators in Minkowski space they are

$$T\phi(x)\phi(0) = 2\sum_a Q^{\chi_a}_{\alpha_1\ldots\alpha_{\ell a}}(x0|i\nabla_z)\, O^a_{\alpha_1\ldots\alpha_{\ell a}}(z)\big|_{z=0} \tag{8.4}$$

If Q^χ is expanded in powers of $i\nabla_z$ one recovers the usual Wilson expansions [21], with coefficients that respect scale invariance as they should.

We have oversimplified matters in some points. The most important qualification is as follows.

Among the poles which the path crosses is one at $-\chi_o$, i.e. $\ell = 0$, $\delta = D-d$. Its contribution looks like coming from a field $\phi^2_{shadow}(x)$ with dimension $D-d$, its coefficient scales with x^{D-3d}. However it is not a physical local field, and in summed up perturbation theory it does not exist*. Thus this unwanted contribution must be cancelled when the Born term (last term on RHS of (2.1) is added to (8.3). It turns out that such a cancellation is achieved for all the n-point amplitudes precisely because of the factorization property (6.2) which, as we saw, is equivalent to the validity of all but one of the dynamical integral equation.

It is most remarkable that the argument can be turned around. Validity of operator product expansions without shadow requires that the contribution from the shadow pole at $-\chi_o$ is cancelled. But this is equivalent to factorization property (6.2) and therefore of the fundamental integral equations albeit with Fig.2 for propagator equation instead of Fig.1e. In this sense we may say that operator product expansions (without shadow) are "the solution" of the fundamental integral equations.

A further oversimplification has been made in deriving (8.4) in that we have ignored the contributions which could come from poles of $N(\chi)Q^{-\chi}$ in χ, we have only included the poles of $N(\chi)^{-1}G^\chi$ so far. Inspection of the explicit expressions (B.4) and (4.11) shows that $N(\chi)Q^{-\chi}$ does have poles in χ, and not all of them are eliminated by zeroes of the Plancherel weight $c(\chi)$. A hasty remedy would be** to postulate zeroes of $N(\chi)G^\chi_{disc}$ for all n

* If we would insist on defining matrix elements of such a field by Eq.(7.1) we would find later that it is a ghost, i.e. its 2-point function is negative definite because the residues of $g(\chi)$ at χ_o and $-\chi_o$ have opposite sign (s.Sec.10).

** Since operator product expansions are valid for the full disconnected Green functions we must in fact consider the partial waves G^χ_{disc} of these; they are easily obtained from those of the connected Green functions by expanding the disconnected parts using (4.12).

at all these χ-values In fact this could be achieved by postulating just zeroes of $b(\chi)^{-1}$. Nevertheless this is almost certainly too strong a demand, it ignores the possibility of cancellations of contributions from poles at different , "partially equivalent"(22), integer points χ. Group theoretically the problem reduces to that of establishing the Paley-Wiener theorem for functions on the homogeneous space $\{(x_1,x_2)\}$ considered in Sec.5. This has not yet been done. In addition there is the physical problem of whether or not some extra "kinematical" contributions to the operator product expansions of Green functions are tolerable (e.g. seagulls); we are not prepared to discuss this question here.

9. - CROSSING SYMMETRY - The problem of solving the infinite set of coupled nonlinear integral equations, figure 1, has been reduced to the much simpler algebraic one of finding meromorphic partial wave amplitudes which satisfy factorization (6.2), and (6.3). However there is a price to pay. Not every solution will be acceptable, we must remember that the Green functions $G(x_1 \ldots x_n)$ must be symmetric in their arguments $x_1 \ldots x_n$. We shall call this crossing symmetry. (Noting that in a local theory the Wightman functions for different ordering of arguments can be obtained from each others by analytic continuation in the differences $x_i - x_j$ one can argue that crossing symmetry is in fact a non-trivial requirement related to locality (23)).

Consider for instance the 4-point function. We can expand in 3 different ways

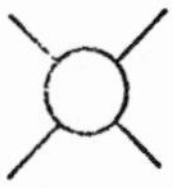 $= \int\!\!\!\!\sum' d\chi\, g(\chi)$ 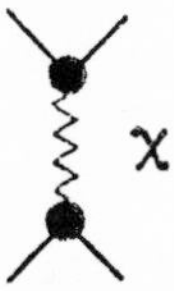 $= \int\!\!\!\!\sum' d\chi\, g(\chi)$ 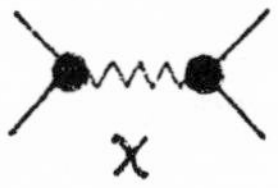 $= \ldots$ (9.1)

with the same $g(\chi)$ because of crossing symmetry. We are using here the expansion (4.14) for the full Green functions.

Now the integrand in (9.1) can itself be expanded in the crossed channel

 $= \int\!\!\!\!\sum d\chi' C(\chi,\chi')$  (9.2a)

with some crossing kernel $C(\chi,\chi')$ that can be explicitly determined from group theory. Inserting this in (9.1) we see that crossing symmetry holds if $g(\chi)$ satisfies the homogeneous linear integral equation[*]

$$g(\chi) = \oint' d\chi g(\chi')C(\chi',\chi) \qquad (9.2b)$$

An explicit expression for the crossing kernel $C(\chi,\chi')$ can be found from (4.12), viz.

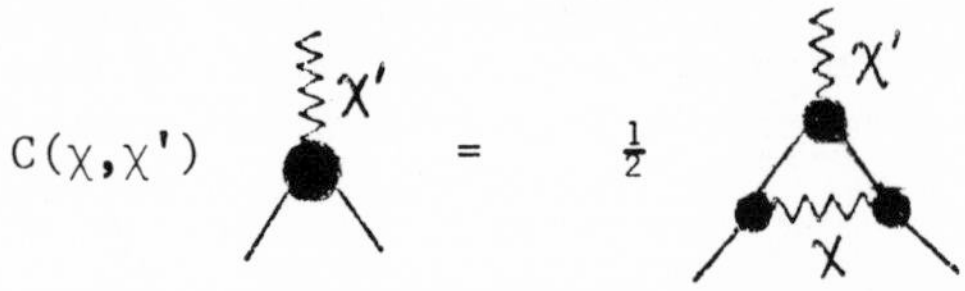

Crossing symmetry for all the higher n-point functions also imposes restrictions. They are again non-coupled, linear integral equations, (one for each n), as opposed to the system of coupled non-linear integral equations from which we started. The problem of constructing a theory has thus been reformulated. It requires that one finds meromorphic (in χ) partial wave amplitude which satisfy factorization (6.2) and (6.3) and in addition solve the linear integral equation (9.2) and higher ones. One hope which one may have is that this new formulation helps in curing the convergence problems of skeleton graph expansions. Another one is that the techniques used in dual resonance models may prove helpful.

Attempts to solve these crossing relations will not be described here[**] except to say that it seems best not to attack the homogeneous integral equations directly, but instead compute the relation between the partial waves and the kernels in the symmetrical integral representation proposed by Symanzik [24]. Meromorphy properties of partial waves should reflect themselves in meromorphy properties of these kernels as were proposed in ref.[25].

It may be interesting to pay special attention to theories for which the requirement of crossing symmetry is sharpened to a kind of duality requirement as follows. Let us make the strong assumption that the path of the δ-integration in equation (8.3) can be closed to the right; this means roughly that operator produced expansions are convergent and not just asymptotic ones. Then the crossing re-

[*] The path of the χ-integration in (9.2a) has to be deformed before the integrations can be interchanged. We leave it to the reader to give a more careful discussion which treats the Born term separately.

[**] Note added in manuscript: In an interesting recent paper, Polyakov has analyzed crossing symmetry relations and related problems. A.M. Polyakov, "Non-Hamiltonian approach to the quantum field theory at small distances", Chernogolovka (June 1973) submitted to JETP.

lation for the 4-point function becomes

$$(9.3)$$

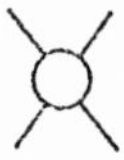 $= \sum_{\substack{\text{poles} \\ \text{in } \chi}} \operatorname{res}_{\chi} g(\chi)$  $= \sum_{\text{poles}} \operatorname{res}\, g(\chi)$ 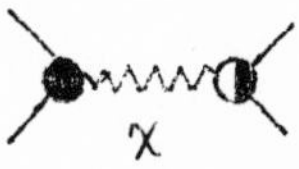 $= \ldots$ etc.

To be more precise, one assumes that the sums converge for some range of the external arguments $x_1 \ldots x_4$, for others the amplitude can then be defined by analytic continuation in the x_i.

This duality type relation can of course also be obtained directly from operator product expansions, assuming they converge, and has also been known to the authors of ref.[5].
There may be a possibility that the above convergence assumptions will be sufficient even if not necessary to ensure validity of the spectrum condition in Minkowski-space. We should acknowledge however that they are rather speculative; it has not been verified that they are consistent with axiomatic positivity (unitarity), etc.

All that has been said so far can be generalized to arbitrary n-point functions, for instance Eq.(9.1) generalizes to

 $= \int \!\!\!\!\!\!\sum{}' d\chi_1 \int \!\!\!\!\!\!\sum{}' d\chi_2 \; g(\chi_1,\chi_2)$ 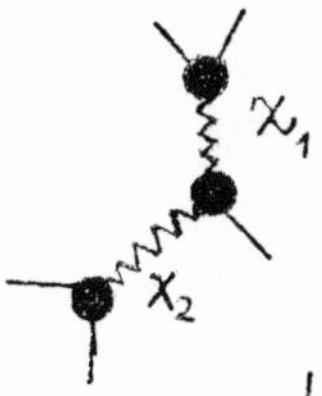

$$(9.4)$$

$= \int \!\!\!\!\!\!\sum{}' d\chi_1 \int \!\!\!\!\!\!\sum{}' d\chi_2 \; g(\chi_1,\chi_2)$  $= \ldots$

etc., in obvious notation (it involves a new Clebsch Gordon kernel).

10. - POSITIVITY - In order that we are dealing with an acceptable quantum field theory, it is necessary that the Wightman functions in Minkowski space satisfy the axiomatic positivity constraint, and also the spectrum condition. Recently, axioms for Euclidean Green functions have been proposed which are supposedly sufficient to guarantee both these properties; the essential ingredient is the Osterwalder-Schrader positivity condition. [31]

Unfortunately it is difficult to exploit this condition systematically. We are therefore at present only able to give some necessary conditions for Wightman positivity.

Let us consider the 2-point functions of the tensor fields $O^a_{\alpha_1 \dots \alpha_{\ell_a}}(x)$ which were considered in Sec.7

$$\langle T\big(O^a_\alpha(x)\; O^a_\beta(y)^*\big)\rangle \equiv \alpha \bullet\!\!-\!\!\!-\!\!\bullet \beta \tag{10.1}$$

[Conformal invariance requires that the tensor fields have real dimensions, as otherwise propagators (10.1) vanish.] They can be computed from the following well known equation [6,26]

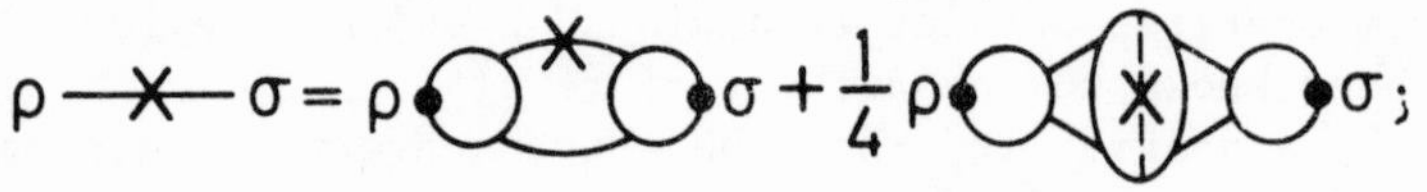

$$\rho -\!\!\times\!\!- \sigma = \rho \bullet\!\!\bigcirc\!\!\!\overset{\times}{\bigcirc}\!\!\bigcirc\!\!\bullet \sigma + \frac{1}{4}\,\rho \bullet\!\!\bigcirc\!\!\!\overset{}{\big(\!\!\times\!\!\big)}\!\!\bigcirc\!\!\bullet \sigma; \tag{10.2}$$

the $\times$-operation is defined as in Fig.2 and Eq.(2.2).
Thus, given the 3-point functions, they are known if the Bethe-Salpeter kernel is known. By uniqueness of conformal invariant 3-point functions,

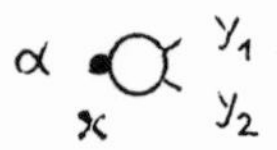

$$\alpha \;\bullet\!\!\bigcirc\!\!\overset{y_1}{\underset{y_2}{}} = g_a \;\Gamma^{X_a}_\alpha(x|y_1 y_2) \equiv g_a \;\underset{-X_\alpha}{\bullet\!\!\!\bullet} \tag{10.3}$$

For the choice of normalization adopted in Sec.7, the proportionality constant is given by

$$g_a = b_a^{-1} \equiv \operatorname*{res}_{\chi=\chi_a} g(\chi) \tag{10.4}$$

due to Eqs.(7.1) and (4.5). We will now proceed to evaluate expression (10.1) in terms of the partial waves $b(\chi)$ of the Bethe Salpeter kernel.

Let $\chi = [\ell,\delta]$ and $\chi' = [\ell,\delta+i\eta] \approx \chi$ in the principal series. From orthogonality relation (4.7) we derive the identity

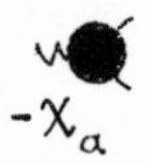

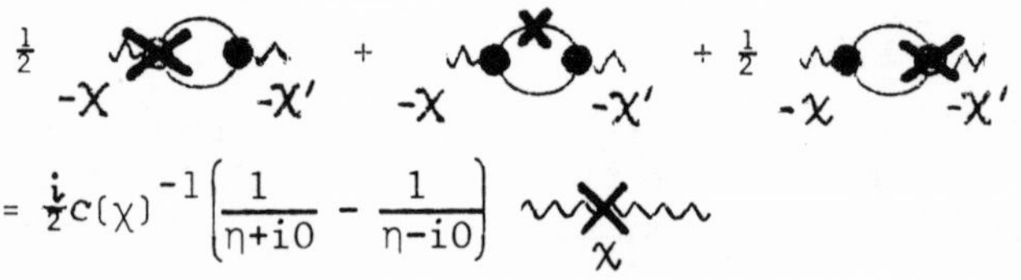

$$\frac{1}{2}\,\underset{-X\quad -X'}{\bullet\!\!\times\!\!\bullet} + \underset{-X\quad -X'}{\bullet\!\!\times\!\!\bullet} + \frac{1}{2}\,\underset{-X\quad -X'}{\bullet\!\!\times\!\!\bullet}$$

$$= \frac{i}{2}\,c(\chi)^{-1}\left[\frac{1}{\eta+i0} - \frac{1}{\eta-i0}\right]\underset{X}{\bullet\!\!\times\!\!\bullet} \tag{10.5}$$

We used Eq.(3.10) for amputation and a familiar representation of the δ-function; the identity (10.5) follows then by distributing crosses over (4.7) in the obvious way; this is possible because of momentum conservation at each vertex (cp. ref.6).

It follows[*] from the convergence criteria of ref.(27) that the first term on the LHS of Eq.(10.5) is analytic for $\mathrm{Im}\,\eta > 0$ while the last term is analytic for $\mathrm{Im}\,\eta < 0$. The middle term is not singular as $\eta \to 0$. Hence

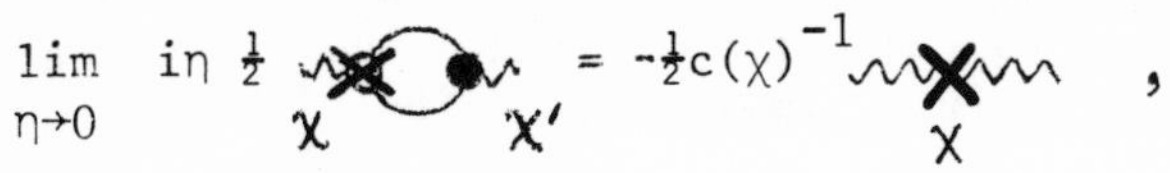

$$\lim_{\eta \to 0}\ \mathrm{in}\ \tfrac{1}{2}\ \ \text{[diagram]} = -\tfrac{1}{2}c(\chi)^{-1}\ \text{[diagram]}\ , \qquad (10.6)$$

$$\chi' = [\ell,\delta+i\eta] \quad (\mathrm{Im}\,\eta > 0)$$

This holds for $\chi = [\ell,\delta]$ in the principal series, viz. $\mathrm{Re}\,\delta = \tfrac{1}{2}D$. However, by uniqueness of analytic continuation in δ, identity (10.6) will then also hold for general χ, both sides being meromorphic functions in δ.

Next we rewrite Eq.(10.2). We distribute crosses $\times$ over the integral equation (7.2b) for the tensor fields 3-point function to obtain

$$\tfrac{1}{2}\ \text{[diagram]} = \text{[diagram]} - \tfrac{1}{2}\ \text{[diagram B]} - \text{[diagram B]}$$

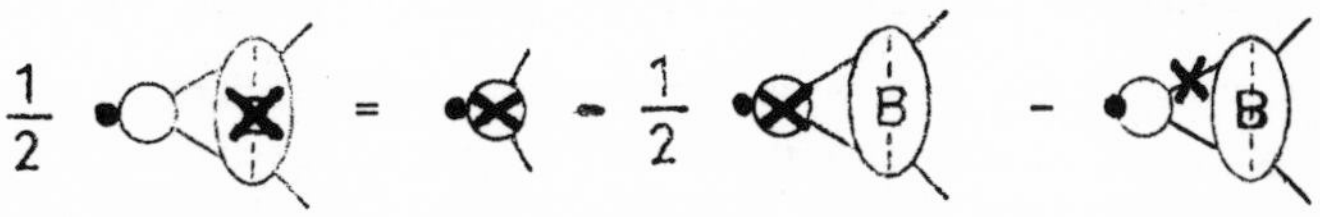

Using this identity and (10.3) we can rewrite Eq.(10.2) as follows

$$\text{[diagram]} = |g_a|^2 \lim_{\chi' \to \chi_a} [1-b(\chi')] \left\{ \text{[diagram]} + \tfrac{1}{2}\ \text{[diagram]} \right\}$$

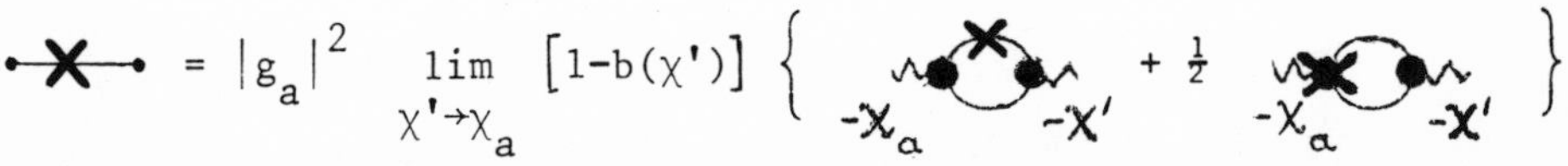

We have also inserted the partial wave expansion (4.6b) of the Bethe Salpeter kernel and used orthogonality relation (4.7) as before in Sec.6. The limit procedure is necessary to avoid expressions of the form $\infty - \infty$. Now we recall that $g(\chi) = b(\chi)\left[1-b(\chi)\right]^{-1}$ whence $b(\chi_a) = 1$ at the pole χ_a of $g(\chi)$ and

$$\langle T\!\left(0_\alpha^a(x)0_\beta^a(0)^*\right)\rangle = \tfrac{1}{2}|g_a|^2 b_a\, \Delta_{\alpha\beta}^{\chi_a}(x) \qquad (10.6)$$

---

[*] When considered in momentum space, all 3 graphs are generalized Feynman integrals in the sense of Speer, i.e. integrals over products of generalized propagators $(p^2)^{-\lambda}$ $\times$ polynomial in p. Their analyticity properties in the λ's are known (28).

by virtue of identity (10.6). Herein $b_a^{-1} = \text{res}\, g(\chi)$ as defined in (10.4).

All the considerations up to now are valid in Euclidean space, however it is a trivial matter to continue analytically to Minkowski space and to go over to the Wightman function by choosing the appropriate $i\varepsilon$-prescription (viz. Im $x^o = -0$) for $\Delta^{\chi}_{\alpha\beta}$.

We will now impose the requirement that the Wightman function $<0^a_\alpha(x)\, 0^a_\beta(y)^* >$ be positive definite. This will be satisfied if

$$b_a^{-1} = \underset{\chi=\chi_a}{\text{res}}\ g(\chi) > 0$$

for all the poles χ_a that are associated with physical fields, and moreover

$$\Delta^{\chi_a}_{\alpha\beta}(x) = n(\chi_a)\,(\tfrac{1}{2}x^2+i\varepsilon x^o)^{-\delta_a} \cdot \{g_{\alpha_1\beta_1}(x)\ldots g_{\alpha_\ell\beta_\ell}(x) - \text{traces}\}$$

is positive definite. [Cp.Eq.(4.9); we use Minkowski metric (+++-)].

The last condition has been analyzed by Rühl for D = 4 space time dimensions, he found the necessary and sufficient conditions (29)

$$\begin{aligned} \delta_a &\geq D - 2 + \ell_a && \text{if } \ell_a \neq 0 \\ \delta_a &\geq \tfrac{1}{2}(D-2) && \text{if } \ell_a = 0 \end{aligned} \qquad (D = 4) \qquad (10.7)$$

In conclusion, Minkowskian positivity requires that the residues of the poles of $g(\chi)$ which are associated with physical fields are positive, moreover their positions must be such that these fields have dimensions bigger or equal to the canonical ones.

The poles $\chi_a = [\ell_a, \delta_a]$ associated with physical fields are those with Re$\delta_a > \frac{1}{2}D$, except the "shadow pole" at $-\chi_o$. Since the pole at $\chi_o = [0, d]$ has a positive residue $b_o^{-1} = \frac{1}{2}g^2$ (it corresponds to the fundamental field) the residue $-b_o^{-1}$ of $g(\chi)$ at $-\chi_o$ is negative [recall that $g(\chi) = g(-\chi)$]. This is why it was crucial that its contribution to the operator product expansion drops out, see the discussion in Sec.8. We see from that discussion that the fundamental integral equations of Fig.1 serve as ghost killers, among other things (Ward identities for $\theta_{\mu\nu}$, see Sec.2).

The result (10.6) shows furthermore how the electromagnetic current's 2-point function $<T(J_\alpha(x) J_\beta(0))>$ is determined in a theory with charged fields, and thereby also the asymptotic magnitude of the total e^+e^--annihilation cross section into hadrons. It is given by the residue b_1 of the pole of $g(\chi)$ at $\chi = \chi_1 = [1,D-1]$, and is thus dynamically determined and not simply by $\Sigma(\text{charges})^2$. The normalization of the current and thereby that of the LHS of Eq.(10.3) is given by the Ward identity for the 3-point functions, this determines g_1 ($\neq b_1^{-1}$ in this case, s.above).

Rühl's result (10.7) also has an interesting consequence which can be tested experimentally.

It is well known that the hypothesis of conformal asymptopia made in Sec.3 plus validity of operator product expansions à la Wilson for the massive theory leads to prediction of, possibly non-canonical, Bjorken scaling of structure function W_2 (q^2,ξ) in electroproduction* [33],

$$\int_0^1 d\xi\ \xi^{s-2} \nu W_2(q^2,\xi) = c_s(q^2)^{-\frac{1}{2}\sigma_s} \quad \text{as} \quad q^2 \to \infty \quad , \quad s=2,4,\ldots \qquad (10.8)$$

where $\xi = q^2/2M\nu$ etc. (usual notation [32] except for metric (+++−)). Therein σ_s is the anomalous part of the dimension of a tensor field of rank s, so that (10.7) implies

$$\sigma_s \geqslant 0 \qquad (s > 0) \qquad (10.9)$$

Further positivity constraints [33] on σ_s follow simply from $W_2 > 0$, e.g. $\sigma_{s+2} \geqslant \sigma_s$. But they do not imply (10.9), which is thus a new result.

Lastly we remark that one can also analyze the implication of the Symanzik-Nelson Euclidean positivity condition for the 4-point function, viz.

$$\int dx_1 \ldots dx_4\ \bar{\varphi}(x_1x_2)\ G_{disc}(x_1\ldots x_4)\ \varphi(x_3x_4) \geqslant 0$$

for arbitrary test functions φ in Euclidean space. The result is that

$$1 + g(\chi) \geqslant 0 \qquad \text{for } \chi \text{ in principal series} \qquad (10.10)$$

viz. $\chi = [\ell,\delta]$, $\text{Re}\delta = \frac{1}{2}D$. In addition the coupling constant g must be real. The extra 1 comes from the disconnected part of the amplitude. Sufficiency of (10.10) follows easily from expansion (4.6),

---

*We assume unsubtracted dispersion relations in ν for the virtual Compton amplitude T_2. If subtractions are necessary, Eq.(10.8) holds for s bigger than some s_o.

symmetry relation (4.10) and $\Gamma^{-\chi} = \bar{\Gamma}^{\chi}$ for principal series χ, and was also observed by Migdal. Necessity follows from the group theoretical version of Bochner's theorem. We can only sketch the argument. Let us use the manifestly covariant formalism and define

$$F(\Lambda) = G_{disc}\ (\hat{\xi}\ \check{\xi}\ \Lambda\hat{\xi}\ \Lambda\check{\xi}) \tag{10.11}$$

where $(\hat{\xi},\check{\xi})$ is an arbitrary standard pair with $\hat{\xi}\cdot\check{\xi} \neq o$. Then (10.10) implies that for arbitrary test function $\varphi(\Lambda)$ on the Euclidean conformal group

$$\int d\Lambda d\Lambda'\ \bar{\varphi}(\Lambda)\ F(\Lambda^{-1}\Lambda')\varphi(\Lambda') \geqslant 0 \tag{10.12}$$

That is, $F(\Lambda)$ is a positive function on the group [18]. By Bochner's theorem its partial waves must then be positive measures on the space $\{\chi\}$ of unitary representations.

Appendix A Decomposition of the Kronecker product of two scalar supplementary series representations of $SO_e(D+1,1)$, $D > 2$.

The question to be asked here is which unitary representations of $SO_e(D+1,1)$ may occur in this decomposition.

To study this we first consider the decomposition of the Kronecker product Ξ of two scalar principal series representations $\chi_1 = [0,d_1]$ and $\chi_2 = [0,d_2]$ with $\mathrm{Re}\,d_i = \frac{1}{2}D$, i=1,2. The representation space to be decomposed consists of functions $f(x_1x_2)$ of a pair of points on compactified Euclidean space. The stability subgroup of such a pair is conjugate to UA in the notation of Sec.5; looking at the transformation law of f under the stability subgroup we see that we are dealing with a unitarily induced representation on the homogeneous space G/UA. (This also follows from a general theorem of Mackey [35]). Moreover the regular representation [18] of G can be decomposed uniquely into unitarily induced representations Ξ on G/UA (they are of course still reducible) by carrying out a harmonic analysis over UA on each coset.* Therefore (almost) every Ξ is contained in the regular representation and can therefore only contain irreducible components which are contained in the regular re-

*The harmonic analysis is easy, as far as the noncompact factor A is concerned it amounts simply to a Mellin transform.

presentation - viz. principal series representations, and discrete series representations if they exist [17,19] .

From now on we restrict our attention to even D = 2,4,6,... . In this case a discrete series does not exist (because the group has no compact Cartan - subalgebra [17,19]). Repeating the discussion of Sec.5 we may therefore write a Plancherel-theorem

$$f(x_1x_2) = \sum\!\!\!\!\!\!\int d\chi \int dx\ \bar{\Gamma}^{\chi}_{\alpha}(x_1x_2|x) \int dy_1dy_2\ \Gamma^{\chi}_{\alpha}(y_1y_2|x)f(y_1y_2) \tag{A.1}$$

with integration over those representations χ of the principal series for which a non-vanishing Clebsch-Gordon kernel exists, they were given in Sec.4. The correct normalization of Clebsch-Gordon kernels must be determined by explicit computation, the result is given by (4.11) with $d_{\pm} = \frac{1}{2}(d_1 \pm d_2)$ and obvious generalization of (4.8). Moreover, the normalization factors $N(\chi)$ given in (4.11) are chosen such that

$$\bar{\Gamma}^{\chi\, d_1 d_2}_{\alpha}(x_1x_2|x) = \Gamma^{-\chi,\bar{d}_1\bar{d}_2}_{\alpha}(x_1x_2|x) = \Gamma^{-\chi d_1 d_2}_{\alpha}(\underline{x}_1\underline{x}_2|x) \tag{A.2}$$

where underlining means amputation with the propagator $G^{-1}(x',x)$ of (3.3b).

Finally we want the decomposition of the Kronecker product of two representations $\chi_o = [0,d]$ of the supplementary series, with

$$\tfrac{1}{2}D-1 < d < \tfrac{1}{2}D \tag{A.3}$$

on physical grounds. We do this heuristically by continuing Eq.(A.1) analytically in d_1, d_2 to $d_1=d_2= d$. After inserting (A.2) the integrand is a meromorphic function* in the d_i, with poles coming in particular from the product $N(\chi)N(-\chi)$ of normalization factors in the Γ's. [c.p Eq.(4.11)] . We do not wish to deform the path of the δ-integration, thus we must watch out for poles of the integrand that may cross the path of the δ-integration ($\chi = [\ell,\delta]$). One finds by inspection that 2 cases have to be distinguished for d in the range (A.3).

i) $D > 2$, or $D = 2$ and $d > \frac{1}{2}$. No poles cross the path of the δ-integration in this case, so (A.1) with (A.2) inserted

---

*We are referring to the notion of a distribution analytic in a parameter [36], from Speers work [28] it is known that $N(\chi)^{-1}\Gamma(\chi)$ are meromorphic in d_i and δ in this sense.

remains valid for $d_1 = d_2 = d$.

ii) $D = 2$ and $d < \frac{1}{2}$. We get an extra contribution from a pole of $N(\chi)N(-\chi)$ which crosses the path of the δ-integration.

The case $D = 2$ has already been completely solved by Naimark [37], therefore we do not need to further discuss ii).

In conclusion we find that for $D > 2$ and even, the Kronecker product of two supplementary series representations $\chi = [0,d]$ with d in the range (A.3) is decomposable into principal series representations only.

Appendix B Representation functions of the second kind.

An explicit expression for the functions Q_α^χ which effect the split (8.1) can be derived from the explicit expression (4.8) for the Clebsch-Gordon kernels Γ_α^χ. It may evidently be rewritten as

$$\Gamma_\alpha^\chi(x_1x_2|x) = N(\chi)\, 2^{2d}\, x_{12}^{-2d+\delta-\ell}\, \vec{\nabla}_\alpha^{12}\, (x_{13}^2 x_{23}^2)^{-\frac{1}{2}(\delta-\ell)} \tag{B.1}$$

where $\vec{\nabla}_\alpha^{12}$ is an ℓ-th order differential operator acting to the right on arguments x_1 and x_2. The Fourier transform of this is according to Ferrara et al. [5]

$$\Gamma_\alpha^\chi(x_1x_2|p) = (2\pi)^p N(\chi)\Gamma\left(\frac{\delta-\ell}{2}\right)^{-2} 2^{2d} x_{12}^{-2d+\delta-\ell}\, \vec{\nabla}_\alpha^{12} \tag{B.2}$$

$$\left\{x_{12}^\nu \int_0^1 du\, e^{-ip[ux_1+(1-u)x_2]} [u(1-u)]^{\frac{1}{4}D-1} \left(\frac{p^2}{4}\right)^{-\frac{1}{2}\nu} K_\nu\left([p^2x_{12}^2u(1-u)]^{\frac{1}{2}}\right)\right\}$$

where $\nu = p - \delta + \ell$ for $\chi = [\ell,\delta]$, $p \equiv D/2$, and $x_{12} = |x_1-x_2|$. K_ν is the modified Besselfunction. It allows for a split

$$K_\nu = -\tfrac{1}{2}\pi\,[\sin\pi\nu]^{-1}\,\{I_\nu(z)-I_{-\nu}(z)\} \tag{B.3}$$

We use this to define

$$Q^{\chi}_{\alpha}(x_1x_2|p) = - N(\chi)\frac{\pi}{2\sin\pi(p-\delta+\ell)}\Gamma\left(\frac{\delta-\ell}{2}\right)^{-2}(2\pi)^{p_2}2^{2d}x_{12}^{-2d+\delta-\ell}\vec{\nabla}^{12}_{\alpha} \tag{B.4}$$

$$\cdot\left\{x^{\nu}_{12}\int_0^1 du\, e^{-ip[ux_1+(1-u)x_2]}\left[u(1-u)\right]^{\frac{1}{4}D-1}\left(\frac{p^2}{4}\right)^{-\frac{1}{2}\nu} I_{\nu}\left(\left[p^2x^2_{12}u(1-u)\right]^{\frac{1}{2}}\right)\right\}$$

It can be readily verified that Q^{χ} have good asymptotic properties in the sense explained in Sec.8; therefore the split (B.3) achieves what it should. The correct factors in (8.1) are found from the symmetry property (4.10) of Γ^{χ} (which is valid for our choice of normalization factors) , viz.

$$\Gamma^{\chi}_{\alpha}(x_1x_2|p) = Q^{\chi}_{\alpha}(x_1x_2|p) + \Delta^{\chi}_{\alpha\beta}(p)\, Q^{-\chi}_{\beta}(x_1x_2|p) \tag{B.5}$$

with $\Delta^{\chi}_{\alpha\beta}(p)$ the Fourier transform of the intertwining kernel (4.9).

ACKNOWLEDGEMENTS : The author is much indebted for helpful discussions to Professors H. Leutwyler, K. Symanzik, I. Todorov and Dr. K. Koller.

REFERENCES :

1. K. Symanzik, lecgures presented at this school.

2. K. Wilson, Phys. Rev. D3, 1818 (1971)

3. G. Mack, in: Lecture Notes in Physics, vol.17, W. Rühl and A. Vancura (eds.) Springer Verlag, Heidelberg 1973.

   A.M. Polyakov, JETP Letters 12, 381 (1970); 32, 296 (1971).

   A.A. Migdal, Phys. Letters 37B, 98, 386 (1971).

   G. Parisi and L. Peliti, Lettere Nuovo Cimento 2, 867 (1971).

4. K. Symanzik, in: Lectures on High Energy Physics, Jaksic (ed) Zagreb 1961, Gordon and Breach, New York 1965.

5. S. Ferrara, R. Gatto and A.F. Grillo, Lettere Nuovo Cimento 2, 1363 (1971). A12, 952 (1972); Nucl. Phys. B34, 349 (1971).

S. Ferrara, R. Gatto, A.F. Grillo and G. Parisi, Nucl. Phys. B49, 77 (1972); Lettere Nuovo Cimento 4, 115 (1972).

6. G. Mack and K. Symanzik, Commun. Math. Phys. 27, 247 (1972).

7. C.G. Callan, S. Coleman, R. Jackiw, Ann. Phys. 59, 42 (1970).

K. Symanzik, in: Cargèse Lectures in Physics, vol.6, J.D. Bessis (ed.) Gordon and Breach, New York 1971.

8. G. Mack (unpublished). The proof cannot be included here because it would amount to a whole picture-book all by itself. A proof using iterative techniques was given in ref.6.

9. G. Mack and Abdus Salam, Ann. Phys. (N.Y.) 53, 174 (1969).

10. M. Hortacsu, R. Seiler, B. Schroer, Phys. Rev. D5, 2518 (1972).

11. M. Gell-Mann and F.E. Low, Phys. Rev. 95, 1300 (1954).

N.N. Bogolubov and D.V. Shirkov, Introduction to the Theory of Quantized Fields, Interscience Publ., New York 1959.

K. Wilson, ref.2.

12. C.G. Callan, Phys. Rev. D2, 1451 (1970).

K. Symanzik, Commun. Math. Phys. 23, 49 (1971).

13. M.A. Naimark, Linear representations of the Lorentz group, Trudy Moscov. Mat Obšč. 10, 181 (1961); engl. transl.: Am. Math. Soc. Transl. 36, 189 (1964).

14. A.M. Polyakov and A.A. Migdal (unpublished) as cited in ref.15.

15. A.A. Migdal, 4-dimensional soluble models of conformal field theory. Chernogolovka 1972, to appear in Nucl. Phys.

16. V.K.Dobrev, V.B. Petkova, S.G. Petrova, I.T.Todorov and G. Mack, to be submitted to Reports Math. Phys.

17. G. Warner, Harmonic Analysis on Semi-Simple Lie Groups, vol.1. Springer Verlag, Heidelberg 1972; Theorem 5.5.1.5 of Sec. 5.5.1.

18. For group theoretical background see
E.M. Stein, in: High Energy Physics and Elementary Particles, p. 563, IAEA, Vienna 1965.

19. P. Hirai, Proc. Japan Acad. 38, 83, 258 (1962), 42, 323 (1965).

20. G. Parisi, Lett. Nuovo Cimento 4, 777 (1972).

21. K. Wilson, Phys. Rev. 179, 1499 (1969).

22. I.M. Gel'fand, M.I. Graev and N.Ya. Vilenkin, Generalized Functions, vol.5, Academic Press, New York 1966; esp. Ch.IV, Sec.5.

23. R. Jost, The general theory of quantized fields, Amer. Math. Soc. Publ., Providence R.I. 1965.

24. K. Symanzik, Lettere Nuovo Cimento 3, 734 (1972).

25. G. Mack, in: Scale and Conformal Symmetry in Hadron Physics, R. Gatto (Ed), John Wiley, New York 1972.

26. K. Symanzik, ref.7, Sec.VI.3.

27. G. Mack and I.T. Todorov, Phys. Rev. D5, 1764 (1973).

28. E.R. Speer, "Generalized Feynman Amplitudes", Princeton University Press, Princeton 1969.

29. W. Rühl, Commun. Math. Phys. 30, 287 (1973).

30. S. Weinberg, Phys. Rev. 118, 838 (1960).

31. K. Osterwalder and R. Schrader, Commun. Math. Phys. 31, 83 (1973)

32. H. Leutwyler and J. Stern, Nucl. Phys. B20, 77 (1970).

33. G. Mack, Nucl. Phys. B35, 592 (1971).
The claim of necessarily canonical dimension made in Sec.4 of this paper is not correct.

34. e.g. I.M. Gel'fand, M.I.Graev and N.Ya.Vilenkin, Generalized functions, vol.5, Academic Press, New York 1966.

35. ref.17, Theorem 5.3.4.3 combined with Proposition 1.2.3.4; $H_i = P$.

36. I.M. Gel'fand and Shilov, Generalized functions, vol.1, Academic Press, New York 1964.

37. K. Symanzik, J. Math. Phys. 7, 510 (1966).

E. Nelson, J. Funct. Analysis 12, 97, 211 (1973).

USE OF INVARIANT TRANSFORMATIONS IN PROBLEMS OF SUPERCONDUCTIVITY

F. Mancini

Istituto di Fisica, Università di Salerno
Via Vernieri 42
84100 Salerno, Italy

INTRODUCTION

In many-body problem various methods of quantum field theory have been frequently used and have led to successful results. The purpose of this lecture is to show how, by using ideas from quantum field theory, it is possible to formulate the theory of superconductivity from a different point of view which offers some advantages over the usual formulation. By applying the "self consistent quantum field theory"(1) we have been developing for some years a formulation of superconductivity which has been called boson formulation of superconductivity.(2) In the course of this application we have introduced the concept of boson transformations(3,4) which are invariant transformations in the sense that they leave invariant the equations of motion but may modify the ground-state expectation values of some of the observables. The use of these invariant transformations has turned out to be a very powerful tool in the study of all those situations related to the presence of a non-homogeneous structure of the ground state. The introduction of the concept of boson transformations in the framework of the self-consistent method provides a technique of computation which has been called "the boson method".[2] Application of this method to the study of superconductivity enables us to describe in a systematic and unified way many different phenomena; besides, the study of these phenomena is made much more simple.

This lecture is divided in two parts: in the first one we shall derive the basic equations of the boson formulation of superconductivity; in the second one we shall illustrate the theory by applying it to the study of magnetic properties of type-II superconductors.

PART I. DERIVATION OF THE BASIC EQUATIONS

1.1 Self-Consistent Solution of the BCS Hamiltonian

For the sake of simplicity we shall give the outline of the formulation in the case of zero temperature and in the case in which the Coulomb interaction among the electrons is not taken into account. The final results will be, however, given in the general form, valid at any temperature below the critical temperature and when the Coulomb interaction is considered.

Let us consider the well-known BCS Hamiltonian:[5]

$$H = \int d^3x\ [\psi_\uparrow^\dagger \varepsilon(\partial)\psi_\uparrow + \psi_\downarrow^\dagger \varepsilon(\partial)\psi_\downarrow] + \lambda \int d^3x\ \psi_\uparrow^\dagger \psi_\downarrow^\dagger \psi_\downarrow \psi_\uparrow\ , \tag{1}$$

where $\psi_\uparrow$ and $\psi_\downarrow$ are the Heisenberg fields for the electrons; λ is the coupling constant of the electron-phonon-electron interaction and

$$\varepsilon(\partial) = -\frac{1}{2m}(\vec{\nabla}^2 + k_F^2)\ ; \tag{2}$$

where m is the electron mass and k_F is the Fermi momentum. It should be noted that this is not the reduced Hamiltonian used by the BCS formulation, but the fully gauge invariant Hamiltonian. Following the scheme of the self-consistent method, let us assume that the equilibrium state of the system can be described in terms of a set of free fields $\{\phi_\uparrow, \phi_\downarrow, B\}$; $\phi_\uparrow$ and $\phi_\downarrow$ correspond to quasi-electron fields; B corresponds to a boson field. The properties of these fields, which are the result of the dynamics of the system described by the Hamiltonian (1), will be determined in a self-consistent way. Introducing the corresponding set of creation and annihilation operators $\alpha_{k\uparrow,\downarrow}$, $\alpha^{\dagger}_{k\uparrow,\downarrow}$ and $B_k, B^{\dagger}_k$ which satisfy canonical commutation relations:

$$[\alpha_{kr},\alpha^{\dagger}_{qs}]_{+} = \delta_{rs}\,\delta(\vec{k}-\vec{q}), \qquad [\alpha_{kr},\alpha_{qs}]_{+} = [\alpha^{\dagger}_{kr},\alpha^{\dagger}_{qs}]_{+} = 0, \tag{3}$$

$$[B_k,B^{\dagger}_q]_{-} = \delta(\vec{k}-\vec{q}), \qquad [B_k,B_q]_{-} = [B^{\dagger}_k,B^{\dagger}_q]_{-} = 0, \tag{4}$$

$$[B_k,\alpha_{qr}]_{-} = [B_k,\alpha^{\dagger}_{qr}]_{-} = 0, \tag{5}$$

we shall construct the Fock space by defining vacuum state $|0\rangle$ as

$$\alpha_{kr}|0\rangle = 0 \qquad B_k|0\rangle = 0\ , \tag{6}$$

and by acting with creation operators on the vacuum. We shall not introduce any other Fock space; every other operator present in the theory should be expressed in terms of the free fields. This means that the Heisenberg fields $\psi_{\uparrow}$ and $\psi_{\downarrow}$ must be written in terms of the set of free fields:

$$\psi_{\uparrow,\downarrow} = g_{\uparrow,\downarrow}\,(\phi_{\uparrow},\phi_{\downarrow},B)\ . \tag{7}$$

Such an expression, which has been called a "dynamical map", must be computed by requiring that the Heisenberg fields satisfy the field equations derived from the Hamiltonian (1); furthermore, expression (7) must satisfy asymptotic conditions and the boundary conditions appropriate to the problem under study. In the present case we shall look for a solution of the Heisenberg equations such that

$$\Delta = \lambda\langle 0|\psi_{\uparrow}\psi_{\downarrow}|0\rangle \neq 0\ . \tag{8}$$

However, since we are interested in the physical interpretation of the theory, instead of computing the dynamical map for $\psi_{\uparrow,\downarrow}$, we shall calculate the expressions of bilinear products of the Heisenberg operators, because only such products appear in the observable quantities. We shall thus write:

$$T[\psi_{\uparrow}(x)\psi_{\downarrow}(y)] = \chi^{(1)}(x-y) + \sum_{p,q} [F^{(1)}(x,y;p,q)\alpha_{q\uparrow}\alpha_{p\downarrow}$$
$$-F^{(2)*}(x,y;p,q)\alpha^{\dagger}_{p\downarrow}\alpha^{\dagger}_{q\uparrow}] + \sum_{\ell}[G^{(1)}(x,y;\ell)B - G^{(2)*}(x,y;\ell)B^{\dagger}_{\ell}]+\ldots \tag{9}$$

where T is the chronological-ordering operator and the dots stand for higher-order normal products; the conservation of spin of quasi-fermions is assumed. Similar expressions can be written for all the other bilinear products. It is easy to see that the coefficients $\chi^{(1)},\chi^{(2)},\ldots,F^{(1)},F^{(2)},\ldots$ and $G^{(1)},G^{(2)},\ldots$ appearing in these expansions are related to the Bethe-Salpeter amplitudes:

$$\chi(x-y) = \langle 0|T[\psi(x)\psi^{\dagger}(y)]|0\rangle , \tag{10}$$

$$F(x,y;p,q) = \langle 0|T[\psi(x)\psi^{\dagger}(y)]|\alpha_{p\downarrow}\alpha_{q\uparrow}\rangle , \tag{11}$$

$$G(x,y;\ell) = \langle 0|T[\psi(x)\psi^{\dagger}(y)]|B_{\ell}\rangle , \tag{12}$$

where ψ is the doublet field:

$$\psi = \begin{pmatrix} \psi_{\uparrow} \\ \psi^{\dagger}_{\downarrow} \end{pmatrix} . \tag{13}$$

$|\alpha\alpha\rangle$ in (11) denotes a state of two quasielectrons: $|\alpha_{p\downarrow}\alpha_{q\uparrow}\rangle = \alpha^{\dagger}_{p\downarrow}\alpha^{\dagger}_{q\uparrow}|0\rangle$; $|B\rangle$ in (12) denotes a state of single boson: $|B_{\ell}\rangle = B^{\dagger}_{\ell}|0\rangle$.

From the Heisenberg equations for the field ψ we can derive equations for the BS amplitudes; these equations can be solved in the pair approximation. Once the solutions are known, we obtain expressions of the observables in terms of free fields. The results are the following.[6] The Hamiltonian $H(\psi)$ becomes the free

Hamiltonian of the fields B and $\phi_{\uparrow,\downarrow}$:

$$H(\psi) = \text{c-number} + H^F + H^B , \tag{14}$$

$$H^F = \int d^3k\ E_k\ [\alpha^\dagger_{k\uparrow}\alpha_{k\uparrow} + \alpha^\dagger_{k\downarrow}\alpha_{k\downarrow}] \tag{15}$$

$$H^B = \int d^3\ell\ \omega_B(\ell)B^\dagger_\ell B_\ell\ . \tag{16}$$

E_k is the energy spectrum of quasielectron fields and has the well-known expression:

$$E_k = \sqrt{\varepsilon_k^2 + \Delta^2}\ . \tag{17}$$

$\omega_B(\ell)$ is the energy spectrum of the boson field and it is fixed by a given equation:[6]

$$W(\ell,\omega_B) = 0\ . \tag{18}$$

The current and charge densities expressed in terms of quasi-particles take the form:

$$\vec{j} = \vec{j}^F + \vec{j}^B , \tag{19}$$

$$\rho = \rho^F + \rho^B\ . \tag{20}$$

The explicit forms of ρ^B and $\vec{j}^B$ are:

$$\rho^B(x) = -\eta(i\vec{\nabla})\pi(x), \quad \vec{j}^B(x) = d(i\vec{\nabla})\vec{\nabla}B(x)\ , \tag{21}$$

where B is the boson field and π its canonical conjugate. The derivative operators $d(i\vec{\nabla})$ and $\eta(i\vec{\nabla})$ are defined by

$$d(\ell) = -i\ \frac{\omega_B}{\ell^2}\sqrt{2(\omega_B^2-\bar{\omega}_\ell^2)}\ , \quad \eta(\ell) = -\ \frac{i}{\omega_B}\ \sqrt{2(\omega_B^2-\bar{\omega}_\ell^2)}\ , \tag{22}$$

where $\bar{\omega}_\ell$ is a given function of the boson energy. From Eqs. (21) and (22) it can be seen that the boson contribution to the current is conserved:

$$\vec{\nabla}\cdot\vec{j}^{B} + \frac{\partial}{\partial t}\rho^{B} = 0 . \tag{23}$$

For what concerns the fermion contributions we do not report the explicit expressions; we only mention the important results that the spatial integral of the charge is identically zero:

$$\int d^3x\ \rho^F(x) = 0 , \tag{24}$$

and that the current is conserved:

$$\vec{\nabla}\cdot\vec{j}^{F} + \frac{\partial}{\partial t}\rho^{F} = 0 . \tag{25}$$

The fact that the current is conserved and that the Hamiltonian is effectively diagonal when expressed in terms of free fields shows that the pair approximation is a reasonable approximation.

1.2 Dynamical Rearrangement of Symmetry[1,8]

The original Hamiltonian (1) is invariant under the constant phase transformation of the electron field:

$$\psi \rightarrow \psi' = e^{i\theta}\ \psi, \tag{26}$$

where θ is a real constant parameter. In a formal sense[9], the transformation (26) is induced by the generator

$$N = \int d^3x\ \rho(x) , \tag{27}$$

where $\rho(x)$ is the electron charge density, as

$$\psi' = e^{-iN\theta}\ \psi\ e^{iN\theta} . \tag{28}$$

Let us now examine how the physical fields ϕ and B change under the transformation induced by (27). From the results of the preceding section the generator N, expressed in terms of physical fields, takes the form:

$$N = \int d^3x\ \rho^B(x) = -\int d^3x\ \eta(i\vec{\nabla})\pi(x) . \tag{29}$$

It is then immediately seen that the physical fields transform as

$$\phi \to \phi, \qquad B \to B' = B + \theta\,\eta(0), \qquad \pi \to \pi \tag{30}$$

The quasielectron fields remain invariant under the phase transformation. The original transformation of the electron field is rearranged at the level of the free fields in the shape of a linear transformation of the boson field B (the Goldstone particle). The original phase of the electron field is completely detached from the quasielectrons, which remain frozen, and is completely carried by the boson field which transports all the symmetry attributes.[7] The transformation $B \to B'$ induces a condensation of bosons which organize the system fixing the phase of the ground state. Since $\theta\eta(0)$ is a space- and time-independent quantity, the equations of motion of the boson field (massless) are obviously invariant under the transformation: the original phase invariance of the theory is not lost but is recovered by the presence of the Goldstone particle.

1.3 Boson Transformations

The theory formulated so far has been obtained by solving the equations of motion in the case that the ground state is constant in space and time. The most important characteristic of the system is that the ground state exhibits the phenomenon of condensation which fixes the properties of the system. We can describe different physical situations by regulating the boson condensation. A simple way of achieving this is to introduce an operator transformation which leaves the equations of motion invariant but modifies the ground state. Such transformations which will regulate the boson condensation need to be expressed only in terms of the boson field, and have been called boson transformations.[3,4] By means of these transformations the spatial and temporal dependence is introduced in the theory.

It is well known in the theory of superconductivity that to move from constant solutions to space- and time-dependent solutions one is faced with great difficulties. In the framework of the BCS theory the Gor'kov equations (or the Bogoliubov equations) are supposed to be able to describe both stationary and non-stationary situations. Unfortunately, owing to their non-local nature, these equations are not easy to solve and the general attitude has been to reduce them, in some appropriate regimes, to the form of local differential equations (Ginzburg-Landau equations and all their generalizations). These difficulties limit the analysis to very small regions in the H,T plane.

In the boson method we are free from these difficulties: the space and time dependence is introduced in the theory by means of the boson transformations. These transformations can induce a change of the ground state of the system, permitting us to describe different physical situations. The different states of a system correspond to the same equations of motion, but solved under different boundary conditions. The advantage presented by the use of the boson transformations is that we do not have to apply the boundary conditions to the original equations of motion but only to the choice of the particular transformation which will induce the ground state characteristic of the physical situation under study.

In the case in which one is interested in studying stationary problems[10] the boson transformation is defined as that transformation which is induced by the following generator:

$$N_f = \int d^3x \; f(\vec{x})\rho^B(\vec{x}) \;, \tag{31}$$

where $f(\vec{x})$ stands for a solution of the Laplace equation

$$\vec{\nabla}^2 f(\vec{x}) = 0 \;. \tag{32}$$

Under this transformation the fields transform as:

$$\phi \to \phi \;, \qquad \pi \to \pi \;, \tag{33}$$

$$B \to B_f = B(\vec{x}) + \eta(i\vec{\nabla})f(\vec{x}) \;, \tag{34}$$

$$\psi \rightarrow \psi_f = \exp\left[i \frac{1}{\eta(i\vec{\nabla})} B_f\right] F(\phi, \vec{\nabla} B_f, \pi) \ . \tag{35}$$

In Eq. (35) F is a certain function of $\phi, \vec{\nabla}B, \pi$ and of their derivatives. As we see, the transformation induced by N_f generates a complicated nonlocal transformation of the Heisenberg field ψ; such transformation will leave invariant the equations of motion of the electron fields since $\dot{N}_f = 0$. It is remarkable that there exist space- and time-dependent transformations[(4)] which keep the electron equations invariant. Such an invariance is not explicitly seen from the original field equations because the boson transformations induce highly nonlocal transformations of ψ. The invariance becomes explicit when we focus our attention on the boson field.

1.4 Basic Equations

The formulation presented so far has been restricted to the analysis of an hypothetical uncharged Fermi system with an attractive short-range interaction at T = 0°K. The general lines of the method are not modified when the Coulomb interaction is taken into account; the Bethe-Salpeter equations are modified[(2)] by the presence of additional terms which represent the contribution of the electromagnetic interaction. The situation is different when we want to study problems at finite temperature. The usual techniques which permit the application of the methods of quantum field theory to statistical mechanics are not suitable for our purposes since they do not allow us to introduce those invariant operator transformations, boson transformations, which play the central role in our formulation. The boson method is based on the observation that different states of the system correspond to different realizations of the field equations; when the theory is formulated in operational form it is very easy to pass to different Hilbert spaces by making use of appropriate canonical transformations. On the other hand, this feature disappears when the theory is formulated in terms of Green functions which refer to a specific choice of the Hilbert space.

In order to incorporate the transformation method in the many-body problem at finite temperature we need to realize quantum statistical mechanics as a quantum field theory. We have proposed[2] a technique which offers the possibility, through the use of canonical transformations, of mapping the entire structure of quantum statistical mechanics to quantum field theory. More precisely, in the present study of superconducting systems, the Hilbert space in which the theory is realized has been constructed as the Fock space of the free fields $\{\phi_\uparrow,\phi_\downarrow,B\}$; let us call this space the zero realization of the fields $\{\phi_\uparrow,\phi_\downarrow,B\}$ and let us denote it by G_o. The field equations for the Heisenberg operators $\psi_\uparrow,\psi_\downarrow$ do not depend on the temperature; the temperature is introduced in the theory by requiring that observable results are given by statistical averages. Given an observable $A(\psi)$, the ground-state expectation value of A is replaced by the quantum-statistical average of $A(\psi)$ over the grand canonical ensemble at temperature T:

$$\langle 0|A(\psi)|0\rangle \to \langle A(\psi)\rangle = T_r\left\{e^{-\beta H}A(\psi)\right\}\Big/T_r\left\{e^{-\beta H}\right\} , \tag{36}$$

where the traces are performed over the state vectors $|n\rangle$ which form an orthonormal basis in G_o. In Eq. (36) β is defined as: $\beta = 1/k_BT$; k_B is Boltzmann's constant. According to Ref. 2 we can construct another realization, let us say $G(T)$, of the fields $\{\phi_\uparrow,\phi_\downarrow,B\}$ such that the vacuum expectation value of $A(\psi)$ in the Hilbert space $G(T)$ is equal to the quantum-statistical average of $A(\psi)$ in the Hilbert space G_o:

$$\langle 0,T|A(\psi)|0,T\rangle = \langle A(\psi)\rangle . \tag{37}$$

In (37) $|0,T\rangle$ is the vacuum state in the Hilbert space $G(T)$. Once we have constructed the space $G(T)$, which satisfies the requirement (37), our approach to finite temperature is the following:

we solve the dynamical problem in the space G(T); this means that the Heisenberg equations for the electron operators $\psi_\uparrow$ and $\psi_\downarrow$ must be read and solved in the representation G(T). When we have found the operator form of the observables, we take the vacuum-expectation values in the space G(T); by means of relation (37) the values so obtained will correspond to observed results. We can then study space- and time-dependence problems at finite temperature by inducing invariant transformations of the fields, as described in the preceding section.

All the details relative to the extension of the boson formulation to finite temperature and to the case in which the electromagnetic interaction is considered can be found in Ref. 2. We shall present here only the final results. As we have already mentioned, under the boson transformation the ground-state expectation value of some of the observables may change due to the fact that the ground state is not invariant under this transformation. In particular, in the case of stationary situations, the ground-state current, the order parameter and the ground-state energy take the following expressions:

$$\vec{J}(\vec{x}) = \frac{1}{4\pi e\, \lambda_L^2} \int d^3y \; c(\vec{x}-\vec{y})\, [\vec{\nabla} f(\vec{y}) - e\vec{A}(\vec{y})] \; , \tag{38}$$

$$\Delta(\vec{x}) = e^{2if(\vec{x})} \, |\Delta(\vec{x})| \; , \tag{39}$$

$$W = \frac{1}{2e} \int d^3x \; [\vec{\nabla} f(\vec{x}) - e\vec{A}(\vec{x})] \cdot \vec{J}(\vec{x}) \; . \tag{40}$$

$\vec{A}$ is the transverse vector potential of the electromagnetic field. Here the electromagnetic field is the total field, which contains the external fields and the fields due to supercurrents. λ_L is the temperature-dependent London penetration depth; $c(\vec{x})$ is the boson characteristic function, whose Fourier transform has the expression

$$c(\ell) = \frac{3[\omega_B^2(\ell)-\bar{\omega}_e^2]}{\lambda N(0)\ell^2 v_F^2} . \tag{41}$$

In Eq. (41) N(0) is the state density at the Fermi level and v_F is the Fermi velocity. Eq. (39) shows the important result that the phase of the order parameter is created by the boson transformation and is given by 2f, where f is a solution of the Laplace equation (32).

The previous expressions represent the basic equations of the boson formulation of superconductivity in the case of time-independent problems. Combining Eq. (38) with the Maxwell equation

$$4\pi \vec{J} = \vec{\nabla}\times \vec{H} , \tag{42}$$

we get the following equation for the vector potential:

$$\vec{\nabla}^2\vec{A}(\vec{x}) - \frac{1}{\lambda_L^2}\int d^3y\ c(\vec{x}-\vec{y})\vec{A}(\vec{y}) + \frac{1}{e\lambda_L^2}\int d^3y\ c(\vec{x}-\vec{y})\vec{\nabla}f(\vec{y}) = 0. \tag{43}$$

The general solution of this equation is given by

$$\vec{A} = \vec{A}_o + \vec{A}_1 , \tag{44}$$

where $\vec{A}_o$ is the solution of the generalized London equation:

$$\vec{\nabla}^2\vec{A}_o(\vec{x}) - \frac{1}{\lambda_L^2}\int d^3y\ c(\vec{x}-\vec{y})\vec{A}_o(\vec{y}) = 0 , \tag{45}$$

and

$$\vec{A}_1(\vec{x}) = \frac{1}{e}\int \frac{d^3k}{(2\pi)^3}\frac{c(k)\vec{g}(\vec{k})}{c(k)+\lambda_L^2k^2}\ e^{i\vec{k}\cdot\vec{x}} . \tag{46}$$

$\vec{g}(\vec{k})$ in Eq. (46) is the Fourier transform of the quantity $\vec{\nabla}f$:

$$\vec{\nabla}f(\vec{x}) = \int \frac{d^3k}{(2\pi)^3}\ \vec{g}(\vec{k})\ e^{i\vec{k}\cdot\vec{x}} . \tag{47}$$

Summarizing, the computation procedure in the case of a space-dependent problem involves the following steps:

1) Look for a function $f(\vec{x})$, which satisfies the Laplace equation

$$\vec{\nabla}^2 f(\vec{x}) = 0 \tag{48}$$

and the boundary conditions appropriate to the problem under consideration.

2) The persistent currents and the magnetic field associated with them are then given by

$$\vec{J} = \frac{1}{4\pi} \vec{\nabla}x\vec{\nabla}x \vec{A} \qquad \vec{h} = \vec{\nabla}x\vec{A} \tag{49}$$

where

$$\vec{A}(\vec{x}) = \frac{1}{e} \int \frac{d^3k}{(2\pi)^3} \frac{c(k)\vec{g}(\vec{k})}{c(k)+\lambda_L^2 k^2} e^{i\vec{k}\cdot\vec{x}} \tag{50}$$

3) The energy is given by

$$E = W + \frac{1}{8\pi} \int d^3x \, \vec{h}\cdot\vec{h} = \frac{1}{8\pi e} \int d^3x \, \vec{h}(x)\cdot\vec{\nabla}x\vec{\nabla}f(\vec{x}) \;; \tag{51}$$

in this expression we have taken into account the energy associated with the electromagnetic field. It should be noted that the quantity $\vec{\nabla}x\vec{\nabla}f$ does not necessarily vanish when f is not continuous (e.g. in cases of the Josephson functions) or is not simply connected in the topological sense (e.g. in the case of flux lines in type-II superconductors). In expression (51) we have not taken into account the condensation energy, which is given by the c-number part in Eq. (14).

4) The magnetic field induced solely by external effects, the Meissner field, and the screening currents are given by

$$\vec{H}_o = \vec{\nabla}x\vec{A}_o \qquad \vec{J} = \frac{1}{4\pi} \vec{\nabla}x\vec{\nabla}x\vec{A}_o \tag{52}$$

where $\vec{A}_o$ is the solution of the generalized London equation (45).

PART II. APPLICATION TO THE STUDY OF THE MAGNETIC PROPERTIES OF TYPE-II SUPERCONDUCTORS

2.1 Type-II Superconductivity

One of the most interesting arguments in the theory of superconductivity is the study of the mixed state in superconductors of the second kind[11]. All superconductors can be divided in two classes according to their behavior in the presence of an applied magnetic field.

Type-I superconductors with a positive surface energy which leads to a complete Meissner effect: complete expulsion of the magnetic flux for $H < H_c$ and transition to the normal state at $H = H_c$. H is the external magnetic field and H_c is the critical magnetic field.

Type-II superconductors with a negative surface energy which leads to a partial Meissner effect in the sense that the transition from the superconducting state to the normal state occurs through a range of values of the magnetic field. More precisely, in type-II superconductors there are two critical magnetic fields: H_{c1} and H_{c2}. When H is less than H_{c1} the sample exhibits a complete Meissner effect; when $H > H_{c1}$, since the surface energy is negative, penetration of the magnetic flux in the form of quantized flux lines parallel to the direction of the applied field is energetically favored. These filaments are usually interpreted in terms of a phenomenological model which describes the flux lines as made up of an internal region, core, of radius of the order of the coherence length ξ, where the superconducting effects are completely suppressed, and of an electromagnetic region, of radius of the order of the penetration depth λ, where there are persistent currents which circulate around the axis of the flux line. When the magnetic field increases, the density of flux lines increases until the internal regions start to overlap and all the sample becomes normal; this transition defines the upper critical field H_{c2}. For values of the magnetic field between H_{c1} and H_{c2} the system is in a mixed state in which both the superconducting and normal phases are present. This state corresponds to a nonhomogeneous solution in which the order parameter is not constant but varies rapidly in space. In all the internal regions of the flux lines the order parameter must go to zero while it must be nearly constant sufficiently far from the centers. A complete and correct description of this situation can be reached only by knowing the spatial variation of the order parameter over a distance of the order of ξ.

In the framework of the BCS theory the only way to get information about the structure of the flux lines is to solve the Gor'kov equations(12) or the Bogoliubov equations.(13) The difficulty of solving these equations has generated the actual theoretical situation where a complete and general description of the mixed state does not exist, but only small regions in the H,T plane can be approximately described. These regions correspond to those situations where it is possible to make expansions in power series of the order parameter. For $T \simeq T_c$ the Gor'kov equations reduce to local differential equations (Ginzburg-Landau equations(14) and their generalizations(15)); for $H \simeq H_{c2}$ it is possible to get from the Gor'kov equations a soluble integral equation for the order parameter.(16)

Recently, there have been some attempts to compute the structure of a single flux line for arbitrary temperature. The Bogoliubov equations have been considered in the WKBJ approximation; the method(17) involves the use of test functions and only numerical calculations can be performed. Information about the asymptotic behavior of magnetic field and order parameter has been obtained(18,19) by studying the Gor'kov equations in the Eilenberger form.(20)

In this second part we shall apply the boson method in

superconductivity to the study of type-II superconductors. The analysis will be developed for any temperature below T_c and will be restricted to the case of pure materials.

2.2 Flux Lines

As we have shown in Ref. 21 the solution of the Laplace equation (48) which corresponds to a single flux line is given by

$$f(\vec{x}) = \frac{1}{2}\,\nu\phi(\vec{x}) \ . \tag{53}$$

$\vec{x}$ is a two-dimensional vector lying in a plane normal to the external magnetic field; $\phi(\vec{x})$ is the polar angle of the vector $\vec{x}$ expressed in cylindrical coordinate; the constant ν has to be an integer in order that the order parameter be single valued. Using the solution (53) we can compute by means of the formulas (49) and (50) the magnetic field and the current of an isolated flux line.

The result is:

$$\vec{h}^{(s)}(\vec{x}) = \vec{e}_3\,\frac{\nu\Phi_o}{2\pi}\int_0^\infty kF(k)J_o(kr)\,dk \ , \tag{54}$$

$$\vec{J}(\vec{x}) = \vec{e}(\phi_x)\,\frac{\nu\Phi_o}{8\pi^2}\int_0^\infty k^2F(k)J_1(kr)\,dk \ , \tag{55}$$

where $\vec{e}_3$ is the unit vector in the direction of the applied field and $\vec{e}(\phi_x)$ is the unit vector in the azimuthal direction; r is the modulus of the vector $\vec{x}$. The function F(k) is defined in terms of the boson characteristic function as

$$F(k) = \frac{c(k)}{c(k)+\lambda_L^2k^2} \ ; \tag{56}$$

J_o and J_1 are Bessel functions; Φ_o is the flux quantum

$$\Phi_o = \frac{\pi}{e} \ . \tag{57}$$

The total magnetic flux Φ can be obtained by integrating (54) over a surface of radius $R >> \lambda_L$;

$$\Phi = \int \vec{h}(\vec{x}) \cdot \vec{dS} = \nu \Phi_o \quad . \tag{58}$$

We thus find that the total magnetic flux is quantized in units of the flux quantum Φ_o. It is interesting to note that the solution of the Abrikosov theory[11] are obtained by putting $c(k) = 1$ and for distances far from the axis center (i.e. $r >> \xi_o$):

$$h^{(s)}(\vec{x}) = \frac{\Phi}{2\pi\lambda_L^2} K_o(\frac{r}{\lambda_L}) \quad , \qquad J(\vec{x}) = \frac{\Phi}{8\pi^2\lambda_L^3} K_1(\frac{r}{\lambda_L}) \quad ; \tag{59}$$

K_o and K_1 are modified Bessel functions.

Let us now consider N flux lines with centers at positions $\vec{r}_1, \vec{r}_2, \ldots \vec{r}_N$; since the Laplace equation (48) is linear in f the solution which describes this situation is given by:[22]

$$f(\vec{x}) = \frac{1}{2} \nu \sum_i \phi(\vec{r} - \vec{r}_i) \quad , \tag{60}$$

where the sum is over all the vortex positions. Inserting this solution in (51) we find that the energy density is

$$E = \frac{n\Phi}{8\pi} \sum_i h^{(s)}(\vec{r}_i) \quad , \tag{61}$$

where n is the density of flux lines and $h^{(s)}(\vec{r}_i)$ is the magnetic field due to a single vortex situated at $\vec{r}_i$. Let us assume that the vortices form a periodic lattice, in which the unit cell is a parallelogram. We shall denote the lengths of sides by a and b and the angle between them by θ. By inserting (54) in (61) we have

$$E = \frac{B^2}{8\pi} \sum_{\ell,m} F(g) \quad , \tag{62}$$

where $B = n\Phi$ is the induction field, ℓ and m are integers running from $-\infty$ to $+\infty$ and g is the magnitude of a vector of the reciprocal lattice:

$$g = 2\pi n\ [b^2\ell^2 + a^2m^2 - 2ab\ell m\cos\theta]^{\frac{1}{2}}\ . \tag{63}$$

The density n is fixed by the equation

$$\partial G/\partial n = 0 \tag{64}$$

where G is the Gibbs free energy per unit volume:

$$G = E - \frac{1}{4\pi}\ BH\ .$$

From (64) we get the constitutive equation B = B(H), where H is the thermodynamic field,

$$H = B \sum_{\ell,m} \left[F(g) + \frac{1}{4}\ g\ \frac{\partial F(g)}{\partial g}\right]\ . \tag{66}$$

From (62) and (66) we also obtain that the Gibbs free energy attached to flux lines has the expression:

$$G = -\ \frac{B^2}{8\pi} \sum_{\ell,m} \left[F(g) + \frac{1}{2}\ \frac{\partial F(g)}{\partial g}\right]\ . \tag{67}$$

Formulae (66) and (67) show that all the magnetic properties of the mixed state of type-II superconductors are expressed in terms of the boson characteristic function. This is not surprising, since the structure of a flux line is mainly controlled by the function c(k), as it can be seen from Eq. (54).

2.3 Neutron Diffraction by Flux Lines and Magnetization Curves

The distribution of the magnetic field of a flux line can be experimentally studied by means of scattering of neutrons by the periodic lattice. More precisely, the Fourier amplitudes of the magnetic field can be measured by means of scattering of neutrons. When the neutrons are scattered mainly by a single vortex, the scattering form factor is given[2,22] by the function F(k), defined in (56).

The boson characteristic function has been computed from the microscopic formulation for small values of k; the result is[23]

$$c(k) = 1 - \alpha(T)\xi_o^2(T)k^2 + O(k^4) \ , \tag{68}$$

where ξ_o is the BCS temperature-dependent coherence length, $\xi_o(T) = v_F/\pi\Delta(T)$, and α is a function of temperature. At T = 0°K:

$$\alpha = \frac{2\pi^2}{45} \tag{69}$$

On the other hand expressions (54) and (55) show that in order to have a stable vortex lattice c(k) must decrease at least as $1/k^2$ when $k \to \infty$. We shall thus write

$$c(k) = \frac{1}{1+\alpha k^2\xi_o^2+\ldots} \tag{70}$$

where dots stand for higher order terms in k^2.

The single vortex form factor in the case of scattering of neutrons by $Nb_{0.27}\,Ta_{0.73}$ has been measured by Schelten et al. (Ref. 24). The solid curve in Fig. 1 shows the experimental result. The Abrikosov approximation is obtained by using c(k) = 1.

This approximation leads to $F(k) = 1/(1+\lambda_L^2 k^2)$ which is represented by the dotted curve in Fig. 1: this curve is quite far from the experimental one. To try the next approximation we shall ignore the terms represented by dots in (70) and we shall take

$$c(k) = \frac{1}{1+\alpha\xi_o^2 k^2} \ . \tag{71}$$

The circles in Fig. 1 represent[25] values of F(k) obtained by using the expression (71). These points agree very well with the experimental data; the agreement is not a trivial one because α in (71) is a purely theoretical result of the microscopic formulation. We have tried to improve the agreement by taking[25]

$$c(k) = \frac{1}{1+\alpha\xi_o^2 k^2+\delta\xi_o^4 k^4} \ , \tag{72}$$

where δ is a parameter which should be phenomenologically fixed. The result with $\delta = 0.025$ is given by crosses in Fig. 1: the existence of the δ-term does not change F(k) very much, though it further improves the form factor.

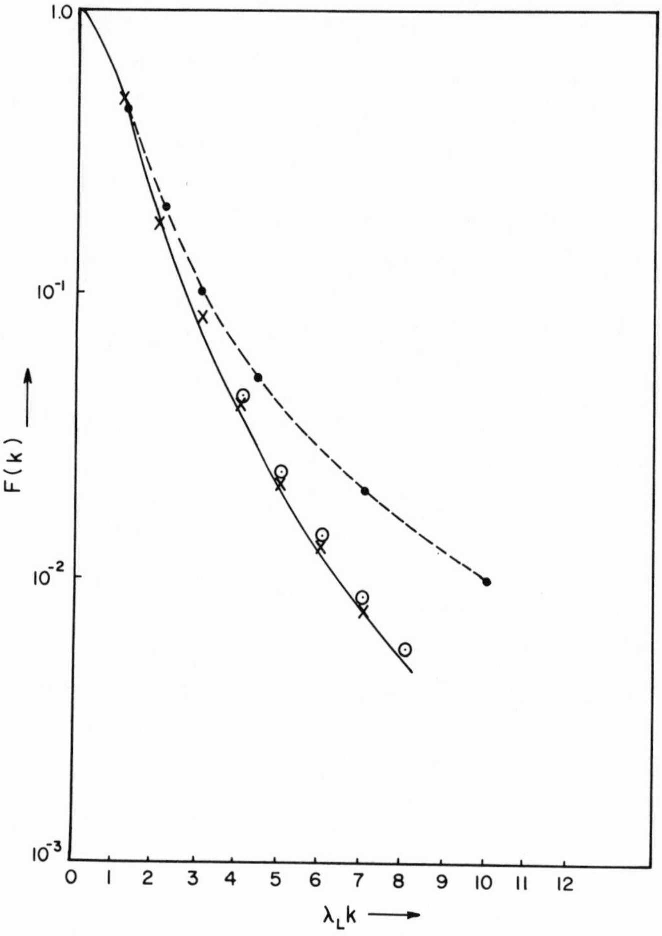

Fig. 1 The single vortex neutron scattering form factor of $Nb_{0.27}Ta_{0.73}$. The solid curve is the experimental data. The circles denote theoretical results of the boson method with $\delta = 0$ and the crosses the ones with $\delta = 0.025$. The dashed curve is the result of the Abrikosov approximation.

Using the expression (72) for the function c(k) with $\delta = 0.025$ we have computed[25] the magnetization of pure type-II superconductors. By means of the formula (66) we can compute the magnetization, $M = (B-H)/4\pi$, in the entire domain of H. Assuming a triangular structure for the vortex lattice ($a = b = \sqrt{2/n\sqrt{3}}$, $\theta = 60°$) we have computed the magnetization of pure Vanadium and Niobium at zero temperature. The summation in (66) has been performed numerically by means of a computer. The results are presented in Fig. 2 for Vanadium and in Fig. 3 for Niobium.

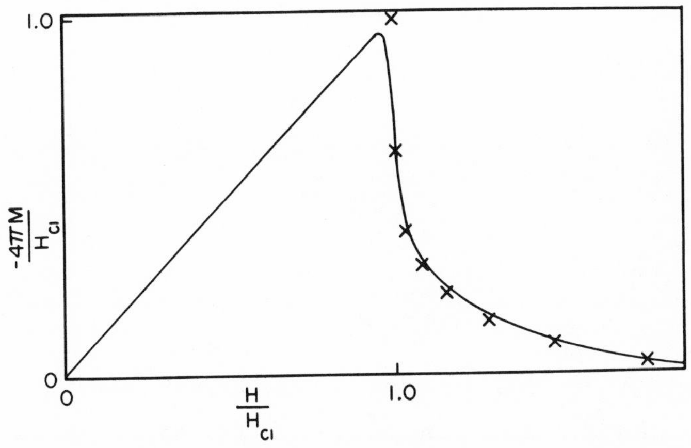

Fig. 2 Magnetization curve of Vanadium normalized by H_{c1}. The solid curve is the experimental data and the crosses denote theoretical results.

In these figures the solid curves are experimental data for $-4\pi M/H_{c1}$ which are taken from Ref. 26. Agreement between theory and experiment is quite good. At H around H(0), the value of H at $n = 0$, there is a rapid drop in the magnetization curve, this shows that H_{c1} is almost equal to H(0). The experimental values,

H_{c1} = 1150 G for V and H_{c1} = 1735 G for Nb together with the relation (66) with n = 0 lead to λ_L = 305 Å for V and λ_L = 250 Å for Nb which are smaller than the experimental values (375 Å for V and 350 Å for N_b).

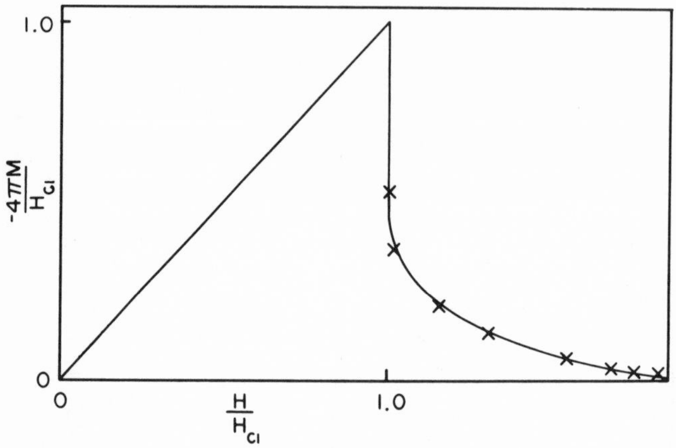

Fig. 3 Magnetization curve of Niobium normalized by H_{c1}.

2.4 First-Order Phase Transition in Type-II Superconductors

It is well known that for type-I superconductors the transition between the superconducting phase and the normal phase is first order; the situation, however, is not very well established in the case of type-II superconductors. The Ginzburg-Landau theory predicts a second-order phase transition at H_{c1}; but this theory is strictly valid only at $T = T_c$. On the other hand, recent experiments(27) have shown that there is a range of values of $\kappa = \lambda_L/\xi_o$, the Ginzburg-Landau parameter, such that the transition at H_{c1} is first order. This result has been theoretically interpreted as being due to the fact that for values of κ less than a critical value there is an attractive interaction among the flux lines. At $H = H_{c1}$ the induction field B will suddenly jump from the zero value of the Meissner state to a finite value B_o which corresponds to a density n_o of flux lines such that the distance d_o between the vortices is the distance where the interaction energy has a minimum:

$$B_o = n_o \Phi = \frac{2}{\sqrt{3}} \frac{\Phi}{d_o^2} \tag{73}$$

The existence of an attractive interaction between the flux lines has been intuitively related to the fact that the magnetic field of a single vortex changes direction at a certain distance r_o; in correspondence the circulating persistent currents will flow in the opposite direction and an attractive interaction will occur.

The first theoretical derivation of the possibility of the existence of an attractive interaction was obtained in Ref. 18 by means of the Gor'kov equations and, independently, in Ref. 21 by means of the boson formulation.

According to the expression (61) the interaction energy per unit length between two vortices is given by:

$$E(d) = \frac{\Phi}{4\pi} h(d) , \tag{74}$$

where d is the distance between centers. This result immediately relates the presence of an attractive interaction to the phenomenon of field reversal.

Let us now compute the magnetic field of a single isolated vortex using the expression (71) for the function c(k). Let us distinguish two cases:[28] i) $\kappa > \kappa_c$; ii) $\kappa < \kappa_c$; κ_c is defined by:

$$\kappa_c = 2 \sqrt{\alpha} . \tag{75}$$

In the case i) we obtain:

$$h(r) = \frac{\Phi\kappa}{2\pi\lambda_L^2\kappa} [K_o(r/\lambda_1) - K_o(r/\lambda_2)] , \tag{76}$$

where $\bar{\kappa} = \sqrt{\kappa^2 - \kappa_c^2}$ and

$$\lambda_1 = \lambda_L \left[\frac{1}{2}\left(1 + \frac{\bar{\kappa}}{\kappa}\right)\right]^{\frac{1}{2}}, \qquad \lambda_2 = \lambda_L \left[\frac{1}{2}\left(1 - \frac{\bar{\kappa}}{\kappa}\right)\right]^{\frac{1}{2}}. \tag{77}$$

In the case ii) we obtain

$$h(r) = \frac{i\Phi\kappa}{2\pi\lambda_L^2\tilde{\kappa}} \left[K_o(br) - K_o(b^* r)\right] , \tag{78}$$

where $\tilde{\kappa} = \sqrt{\kappa_c^2 - \kappa^2}$, $b = b_1 + ib_2$, $b^* = b_1 - ib_2$ and

$$b_1 = \frac{1}{\lambda_L} \left[\frac{\kappa}{\kappa_c}\left(1 + \frac{\kappa}{\kappa_c}\right)\right]^{\frac{1}{2}}, \quad b_2 = \frac{1}{\lambda_L} \left[\frac{\kappa}{\kappa_c}\left(1 - \frac{\kappa}{\kappa_c}\right)\right]^{\frac{1}{2}} . \tag{79}$$

In Eqs. (76) and (78) K_o is the modified Bessel function of zero order.

Let us first discuss the case i). Since $\lambda_1 > \lambda_2$, and since $K_o(x)$ is a monotonically decreasing function of x, the magnetic field goes exponentially to zero when the distance from the axis center goes to infinity. For $\kappa > \kappa_c$ the interaction energy is always repulsive and the transition at H_{c1} is second order.

The situation is different in the case ii). Observing that $|b|^2 = 2\kappa/\lambda_L^2 \kappa_c$, we can use for $r >> \lambda_L$ the asymptotic expression for the Bessel function: $K_o(z) = \sqrt{\pi/2z}\, e^{-z}$. We thus obtain the following asymptotic behavior for the magnetic field:

$$h(r) = \frac{\Phi\kappa}{\lambda_L^2\tilde{\kappa}} \left[\frac{1}{2\pi r|b|}\right]^{\frac{1}{2}} \exp(-rb_1)\sin\left(\frac{1}{2}\,\theta + rb_2\right) , \tag{80}$$

where

$$\theta = \mathrm{arctg}\left(\frac{b_2}{b_1}\right) . \tag{81}$$

For $\kappa < \kappa_c$ the magnetic field of a single vortex is an oscillating exponentially damped function; the nodes of this functions are given by the relation

$$r_n = \frac{1}{b_2}\left[n\pi - \frac{1}{2}\theta\right] \qquad n = 1,2\ldots \tag{82}$$

We thus find that materials with $\kappa < \kappa_c$ present an interaction energy which becomes attractive at a distance r_o given by

$$\frac{r_o}{\lambda_L} = \left[\frac{\kappa_c^2}{\kappa(\kappa_c-\kappa)}\right]^{\frac{1}{2}}\left[\pi - \frac{1}{2}\,\mathrm{arctg}\sqrt{\frac{\kappa_c-\kappa}{\kappa_c+\kappa}}\right]. \tag{83}$$

Summarizing, there is a critical value κ_c, such that for $\kappa > \kappa_c$ the transition at H_{c1} is second order, while for $\kappa < \kappa_c$ the transition is first order. This critical value, which separates type II/1 from type II/2 superconductors, according to the nomenclature of Ref. 27, is a function of temperature. We have so far computed its value at T = 0°K and we find $\kappa_c = 1.27$ (in our formulation κ is defined as $\kappa = \lambda_L/\xi_o$, while in the BCS formulation κ is defined as $\kappa = 0.96\ \lambda_L/\xi_o$) which should be compared with the experimental value $\kappa_c = 1.14$ of Ref. 27. The agreement is sufficiently good and the situation can be improved by using a more accurate expression for the function c(k).

By using the expression (72) for the function c(k), it can be proved that the critical value of κ is given by

$$\kappa_c^2 = \frac{\alpha(2\alpha^2-9\delta) + 2(\alpha^2-3\delta)^{3/2}}{\alpha^2-4\delta} \tag{84}$$

In order that this expression agrees with the experimental value of Ref. 27 we should take at T = 0°K $\delta = 0.057$. The expression of the magnetic field of a single flux line can be also analytically computed. For $\kappa < \kappa_c$, where κ is now defined by Eq. (84), we find that there is an attractive interaction among the flux lines. All the detailed calculations will be published elsewhere.

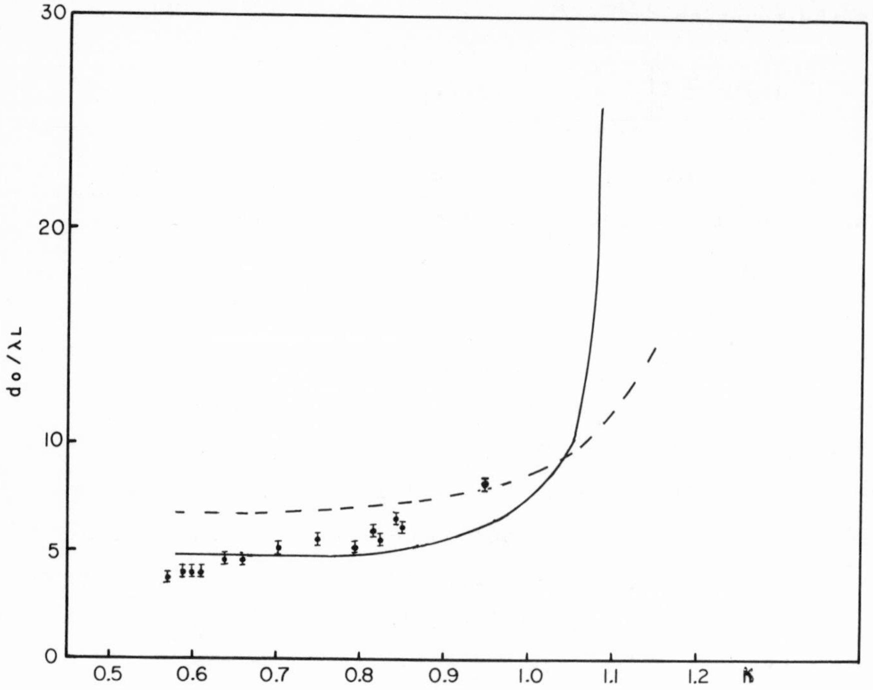

Fig. 4 The flux-line parameter d_o as a function of the Ginzburg-Landau parameter κ. The points are the experimental data taken from Ref. 27. The dashed curve denotes the theoretical results with $\delta = 0$. The solid curve denotes the theoretical results with $\delta \neq 0$ at $t = 0.3$.

In Fig. 4 we compare the theoretical values of d_o/λ_L with the experimental data of Ref. 27. The dashed curve represents the theoretical results obtained with $\delta = 0$. The solid curve represents the results with $\delta \neq 0$; this curve has been computed at $t = T/T_c = 0.3$ by fitting the theoretical value of κ_c, given by Eq.(84), to the experimental value $\kappa_c = 1.09$ at $t = 0.3$ of Ref. 27. We have assumed the reasonable approximation that the quantity δ/α^2 does not appreciably depend on temperature.

To study in more detail the situation when $\kappa < \kappa_c$, the magnetization curves for $\kappa = 0.67$ and $\kappa = 0.35$ have been computed.[29] The results are shown in Fig. 5 and 6.

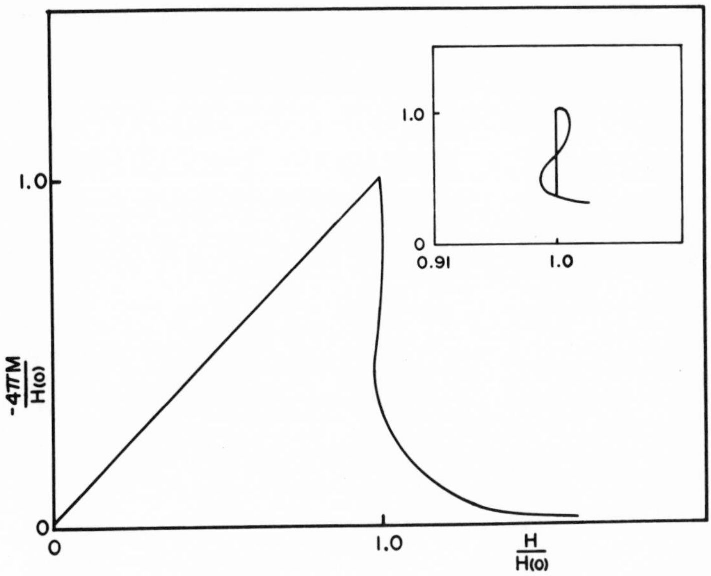

Fig. 5 Magnetization curve of a pure superconductor with $\kappa = 0.67$ normalized by H(0). Also in the box is the enlarged version of the inverted S form.

These curves have been obtained by assuming that the flux lines form a triangular lattice for all values of H. The results show that around H_{c1} the magnetization curves have an S-shape; this effect is more pronounced for smaller values of κ and disappears when κ is bigger than κ_c. This characteristic S-shape shows that for some values of H the triangular lattice is not permitted; this happens because the stability condition

$$\frac{\partial^2 G}{\partial n^2} \geq 0 \tag{85}$$

is not satisfied; in fact it can be shown that

$$\frac{\partial^2 G}{\partial n^2} = \frac{\Phi^2}{4\pi} \frac{\partial H}{\partial B} ; \tag{86}$$

the system is unstable in the region where $\partial H/\partial B < 0$. For those values of H which correspond to a distance among the vortices $d > r_o$ the interaction is attractive; in this case the triangular lattice is unphysical. We are faced with a typical situation for a first order phase transition.

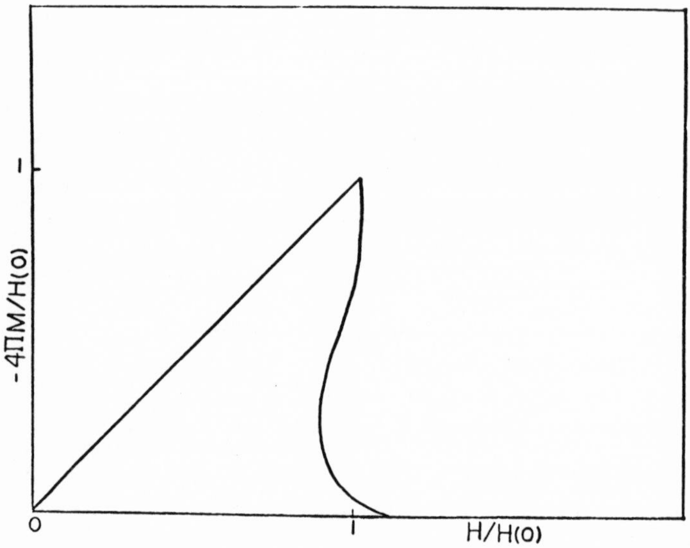

Fig. 6 Magnetization curve for $\kappa = 0.35$. (Ref. 29).

It is interesting to note that the experimental difference[27] of H for going and coming back curves is of the same order of magnitude as the width of the inverted S-form in the theoretical curves.

References

1. H. Umezawa, Acta Phys. Hung. 19, 9 (1965); H. Umezawa, Nuovo Cimento 40, 450 (1965).

2. L. Leplae, F. Mancini, and H. Umezawa, preprint UWM-4867-73-4 (1973); F. Mancini, Ph.D. Thesis, University of Wisconsin-Milwaukee (1972).

3. L. Leplae, F. Mancini, and H. Umezawa, Phys. Rev. B2, 3594 (1970).

4. L. Leplae, F. Mancini, and H. Umezawa, Phys. Rev. B5, 884 (1972).

5. J. Bardeen, L.N. Cooper, and J.R. Schrieffer, Phys. Rev. 108, 1175 (1957).

6. The Bethe-Salpeter equations for the amplitudes (10)-(12) were first solved in Ref. 7 in the small-momentum limit; in Ref. 2. the calculations were extended to higher-order powers in the momentum expansion. Recently we have solved the Bethe-Salpeter equations in complete generality without making any expansion; the expressions reported in this lecture are these last results. All the details calculations will be published elsewhere.

7. L. Leplae, and H. Umezawa, Nuovo Cimento 44, 410 (1966).

8. L. Leplae, R.N. Sen, and H. Umezawa, Suppl. Progr. Phys. Extra Number, 637 (1965).

9. It is well known that in any case of spontaneous breakdown of symmetry the generator is not defined as an operator acting on the physical representation. This fact does not cause any difficulty since in all the practical calculations one is concerned only with commutators of N with other operators; the generator does not need to be defined; what is important is the algebra that it satisfies. However, we can still use the notion of generator if we adopt the following definition. Let $\{g_n\}$ be a sequence of square-integrable functions uniformly convergent to 1 as $n \to \infty$. Then let us introduce the operator

$$Ng_n = \int d^3x \; g_n(\vec{x}) \; \rho(x)$$

which is always defined. In all the cases in which we shall use the generator this must be intended as $N = \lim_{n\to\infty} Ng_n$, where the limit $n \to \infty$ must be taken once all the calculations have been performed.

10. The general case of nonstationary situations is discussed in Ref. 4.

11. A.A. Abrikosov, Zh. Eksp. i Teor. Fiz. 32, 1442 (1957); Sov. Phys. JETP 5, 1174 (1957). For a review of the theory of type-II superconductors see A.L. Fetter and P.C. Hohenberg,

in Superconductivity, ed. by R.D. Parks (Dekker, New York, 1969), Chap. 14.

12. L.P. Gor'kov, Zh. Eksp. i Teor. Fiz. 36, 1918; 37, 1407 (1959) [Sov. Phys. JETP 9, 1364 (1959); 10, 998 (1960)].

13. N.N. Bogoliubov, V.V. Tolmachev, and D.V. Shirkov, A New Method in the Theory of Superconductivity, Consultants Bureau, New York, 1959.

14. V.L. Ginzburg, and L.D. Landau, Zh. Eksp. i Teor. Fiz. 20, 1064 (1950).

15. N.R. Werthamer, Phys. Rev. 132, 663 (1963); L. Tewordt, Phys. Rev. 132, 595 (1963).

16. E. Helfand, and N.R. Werthamer, Phys. Rev. 147, 288 (1966); K. Maki, and T. Tsuzuki, Phys. Rev. 139, A 868 (1965).

17. J. Bardeen, R. Kümmel, A.E. Jacobs, and L. Tewordt, Phys. Rev. 187, 556 (1969).

18. G. Eilenberger, and H. Büttner, Z. Physik 224, 335 (1969).

19. A.E. Jacobs, J. Low Temp. Phys. 10, 137 (1973).

20. G. Eilenberger, Z. Physik 214, 195 (1968).

21. L. Leplae, F. Mancini, and H. Umezawa, Lettere al Nuovo Cimento, Serie I, Vol. 3, 153 (1970).

22. F. Mancini, and H. Umezawa, Phys. Letters 42A, 287 (1972).

23. F. Mancini, and H. Umezawa, Lettere al Nuovo Cimento 7, 125 (1973).

24. J. Schelten, H. Ullmaier, and W. Schmatz, Phys. Stat. Sol. (b) 48, 619 (1971).

25. F. Mancini, G. Scarpetta, V. Srinivasan, and H. Umezawa, Phys. Rev., in press.

26. S. T. Sekula, and R.H. Kernohan, Phys. Rev. B5, 904 (1972).

27. J. Auer, and H. Ullmaier, Phys. Rev. B7, 136 (1973).

28. F. Mancini, Preprint University of Salerno (1973).

29. L. Leplae, V. Srinivasan, and H. Umezawa, Preprint University of Wisconsin-Milwaukee (1973).

ANALYTIC RENORMALIZATION OF THE σ-MODEL

M.MARINARO - L.MERCALDO - G.VILASI

Istituto di Fisica dell'Università di Salerno

Via Vernieri n.42, 84100 SALERNO

ABSTRACT

The renormalization of the σ-model is studied in the scheme of the analytic regularization. It is shown that by using a suitable analytic regularization it is possible to verify the Ward-Takahashi identities at every stage of the computation.

1-INTRODUCTION

The study of the renormalization problem for field theories exhibiting internal symmetries, such as gauge invariance of first and second kind, chiral invariance and so on, has been extensively developed in the recent literature. Many approaches have been introduced in order to find a renormalization scheme which preserves the invariance of the theory at every stage of the computation. It has been shown that, in the frame of subtractive renormalization [1,2], the so called dimensional regularization [3] satisfies this property. Nothing has been proved, at our knowledge, in the scheme of analytic regularization. In this paper we want to study the aforesaid problem. We shall analyze the renormalization of σ-model in the simple case in which only meson fields are present, and we shall show that it is possible to construct a renormalization scheme which preserves the invariance property of the theory at every stage of the computation. This is performed by using a suitable analytic regularization [5] which consists, essentially, in introducing a splitting of coincident variables in the configuration space. In particular we shall prove that the Ward-Takahashi (W-T) identities are satisfied by the regularized perturbative expansion of the Green's functions, and that the rescaling of the parameters of the theory (normalization of field, coupling constants, mass) is the same as in the symmetric theory (i.e. the theory which is obtained when the coupling constant, of the linear term, goes to zero). In section 2. starting from the Lagrangian of the σ-model and introducing a suitable analytic regularization ($\hat{\delta}$-procedure) we construct the renormalized perturbative expansion of the propagators. In section 3. explicit computation of the rescaling of the parameters of the theory is performed. It is shown that the rescaling does not depend on the coupling constant of the linear term and is the same as in the symmetric theory. In section 4. the W-T identities are studied; it is shown that they are satisfied at every order of the regularized and renormalized perturbative expansion. In section 5. a rearrangement of the perturbative series is performed in order to have an expansion in tree, one loop...m-loop diagrams. The equivalence, of this expansion to the one studied in section 3., as far as the renormalization problem is concerned, is proved. In section 6. the W-T identities are proved for rearranged series and the phenomena of the spontaneous breakdown of the symmetry are analyzed. In particular it is shown how it is possible to obtain the anomalous solutions through analytic continuation of the normal solution. Concluding remarks are given in section 7. The appendices A, B and C are devoted to explicit computations of some technical details.

2-RENORMALIZED PERTURBATIVE EXPANSION

In this section we shall discuss the renormalization of the so-called σ-model [4] in the scheme of the analytic regularization [5].

For the sake of semplicity we shall be concerned exclusively with meson fields; the Lagrangian density defining the model is then:

$$(1)\quad L(x) = \frac{1}{2}\,\partial_\mu\,\phi^\alpha(x)\,\partial^\mu\,\phi^\alpha(x) + m^2\,\phi^\alpha(x)\,\phi^\alpha(x) + $$

$$+ g(\phi^\alpha(x)\phi^\alpha(x))^2 + c\phi^4(x)$$

where $\phi^{\alpha}(x)$ (α=1,2,3,4) are the components of an isovector and $\phi^{\alpha}(x)$ (α=1,2,3), $\phi^4(x)$ are respectively pseudoscalar and scalar fields. Here and in the following summations over repeated indices are understood. The Lagrangian (1) is invariant under chiral SU(2) $\otimes$ SU(2) transformation but for the last term. The vector and axial-vector currents corresponding to the Lagrangian (1) are

$$V^{\alpha}_{\mu} = \varepsilon^{\alpha\beta\gamma}\, \phi^{\beta}\, \partial_{\mu}\, \phi^{\gamma} \qquad \alpha,\ \beta,\ \gamma = 1,2,3$$

$$A^{\alpha}_{\mu} = \phi^{\alpha}\, \partial_{\mu}\, \phi^{4} - \phi^{4}\, \partial_{\mu}\, \phi^{\alpha} \qquad \alpha = 1,2,3$$

$\varepsilon^{\alpha\beta\gamma}$ is the completely antisymmetrical tensor.

It is immediate to see that the vector currents are conserved, while the axial currents satisfy the P.C.A.C. (partial conserved axial-current) condition:

$$(2)\quad \partial^{\mu}\, A^{\alpha}_{\mu} = -\, c\, \phi^{\alpha} \qquad \alpha = 1,2,3$$

Our aim is to calculate the renormalized perturbative expansion of the Green's functions corresponding to the Lagrangian (1).

In order to do this we introduce a suitable analytic regularization which, as we shall show, does not destroy the W-T identities verified by the unrenormalized theory. We start by introducing the distribution [5]

$$(3)\quad \hat{\delta}(x;\lambda) = i\pi^{-2}\lambda f(\lambda)\, (x^2 - io^{+})^{\lambda-2}$$

where $x^2 = x_o^2 - \vec{x}^2$, and $f(\lambda)$ is a function analytic at $\lambda = o$ satisfying the condition $\lim_{\lambda\to o} f(\lambda) = 1$. The distribution $\hat{\delta}(x,\lambda)$ verifies the equation

$$(4)\quad \lim_{\lambda\to o}\int \hat{\delta}(x,\lambda)\, F(x) = \int \delta(x) F(x) = F(o) \quad \forall F \text{ regular at } x=o$$

Then by using the distribution (3) we define the regularized propagator as follows:

$$(5)\quad K^{(R)}_{N}\,(x_1^{\alpha_1} \ldots\ldots x_N^{\alpha_N}) =$$

$$= Z^{\frac{N}{2}}\ {}_{o}\!\langle o|\, T(\phi_o^{\alpha_1}(x_1)\ldots\ldots\phi_o^{\alpha_N}(x_N)\ \exp i\!\int \hat{L}_I[\phi_o(x)]dx)\,|o\rangle_{o}$$

where ϕ_o is the free field of mass $\overline{m}$, $|o\rangle$ is the Fock-space vacuum state for ϕ_o, T is the chronological ordering,

$$(6) \qquad \hat{L}_I\left[\phi_o(x)\right] = \overline{g}\int dudvdtdz \;\; \phi_o^\beta(u)\;\phi_o^\beta(v)\;\phi_o^\gamma(t)\phi_o^\gamma(z)\Gamma_x + \overline{c}\phi_o^4(x)$$

with

$$\Gamma_x = \hat{\delta}(x-u,\lambda^u)\;\hat{\delta}(x-v,\lambda^v)\;\hat{\delta}(x-t,\lambda^t)\hat{\delta}(x-z,\lambda^z)$$

$Z, \overline{g}, \overline{c}, \overline{m}$ are arbitrary finite parameters to be determined at the end by the normalization conditions. Comparing the expression (5) with the one for the unregolarized propagator

$$K_N\,(x_1^{\alpha_1}\ldots x_N^{\alpha_N}) =$$

$$\langle o|\,T(\phi_o^{\alpha_1}(x_1)\ldots\phi^{\alpha_N}(x_N)\;\exp\, i\int L_I\left[\phi_o(x)\right]\,dx)|o\rangle$$

$$L_I = g\;\phi_o^\beta(x)\;\phi_o^\beta(x)\;\phi_o^\gamma(x)\;\phi_o^\gamma(x) + c\;\phi_o^4$$

we see that the regularization introduced consists in splitting coincident points. In terms of graphs we have for example

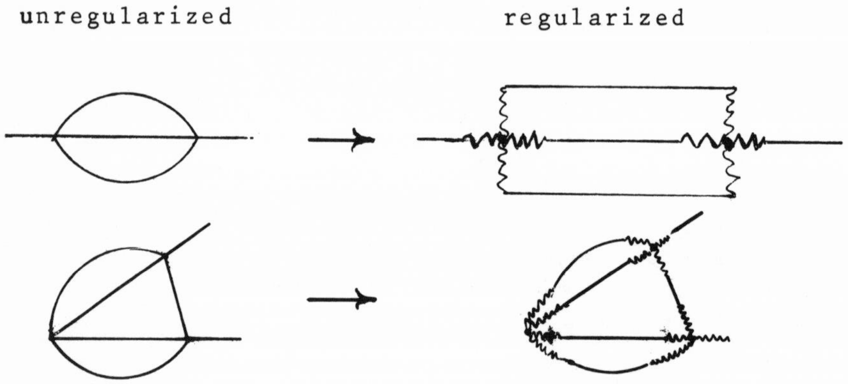

$$\left(\;\underset{x\qquad y}{\overline{\qquad\qquad}} = \delta_{\alpha\beta}\;\Delta c(x-y)\quad;\quad \underset{x\qquad\xi}{\sim\!\sim\!\sim} = \hat{\delta}(x-\xi,\lambda)\;\right).$$

It is convenient to write (5) in more suitable form; to this end we introduce the algorithm $[x_1 \dots x_k]$ (hafnian[6]) defined by development

$$(7) \quad [x_1 \dots x_k] = \sum_{j=2}^{k} f(x_1 \; x_j) \; [x_2 \dots \not{x}_j \dots x_k]$$

where

$$[x_2 \dots \not{x}_j \dots x_k]$$

is the hafnian of k-2 variables obtained from $[x_1 \dots x_k]$ by dropping the variables x_1 and x_j, and

$$f(x_1 \; x_j) = \langle o | T(A(x_1) \; A \; (x_j)) | o \rangle$$

is called "element of the hafnian", in the following it will be denoted by $[x_1 \; x_j]$.
The vacuum expectation value of a product of k boson fields A(x) can be written [6] as follows:

$$(7^1) \quad \langle T(A(x_1) \dots A(x_k)) \rangle = [x_1 \dots x_k]$$

By keeping in mind (7^1), we can write (5)

$$(8) \quad K_N^{(R)} (x_1^{\alpha_1} \dots x_N^{\alpha_N}) =$$

$$= Z^{N/2} \sum_{n=o}^{\infty} \sum_{p=o}^{\infty} \frac{(i\bar{g})^n}{n!} \frac{(i\bar{c})^p}{p!} \int dS_n \int dQ_p \; \Gamma_{\xi_1} \dots \Gamma_{\xi_n} \cdot$$

$$\Big[x_1^{\alpha_1} \dots x_N^{\alpha_N} \; u_1^{\beta_1} \; v_1^{\beta_1} \; t_1^{\gamma_1} \; z_1^{\gamma_1} \dots u_n^{\beta_n} \; v_n^{\beta_n} \; t_n^{\gamma_n} \; z_n^{\gamma_n}$$

$$\eta_1^4 \dots \eta_p^4 \Big]$$

where

$$\int dS_n = \prod_{i=1}^{n} \int \dots \int d\xi_i \; du_i \; dv_i \; dt_i \; dz_i ; \quad \int dQ_p = \prod_{i=1}^{p} \int d\eta_i$$

The element of the hafnian in (8) is given by

$$(9)\quad f_{\alpha\beta}(x\ y) = \langle 0 | T(\phi_0^\alpha(x)\ \phi_0^\beta(y)) | 0 \rangle \equiv \left[x^\alpha\ y^\beta\right] =$$

$$= \delta_{\alpha,\beta}\ \Delta_c\ (x-y,\bar{m})$$

$\delta_{\alpha,\beta}$ is the Kronecker symbol and $\Delta_c(x-y,\bar{m})$ is the free propagator $((\Box_x-\bar{m}^2)\ \Delta_c(x) = i\delta(x))$

The expression at the right handside of (8) is a formal power series in $\bar{g}$ and $\bar{c}$ whose terms define a distribution [5]

$$T(\underline{\xi}\ \underline{\lambda}),\ \xi = \{x_1^{\alpha_1},\ldots.\eta_p^4\},\ \underline{\lambda} = \{\lambda^{u_1}\ldots.\lambda^{t_n}\}$$

as an element of S' holomorphic in a region Ω of $\underline{C}$, where $\underline{C}$ is the tensor product of the complex planes of the parameters λ. This distribution can be continued [5] to a distribution $T(\underline{\xi},\underline{\lambda})$ which is meromorphic in $\underline{C}$. It is therefore possible by means of a general procedure which is based on the use of "generalized evaluators" for $\underline{\lambda} \to 0$ to obtain from (8) the renormalized perturbative expansion of the propagator. Infact we introduce the operation

$$(10)\quad \Theta_n(\underline{\lambda}) = \theta(\lambda^{u_1})\ \theta(\lambda^{v_1})\ldots.\theta(\lambda^{z_n})$$

where

$$(11)\quad \theta(\lambda)\ F(\lambda\ldots) = \lim_{\lambda\to 0} \text{ of the regular part in } \lambda \text{ of the}$$

Laurent expansion of $F(\lambda\ldots)$ and define the renormalized propagator in the following way: [6]

$$\bar{K}_N\ (x_1^{\alpha_1}\ldots x_N^{\alpha_N}) = Z^{\frac{N}{2}} \sum_{n=0}^{\infty} \sum_{p=0}^{\infty} \frac{(i\bar{g})^n}{n!} \frac{(i\bar{c})^p}{p!}\ \Theta_n(\underline{\lambda})\cdot$$

$$\int dS_n \int dQ_p\ \Gamma_{\xi_1}\ldots\Gamma_{\xi_n}$$

$$(12)\quad \left[x_1^{\alpha_1}\ldots x_N^{\alpha_N}\ u_1^{\beta_1}\ v_1^{\beta_1}\ldots\ldots.t^{\gamma_n}\ z^{\gamma_n}\ \eta_1^4\ldots\eta_p^4\right]$$

Note that, as a consequence of the symmetry of the integrand, the order in which the operations $\theta(\lambda)$ are performed is irrelevant. Each term [5] of the expansion (12) is an element in S'.

3-RESCALING OF THE PARAMETERS OF THE THEORY

This section is devoted to the study of the renormalized propagator. We shall show that the expression (12) for the propagator is meaningfull from the physical point of view [5,7]. We mean that the renormalization procedure adopted, does not affect the physical results, i.e. the finite regularization arbitrariness introduced by the regularization rule is incorporated in the form of distributions of point support and therefore give only a change of the parameters (mass and coupling constants). Besides, we shall show that the change of the parameters of the σ-model coincides with that of the symmetric theory ($\bar{c}=o$). The explicit calculation of the counterterms will show that these do not depend on $\bar{c}$ and coincide with the counterterms of the symmetric theory.

In order to see how the renormalized propagator is affected by the renormalization procedure adopted, we have to calculate the variation of

$$K(x_1^{\alpha_1} \dots x_N^{\alpha_N})$$

under a change of the rule. It is sufficient [8] to consider the variation obtained through an infinitesimal change of the function $f(\lambda)$ in the class of the functions analytic near the point $\lambda=o$ and equal to 1 at $\lambda=o$.

Let us indicate by $\delta_f \bar{K}_N (x_1^{\alpha_1} \dots x_N^{\alpha_N})$ the variation of $\bar{K}_N$ for an infinitesimal change of $f(\lambda)$; we have:

$$(13) \quad \delta_f \bar{K}_N (x_1^{\alpha_1} \dots x_N^{\alpha_N}) = Z^{N/2} \sum_{n=o}^{\infty} \sum_{p=o}^{\infty} \frac{(i\bar{g})^n}{n!} \frac{(i\bar{c})^p}{p!}$$

$$\sum_{i=1}^{n} \Theta_n(\underline{\lambda}) \int dS_n \int dQ_p \, \Gamma_{\xi_1} \dots (\delta_f \Gamma_{\xi_i}) \, \Gamma_{\xi_{i+1}} \dots \Gamma_{\xi_n} \cdot$$

$$\Big[x_1^{\alpha_1} \dots x_N^{\alpha_N} \, u_1^{\beta_1} \, v_1^{\beta_1} \, t_1^{\gamma_1} \, z_1^{\gamma_1} \dots u_n^{\beta_n} \, v_n^{\beta_n} \, t_n^{\gamma_n} \, z_n^{\gamma_n}$$

$$\eta_1^4 \dots \eta_p^4 \Big]$$

where

$$\delta_f \Gamma_{\xi i} = \sum_{\{c\}} \left(\delta_f \hat{\delta}(\xi_i - q_i, \lambda_i^q)\right) \hat{\delta}(\xi_i - m_i ; \lambda_i^m)$$

$$\hat{\delta}(\xi_i - n_i ; \lambda_i^n) \; \hat{\delta}(\xi_i - p_i ; \lambda_i^p)$$

$\{c\}$ denotes summation over all the possible ways of extracting the variable q_i from the sequence u_i v_i t_i z_i, and m_i, u_i, p_i are the remaining variables.

Straightforwards, but tedious computations reported in appendix A allow us to express the variation of $\overline{K}_N$ in terms of $\overline{K}_N$ and its derivatives with respect to mass and charges, namely we have:

$$(14) \quad \delta_f \overline{K}_N (x_1^{\alpha_1} \ldots x_N^{\alpha_N}) = (R_o + \frac{N}{2} R_2^{(2)} + 24\, R_4\, B^2\, \Omega)\, \overline{K}_N +$$

$$+ (-\,12\, B\, R_4 + R_2^{(1)})\; 2i \frac{\partial}{\partial \overline{m}^2} \overline{K}_N + (-i\, R_4 + 2\overline{g}\, R_2^{(2)}) \cdot$$

$$\frac{\partial \overline{K}_N}{\partial \overline{g}} + \frac{\overline{C}}{2} R_2^{(2)} \frac{\partial \overline{K}_N}{\partial \overline{c}}$$

where

$$B = \theta(\lambda) \int dy\; \Delta_c\, (x-y)\; \hat{\delta}(x-y, \lambda)\, dy$$

and R_o, $R_2^{(1)}$, $R_2^{(2)}$, R_4 are expressed by series in $\overline{g}$ as follows:

$$R_o = \sum_{r=1}^{\infty} \frac{(i\overline{g})^r}{r!}\, \theta_{\lambda_1, \lambda_2 \ldots \lambda_r} \int dS_r\, (\delta_f \Gamma_{\xi_1})\, \Gamma_{\xi_2} \cdots \Gamma_{\xi_r} \cdot$$

$$\left[u_1^{\beta_1}\; v_1^{\beta_1} \;\ldots\; t_r^{\gamma_r}\; z_r^{\gamma_r} \right]$$

$$(15) \quad R_4 = \sum_{r=1}^{\infty} \frac{(i\overline{g})^r}{r!}\, \theta_{\lambda_1, \lambda_2 \ldots \lambda_r} \int du_1 \ldots dz_1 \int dS_{r-1} (\delta_f \Gamma_{\xi_1})\, \Gamma_{\xi_2} \cdots \Gamma_{\xi_r}$$

$$\sum_{(4)} \left[u_1^{\beta_1}\; v_1^{\beta_1}\; t_1^{\gamma_1}\; z_1^{\gamma_1} \;\ldots\; u_r^{\beta_r}\; v_r^{\beta_r}\; t_r^{\gamma_r}\; z_r^{\gamma_r} \right]$$

$$R_2^{(1)} = \sum_{r=1}^{\infty} \frac{(i\bar{g})^r}{r!} \theta_{\lambda_1,\lambda_2 \cdots \lambda_r} \int du_1 \cdots dz_1 \int dS_{r-1} (\delta_f \Gamma_{\xi_1}) \cdot \Gamma_{\xi_2} \cdots \Gamma_{\xi_r} (A_r + \frac{\bar{m}^2}{2} B_r)$$

$$R_2^{(2)} = i \sum_{r=1}^{\infty} \frac{(i\bar{g})^r}{r!} \theta_{\lambda_1,\lambda_2 \cdots \lambda_r} \int du_1 \cdots dz_1 \int dS_{r-1} (\delta_f \Gamma_{\xi_1}) \cdot \Gamma_{\xi_2} \cdots \Gamma_{\xi_r} B_r$$

with

$$A_r = \sum_{(2)} \left[u_1^{\beta_1} \; v_1^{\beta_1} \; t_1^{\gamma_1} \; z_1^{\gamma_1} \cdots z_r^{\gamma_r} \right]$$

$$B_r = \frac{1}{2} \sum_c \{ (P_1 - \xi_1) \cdot (P_2 - \xi_1) - (P_1 - \xi_1)^2 - (P_2 - \xi_1)^2 \} \left[\zeta_1^{\mu_1} \; \zeta_2^{\mu_2} \cdots \zeta_{4r-2}^{\mu_{4r-2}} \right]$$

Where $\sum_{(q)}$ denotes summations over all the hafnians obtainable from $\left[u_1^{\beta_1} \cdots\cdots z_r^{\gamma_r} \right]$ by deleting any q variables from it. $\sum_c$ denotes summation over all the ways of extracting two variables, P_1, P_2, from the sequence $u_1^{\beta_1} \cdots z_r^{\gamma_r}$, and $\zeta_1^{\mu_1}, \ldots, \zeta_{4r-2}^{\mu_{4r-2}}$ denote the remaining variables.

It is now evident that the variation of $\bar{K}_N$ can be compensed by a variation of $\bar{g}, \bar{m}, \bar{c}$ and Z, but for a factor. In fact let us consider the total variation of K_N for infinitesimal change of rule and parameters, we have:

$$(16) \quad \delta \bar{K}_N = \delta_f \bar{K}_N + \frac{N}{2} \frac{\delta Z}{Z} \bar{K}_N + \delta \bar{m}^2 \frac{\partial \bar{K}_N}{\partial \bar{m}^2} + \delta \bar{g} \frac{\partial \bar{K}_N}{\partial \bar{g}} + \delta \bar{c} \frac{\partial \bar{K}_N}{\partial \bar{c}}$$

Keeping in mind (14) and (15) and putting

$$\begin{aligned}
&\delta\overline{m}^2 = -2i\,(R_2^{(1)} - 12BR_4)\\
&\delta\overline{g} = iR_4 - 2\overline{g}\,R_2^{(2)}\\
(17)\quad &\delta\overline{c} = \frac{\overline{c}}{2}\,R_2^{(2)}\\
&\frac{\delta Z}{Z} = R_2^{(2)}
\end{aligned}$$

(16) becomes

$$(18)\quad \delta\overline{K}_N = (R_o + 24\,R_4 B^2\,\Omega)\,\overline{K}_N$$

From (18) it follows that the Green's functions of N variables:

$$(19)\quad G_N\,(x_1^{\alpha_1} \,..\, x_N^{\alpha_N}) = \frac{\overline{K}_N(x_1^{\alpha_1} \,..\, x_N^{\alpha_N})}{\overline{K}_o}$$

is unchanged under a simultaneous variation of parameters and regularization rule.

Finally it is worthwhile noting that, in spite of the fact that there are more primitive divergences [4] in the σ-model than in the symmetric theory ($\bar{c}=o$), the arbitrariness introduced by the regularization has been eliminated by rescaling (just) the four parameters $Z, \overline{m}^2$, $\overline{g}$ and $\overline{c}$, and that the rescaling is given through equations (17) in terms of the quantities R_o R_4 $R_2^{(1)}$ $R_2^{(2)}$ which do not depend on $\bar{c}$. The rescaling of the parameters is, thus, the same as in the symmetric theory ($\bar{c}=o$). This result is conform to previous results by Symanzik, Lee ..; it is a consequence of the residual chiral symmetry still present [4] in the σ-model, which makes the arbitrary part strictly chiral symmetric.

4-WARD-TAKAHASHI IDENTITIES

In the previous sections we have shown how by using a suitable analytic regularization it is possible to construct a renormalized theory corresponding to the Lagrangian (1). In the following we want show the most remarkable feature of this renormalization procedure, i.e. the W-T identities are preserved at every stage of the computation. The analytic regularization adopted does not destroy the residual chiral symmetry of the σ-model.
In this section we shall show that the W-T identities are satisfied order by order (∀ p,n) by the regularized per-

turbative expansion (8), and therefore by the renormalized expansion (12) too, in the section 6 we shall show that the identities are satisfied also if a suitable partial summations is performed.
The derivation of W-T identities from the regularized perturbative expansion (8) is based on the induction method and make use of the equation (7) for the expansion of the hafnians.
For example in order to derive the relation between the pion propagator and the vacuum expectation value of the σ we first prove by induction (see appendix B) that

$$\int dQ_{2p+1}\left[x_1^4\, u_1^{\beta_1}\, v_1^{\beta_1}\, t_1^{\gamma_1}\, z_1^{\gamma_1} \ldots u_k^{\beta_k}\, v_k^{\beta_k}\, t_k^{\gamma_k}\, z_k^{\gamma_k}\, \eta_1^4 \ldots \eta_{2p+1}^4\right] = (2p+1)\int dQ_{2p}\; dx_1\left[x_1^{\alpha_1}\, x_2^{\alpha_2}\, u_1^{\beta_1}\, v_1^{\beta_1} \ldots\ldots z_k^{\gamma_k}\, \eta_1^4 \ldots \eta_{2p}^4\right] \tag{20}$$

$$\alpha_i = 1,2,3$$

Then, keeping in mind (8), we have

$$Z^{-\frac{1}{2}} i\bar{c}\int K_2^{(R)}(x_1^{\alpha_1}\; x_2^{\alpha_2})\; dx_2 = K_1^{(R)}(x_1^4) \tag{21}$$

Equations(21) means that the perturbative expansions of the left and right handside coincide order by order in powers of $\bar{g}$ and $\bar{c}$, i.e.

$$Z^{-\frac{1}{2}} i\bar{c}\int K_2^{(n,2p)}(x_1^{\alpha_1}\; x_2^{\alpha_2})\; dx_2 = K_1^{(n,2p+1)}(x_1^4) \qquad \forall\; n,p \tag{22}$$

where

$K_2^{(\ell,m)}$ ($K_1^{(\ell,m)}$) is the coefficient of the power $\bar{g}^{\ell}$ $\bar{c}^{m}$ of the perturbative expansion of $K_2^R(K_1^R)$.
From (21) follows that also the renormalized propagators satisfies the relation

$$Z^{-\frac{1}{2}} i\bar{c}\int \bar{K}_2(x_1^{\alpha_1}\; x_2^{\alpha_2})\; dx_1 = \bar{K}_1(x_1^4) \tag{23}$$

The eq.(23) have to be understood in the same way of (21), i.e equality between two formal series.
Finally it is easy to show that the relations (21) (23) remain true if one considers only connected graphs of

both sides. This implies the relation:

$$(24) \quad Z^{-\frac{1}{2}} i\bar{c} \int G_2^{(n,2p)}(x_1^{\alpha_1}\ x_2^{\alpha_2})\, dx_2 = G_1^{(n,2p+1)}(x_1^4) \quad \forall\ n,p.$$

where we have indicated $G_2^{(\ell,m)}$ $(G_1^{(\ell,m)})$ the coefficient of the power $\bar{g}^\ell\ \bar{c}^m$ of the regularized perturbative expansion of

$$G_2^R\ (x_1^{\alpha_1}\ x_2^{\alpha_2})((G_1^R\ (x_1^4)) \text{ with}$$

$$G_2^R = \frac{K^R(x_1^{\alpha_1}\ x_2^{\alpha_2})}{K_o^R}$$

The same procedure can be used to prove the others W-T identities. It is worthwhile noting that relations (23),(24) are meaningfull only in perturbative sense and therefore cannot be used to study the phenomena of the spontaneous breakdown of the symmetry. The study of these phenomena is referred to section 6-

5-REARRANGEMENT OF THE PERTURBATIVE EXPANSION

Until now we have discussed the σ-model studing the perturbative expansions obtained from Lagrangian (1). On the other hands the presence of the linear term in (1) makes the vacuum expectation value of the ϕ^4 component of $\vec{\phi}$ field different from zero. Therefore it would be convenient to operate a shift of the field ϕ^4

$$(25) \quad \phi^4 \to \phi^{4\prime} + F$$

(F is determined so that $\phi^{4\prime}$ has vacuum expectation value equal to zero) and to apply a perturbative method to the new Lagrangian obtained from (1) through the substitution (25). We call this Lagrangian L'. The aim of this section is to show that for what concerns renormalization the two schemes are completely equivalent. The proof is based on the equivalence (see Lee in ref.4) between the perturbative expansion of the Lagrangian L' and the expansion in which one sums all the tree diagrams in the lowest order; diagrams with one loop in the next order; etc.. To reproduce this expansion in our scheme we procede as it follows. Let us confine, for the sake of simplicity, to the study of propagator with only one external ϕ^4 line. The expansion (12) becomes

$$\text{(26}\quad \overline{K}(x_1^4) = Z^{\frac{1}{2}} \sum_{n=o}^{\infty} \sum_{p=o}^{\infty} \frac{(i\overline{g})^n}{n!} \frac{(i\overline{c})^{2p+1}}{(2p+1)!} \Theta_n(\underline{\lambda})\cdot$$

$$\int dS_n \int dQ_{2p+1} \Gamma_{\xi_1} \cdot\cdot \Gamma_{\xi_n} \left[x_1^4 \; u_1^{\beta_1} \; v_1^{\beta_1} \; \ldots\ldots \; \eta_{2p+1}^4\right]$$

note that the sum on p in (12) is an even (odd) numbers if the number of the external ϕ^4 lines is even (odd). It is evident that for n<p there are only disconnected graphs in (26); therefore, if we neglect these we can write (26) as follows

$$\text{(27)}\quad \overline{K}_{con}^{(r)} (x_1^4) = G^{(r)}(x_1^4) = i\overline{c} \sum_{p=o}^{\infty} \sum_{t=o}^{\infty} \frac{(-i\overline{g}\;\overline{c}^2)^p}{(2p+1)!}$$

$$\frac{(i\overline{g})^t}{(p+t)!} \Theta_{t+p}(\underline{\lambda}) \int dS_{p+t} \int dQ_{2p+1} \Gamma_{\xi_1} \cdot\cdot \Gamma_{\xi_{p+t}}$$

$$\left[x_1^4 \; u_1^{\beta_1} \; v_1^{\beta_1} \; \ldots \; \eta_{2p+1}^4\right]_{con}$$

Where the subscript "con" means that we consider, in the development of the hafnian, only terms corresponding to connected graphs, the superscript (r) remember that we are considering the rearranged series. As before one can show that any terms of (27) is an element of S'.(27) is the wanted expansion, infact the sum on p for t=o gives the tree approximation for t=1 the one loop approximation and so on.

In order to study the renormalization problem in the rearranged series, we, as in section 3, consider the variation of $K_N^{(r)}$ under an infinitesimal change of the rule f.

In our scheme is most convenient consider both connected and disconnected graphs. Therefore we consider instead of (27) the expansion

$$\text{(28)}\quad \overline{K}^{(r)}(x_1^4) = i\overline{c} \sum_{p=o}^{\infty} \sum_{t=-p}^{\infty} \frac{(-i\overline{g}\overline{c}^2)^p}{(2p+1)!} \frac{(i\overline{g})^t}{(t+p)!} \cdot$$

$$\Theta_{t+p}(\lambda) \int dS_{p+t} \int dQ_{2p+1} \Gamma_{\xi_1} \cdot\cdot \Gamma_{\xi_{p+t}} \left[x_1^4 \; u_1^{\beta_1} \; v_1^{\beta_1} \dots \eta_{2p+1}^4 \right]$$

(Naturally the relation between (27) and (28) is

$$\overline{K}^{(r)}_{con} (x_1^4) = G^{(r)}(x_1^4) = \frac{\overline{K}_1^{(r)}(x_1^4)}{\overline{K}_o^{(r)}} \Bigg).$$

Following the same procedure and adopting the same notations of the section 3 we have

$$(29) \quad \delta_f \; \overline{K}^{(r)}(x_1^4) = i\overline{c} \sum_{p=o}^{\infty} \; \sum_{t=-p}^{\infty} \; \sum_{i+1}^{p+t} \frac{(-i\overline{g}\overline{c}^2)^p}{(2p+1)!} \; \frac{(i\overline{g})^p}{(t+p)!} \, .$$

$$\Theta_{p+t}(\lambda) \int dS_{p+t} \int dQ_{2p+1} \Gamma_{\xi_1} \cdot\cdot (\delta_f \Gamma_{\xi_i}) \cdot\cdot \Gamma_{\xi_{p+t}} \, .$$

$$\left[x_1^4 \; u_1^{\beta_1} \; v_1^{\beta_1} \dots \eta_{2p+1}^4 \right]$$

The expression (29) can be written, after same manipulations

$$(30) \quad \delta_f \; \overline{K}^{(r)}_N = (R_o + \frac{N}{2} R_2^{(2)} + 24 R_4 \, B^2 \, \Omega) \; \overline{K}^{(r)}_N \; +$$

$$+ \; (-12 \; BR_4 + R_2^{(1)}) 2i \frac{\partial}{\partial \overline{m}^2} \overline{K}^{(r)}_N + (-iR_4 + 2\overline{g}R_2^{(2)}) .$$

$$\frac{\partial \overline{K}^{(r)}_N}{\partial \overline{g}} + \frac{\overline{c}}{2} R_2^{(2)} \frac{\partial \overline{K}^{(r)}_N}{\partial \overline{c}} \, .$$

Where R,B are the same quantities which appear in the equations (15).
Keeping in mind equations (17), we can conclude that the rearrangement of the series does not affect the renormalization of the theory. We have the same rescaling of the parameters in both the cases.
It is worthwhile noting that, in order to obtain the expression (30) from (29) we must shift the series in t. This produces a mixing of the tree one loop..n-loop approximation ; therefore, if, instead of considering the whole series in t, one stops to a fixed value of t,

one obtains approximations which are not renormalizable. The only exception is the trivial case in which there are not U.V. divergences, i.e. the tree approximations (t=o).

6-WARD TAKAHASHI IDENTITIES AND GOLDSTONE THEOREM

In section 4- we have proved that the W-T identities are satisfied order by order in perturbative expansion in powers of $\bar{g}$ and $\bar{c}$.
It is now immediate to see that the relation (24) allow us to prove that the W-T identities are satisfied by tree, one loop n-loop approximations.
Infact if we choose n=p in (24) we have:

$$(31)\quad \overline{G}^{(p,2p+1)}(x_1^4) = Z^{-\frac{1}{2}} i\bar{c} \int \overline{G}^{(p,2p)}(x_1^{\alpha_1}\ x_2^{\alpha_2})\, dx_2$$

$$\alpha_1 = \alpha_2 = (1,2,3)$$

where

$$(32)\quad G^{(p,2p+1)}(x_1^4) = Z^{\frac{1}{2}} i\bar{c}\, \frac{(-i\bar{g}\ \bar{c}^2)^p}{(2p+1)!p!} \int dS_p \int dQ_{2p+1} \cdot$$

$$\Gamma_{\xi_1} .. \Gamma_{\xi_p} \left[x_1^4\ u_1^{\beta_1}\ v_1^{\beta_1} \ldots \eta_{2p+1}^4 \right]_{con}$$

$$(33)\quad G^{(p,2p)}(x_1^{\alpha_1}\ x_2^{\alpha_2}) = Z\, \frac{(-i\bar{g}\ \bar{c}^2)^p}{(2p)!p!} \int dS_p \int dQ_{2p}$$

$$\Gamma_{\xi_1} .. \Gamma_{\xi_p} \left[x_1^{\alpha_1}\ x_2^{\alpha_2}\ u_1^{\beta_1} .. \eta_{2p}^4 \right]_{con}$$

From (31), (32), (33) by using the operation $\Theta(\underline{\lambda})$ and performing the summation over p we obtain:

$$(34)\quad G^{tree}(x_1^4) = Z^{-\frac{1}{2}}\ i\bar{c} \int G^{tree}(x_1^{\alpha_1}\ x_2^{\alpha_2})\, dx_2$$

with

$$(35)\quad G^{tree}(x_1^4) = Z^{\frac{1}{2}} i\bar{c} \sum_{p=o}^{\infty} \frac{(-i\bar{g}\,\bar{c}^2)^p}{p!(2p+1)!} \int d\xi_1 \ldots d\xi_p \cdot$$

$$\int dQ_{2p+1} \left[x_1^4 \quad \xi_1^{\beta_1} \quad \xi_1^{\beta_1} \quad \xi_1^{\gamma_1} \quad \xi_1^{\gamma_1} \ldots\ldots \eta_1^4 \ldots \eta_{2p+1}^4 \right]_{con}$$

$$(36)\quad G^{tree}(x_1^{\alpha_1} \; x_2^{\alpha_2}) = Z \sum_{p=o}^{\infty} \frac{(-i\bar{g}\,\bar{c}^2)^p}{p!(2p)!} \int d\xi_1 \ldots d\xi_p \cdot$$

$$\int dQ_{2p+1} \left[x_1^{\alpha_1} \quad x_2^{\alpha_2} \quad \xi_1^{\beta_1} \quad \xi_1^{\beta_1} \quad \xi_1^{\gamma_1} \quad \xi_1^{\gamma_1} \ldots \eta_1^4 \ldots \eta_{2p}^4 \right]_{con}$$

Equation (34) is the well-known relation which connects, the vacuum expectation value of the σ-field with the π-propagators, in tree approximation. It is worthwhile emphasizing the difference between (23) and (34). The equality (23) has only formal meaning: it is an equality between two asymptotic series; instead the series which appears to left and right hand side of (34) are convergent series and therefore the equality is true also for their sums.

In the same way by taking $n=p+1, \ldots n=p+k$ in (24) we can prove the W-T identities in one loop...k-loop approximation. We can study the phenomena of the spontaneous breakdown of the symmetry. Let us come back to equation (35). It is easy to show (see appendix C) that

$$(37)\quad G^{tree}(x_1^4) = Z^{\frac{1}{2}} \frac{\bar{c}}{\bar{m}^2} \sum_{p=o}^{\infty} \frac{(3p)!}{p!(2p+1)!} \left(\frac{4\bar{g}\bar{c}^2}{\bar{m}^6} \right)^p$$

The series (37) converges for $\frac{27\bar{g}\,\bar{c}^2}{\bar{m}^6} < 1$ to an hypergeometric [9] function.

$$(38)\quad G^{tree}(x_1^4) = \frac{\bar{c}}{m^2}\, F\left(\frac{1}{3}, \frac{2}{3}, \frac{3}{2}, \frac{27\,\bar{g}\,\bar{c}^2}{\bar{m}^6}\right) = \left[-\frac{\bar{c}}{8\bar{g}} + \right.$$

$$\left. + \sqrt{\frac{\bar{c}^2}{4^3\bar{g}^2} - \frac{\bar{m}^6}{4^3 27\bar{g}^2}} \right]^{\frac{1}{3}} + \left[\frac{-\bar{c}}{8\bar{g}} - \sqrt{\frac{\bar{c}^2}{4^3\bar{g}^2} - \frac{\bar{m}^6}{4^3 27\bar{g}^3}} \right]^{\frac{1}{3}}$$

The function to the r.h.s. of (38) is a many-valued function; the branch of this function which is represented by the series (37) is the one that goes to zero, as $\bar{c}$ goes to zero.

On the other hand, the expression (38) can be used to obtain the analytic continuation of $G(x^4)$ outside of the convergence circle, in the complex $\bar{c}$ plane.

Calling $\tilde{G}^{tree}(x^4)$ this analytic continuation we find that $G(x^4)$ takes three values at $\bar{c}=o$

$$\tilde{G}^{tree}(x^4)\Big|_{\bar{c}=o} = o \tag{39}$$

$$\tilde{G}^{tree}(x^4)\Big|_{\bar{c}=o} = \pm\sqrt{\frac{\bar{m}^2}{12\bar{g}}} \tag{40}$$

The solution (39) is called normal: it is representable through the series (38) and gives a vanishing vacuum expectation value of σ in the limit $\bar{c}\to o$. The solutions (40), which give a vacuum expectation value of σ different from zero in the limit $\bar{c}\to o$, are called anomalous solutions. It is immediate to connect the anomalous solutions with the existence of a mass zero particle (Goldstone Theorem). Infact by taking the analytic continuation of the left and right handside of (34) we have:

$$\tilde{G}^{tree}(x_1^4) = Z^{-\frac{1}{2}}\, i\bar{c}\int \tilde{G}^{tree}(x_1^{\alpha_1}\, x_2^{\alpha_2})dx_2;\alpha_1=\alpha_2=(1,2,3) \tag{41}$$

Equation (41) gives in correspondence of the two solutions (39)(40) respectively

$$\int \tilde{G}^{tree}(x_1^{\alpha}\, x_2^{\alpha})\,dx_2\Big|_{\bar{c}=o} = \text{cost} \tag{42}$$

$$\int \tilde{G}^{tree}(x_1^{\alpha}\, x_2^{\alpha})\,dx_2\Big|_{\bar{c}=o} = \infty \tag{43}$$

Thus to the normal solution corresponds a pion with mass different from zero, while to the anomalous correspond a massless pion.

7-CONCLUSIONS

In the previous section has been proved how it is possible to construct, by using a suitable analytic regularization, the renormalized perturbative expansions for the Green's functions of the σ-model. This construction has been performed by considering expansion in $\bar{c}$ and $\bar{g}$ separately, and again by considering rearranged series in powers of $\bar{c}^2\bar{g}$ and $\bar{g}$. Besides we have proved that the W-T identities are satisfied by both the expansions at every stage of the computation.

The use of the rearranged series has been particularly convenient; infact it made possible to treat phenomena like the spontaneous breakdown of the symmetry. Unfortunately, as we have seen, the regularization procedure applied to rearranged series mix terms of different order in $\bar{g}$ and therefore the n-loop approximation is not renormalizable alone.

Hence, in order to go beyond the tree approximation one has to construct a new approximation scheme. This can be made by using the recursive equations among the Green's functions and performing approximations [Lee,4] of the mass and vertex operators.

A detailed study of this problem will be reported in a next paper. It is also interesting to extend the above technique to the study of the regularization of σ-model with both fermion and boson fields.

APPENDIX A

This appendix is devoted to the explicit calculation of the variation of the propagator under variation of regularization rule. We first observe that the integrand in (13) is symmetric in all variables except the ith; it is therefore convenient to arrange the limiting procedures so as to perform the ith limit first.

We use the relation

$$(A.1)\quad \Theta_n(\underline{\lambda})\, F_i = \theta(\underline{\lambda}_1)\ldots\theta(\underline{\lambda}_i)\ldots\theta(\underline{\lambda}_n)\, F_i = \sum_{k=o}^{n-i}\binom{n-i}{k}$$

$$\theta(\underline{\lambda}_1)..\theta(\underline{\lambda}_{i-1})\,\theta(\underline{\lambda}_{i+k+1})..\theta(\underline{\lambda}_n)\,\theta_{\lambda_i,\lambda_{i+1}..\lambda_{i+k}}\, F_i{}^{*}$$

where $\theta(\underline{\lambda}_i) = \theta(\lambda^{u_i})\ldots\theta(\lambda^{t_i})$ and, denoting $[A,B]$ the commutator, we have called

$$\theta_{\lambda_i,\lambda_{i+1}..\lambda_k} = \left[\theta_{\lambda_i,\lambda_{i+1}..\lambda_{k-1}},\theta(\underline{\lambda}_k)\right]$$

$$\theta_{\lambda_i,\lambda_{i+1}..\lambda_{i+k}} = \theta(\lambda_i) \quad \text{for } k=o$$

$F_i = F(\underline{\lambda}_1 \ldots \underline{\lambda}_n)$ is a function symmetric in all the variables $\underline{\lambda}_j = (\lambda^{u_j}\ldots\lambda^{t_j})$ except the ith.

The expression (A.1) can be put in the following form

$$(A.2)\quad \sum_{i=1}^{n}\Theta_n(\underline{\lambda})\, F_i = \sum_{s=1}^{n}\binom{n}{s-1}\theta(\underline{\lambda}_2)..\theta(\underline{\lambda}_s)\,\theta_{\lambda_1,\lambda_{s+1}..\lambda_n}\, F_1$$

Keeping in mind (A.2) equation (13) becomes

$$(A.3)\quad \delta_f\, \overline{K}_N(x_1^{\alpha_1}\ldots x_N^{\alpha_N}) = Z^{\frac{N}{2}}\sum_{s=1}^{\infty}\sum_{p=o}^{\infty}\sum_{n=s}^{\infty}\frac{(i\overline{c})^p}{p!}\,\frac{(i\overline{g})^n}{(s-1)!(n-s-1)!}$$

$$\theta(\underline{\lambda}_2)..\theta(\underline{\lambda}_s)\,\theta_{\lambda_1,\lambda_{s+1}..\lambda_n}\int dS_n\int dQ_p(\delta_f\,\Gamma_{\xi_1})\,\Gamma_{\xi_2}..\Gamma_{\xi_n}\cdot$$

$$\left[x_1^{\alpha_1}..x_N^{\alpha_N}\; u_1^{\beta_1}\; v_1^{\beta_1}\ldots z_n^{\gamma_n}\; \eta_1^4\ldots\eta_p^4\right]$$

---

* (A.1) for k=N-i has to be read $\theta(\lambda_1)..\theta(\lambda_{i-1})\cdot$ $\cdot\theta_{\lambda_1,\lambda_{i+1}..\lambda_n}$

The actual contributions of the terms at the r.h.s. is easily evaluated by means of the extension [6] to the hafnians of Arnaldi's Theorem. We write

$$(A.4)\quad \left[x_1^{\alpha_1} \ldots x_N^{\alpha_N}\, u_1^{\beta_1}\, v_1^{\beta_1} \ldots z_n^{\gamma_n} \ldots \eta_p^4\right] = \sum_{\sigma=o}^{\frac{N}{2}+2(s-1)+\frac{p}{2}} \sum_{c_\sigma}$$

$$\left[\zeta_1^{\mu_1} \ldots \zeta_{2\rho}^{\mu_{2\rho}}\right]\left[\overset{o}{w}{}_1^{\nu_1} \ldots \overset{o}{w}{}_{2\sigma}^{\nu_\sigma}\, u_1^{\beta_1} \ldots z_1^{\gamma_1}\, u_{s+1}^{\beta_{s+1}} \ldots z_{s+1}^{\gamma_{s+1}} \ldots \ldots u_n^{\beta_n} \ldots z_n^{\gamma_n}\right]$$

where $\sum\limits_{c_\sigma}$ denotes summations over all possible ways of extracting, in natural order, the sequence

$$\overset{o}{w}{}_1^{\nu_1}, \ldots, \overset{o}{w}{}_{2\sigma}^{\nu_{2\sigma}}$$

from the sequence

$$x_1^{\alpha_1}, \ldots, x_N^{\alpha_N}, u_2^{\beta_2}, v_2^{\beta_2}, \ldots, u_s^{\beta_s}, v_s^{\beta_s}, t_s^{\gamma_s}, z_s^{\gamma_s}, \eta_1^4, \ldots, \eta_p^4$$

and $\zeta_1^{\mu_1}, \ldots, \zeta_{2\rho}^{\mu_{2\rho}}$ are the remaining variables, so that

$$2\sigma + 2\rho = N + 4(s-1) + p$$

The zeros on the variables $w_i^{\nu_i}$ means that in the development of the hafnian the elements of the type $\left[\overset{o}{w}{}_i^{\nu_i}\ \overset{o}{w}{}_\ell^{\nu_\ell}\right]$ have to be put equal to zero. By substituting (A.4) in (A.3) we have:

$$(A.5)\quad \delta \overline{K}_N (x_1^{\alpha_1} \ldots x_N^{\alpha_N}) =$$

$$= Z^{\frac{N}{2}} \sum_{t=o}^{\infty} \sum_{p=o}^{\infty} \frac{(i\bar{c})^p}{p!} \frac{(i\bar{g})^t}{t!}\, \Theta_t(\underline{\lambda}) \int dS_t \int dQ_p\, \Gamma_{\xi_1} \ldots \Gamma_{\xi_t}$$

$$\sum_{\sigma}^{\frac{N}{2}+\frac{p}{2}+2t} \sum_{(c\sigma)} \left[x_1^{\alpha_1} \ldots x_N^{\alpha_N}\, u_1^{\beta_1}\, v_1^{\beta_1} \ldots t_t^{\gamma_t}\, z_t^{\gamma_t}\, \eta_1^4 \ldots \eta_p^4\right] \cdot$$

$$\cdot R_{2\sigma} (w_1^{\nu_1} \ldots w_{2\sigma}^{\nu_{2\sigma}})$$

where

$$(A.6)\quad R_{2\sigma}(w_1^{\nu_1} \ldots w_{2\sigma}^{\nu_{2\sigma}}) = \sum_{r=1}^{\infty} \frac{(i\bar{g})^r}{r!}\, \theta_{\lambda_1, \lambda_2 \ldots \lambda_r} \int dS_r$$

$$\cdot(\delta_f \Gamma_{\xi_1})\ \Gamma_{\xi_2} \cdot\cdot \Gamma_{\xi_r} \left[\overset{o}{w}{}_1^{\nu_1} \ldots \overset{o}{w}{}_{2\sigma}^{\nu_{2\sigma}} \ u_1^{\beta_1} \ v_1^{\beta_1} \cdot\cdot t_r^{\gamma_r} \ z_r^{\gamma_r} \right],$$

$\sum_{(2\sigma)} \left[x_1^{\alpha_1} \cdot\cdot x_N^{\alpha_N} \ldots \eta_p^4 \right]$ denotes summation over all hafnians obtainable from

$\left[x_1^{\alpha_1} \ldots x_N^{\alpha_N} \ u_1^{\beta_1} \ v_1^{\beta_1} \ldots \eta_p^4 \right]$ by deleting any 2σ variable from it.

We want to analyze the quantities $R(w_1^{\nu_1} \ldots w_\sigma^{\nu_\sigma})$. To this end we use the property that the operation $\theta_{\lambda_1, \lambda_2 \cdot\cdot \lambda_r}$ yields a non vanishing result only when acting on connected graphs exhibiting "superficial divergences" i.e. overalldivergence according to power counting. This imply that $R(w_1^{\nu_1} \cdot\cdot w_\sigma^{\nu_\sigma})$ is equal to zero for $\sigma > 4$ and $R_4(w_1^{\nu_1} \ldots w_4^{\nu_4})$, $R_2(w_1^{\nu_1} \ w_2^{\nu_2})$ can be expressed as follows:

$$\text{(A.7)} \quad R_4(w_1^{\nu_1} \cdot\cdot w_4^{\nu_4}) =$$

$$R_4^{(1)} \int d\xi_1 \left[\overset{o}{w}{}_1^{\nu_1} \ \overset{o}{w}{}_2^{\nu_2} \ \overset{o}{w}{}_3^{\nu_3} \ \overset{o}{w}{}_4^{\nu_4} \ \xi_1^{\beta_1} \ \xi_1^{\beta_1} \ \xi_1^{\gamma_1} \ u_1^{\gamma_1} \right]$$

$$\text{(A.8)} \quad R_2(w_1^{\nu_1} \ w_2^{\nu_2}) =$$

$$= R_2^{(1)} \int d\xi_1 \left[\overset{o}{w}{}_1^{\nu_1} \ \overset{o}{w}{}_2^{\nu_2} \ \xi_1^{\delta_1} \ \xi_1^{\delta_1} \right] + R_2^{(2)} \left[w_1^{\nu_1} \ w_2^{\nu_2} \right]; \nu_1 = \nu_2$$

In order to obtain (A.7) we develop the hafnian with respect to the variables w_i:

$$\text{(A.9)} \quad \left[\overset{o}{w}{}_1^{\nu_1} \cdot\cdot \overset{o}{w}{}_4^{\nu_4} \ u_1^{\beta_1} \ v_1^{\beta_1} \cdot\cdot z_r^{\gamma_r} \right]$$

$$\sum_c \left[\overset{o}{w}{}_1^{\nu_1} \ldots \overset{o}{w}{}_4^{\nu_4} \ p_1^{\delta_1} \ p_2^{\delta_2} \ p_3^{\delta_3} \ p_4^{\delta_4} \right] \left[\zeta_1^{\mu_1} \ldots \zeta_{4r-4}^{\mu_{4r-4}} \right] =$$

$$= \sum_p \sum_c \left[w_1^{\nu_1} \ p_i^{\nu_1} \right] \left[w_2^{\nu_2} \ p_j^{\nu} \right] \left[w_3^{\nu_3} \ p_k^{\nu_3} \right] \left[w_4^{\nu_4} \ p_s^{\nu_4} \right] \left[\zeta_1^{\mu_1} \cdot\cdot \zeta_{4r-4}^{\mu_{4r-4}} \right]$$

$\sum_c$ denotes summations over all the possible ways of extracting the $p_1^{\delta_1} .. p_4^{\delta_4}$ variables from the sequence $u_1^{\beta_1}, v_1^{\beta_1}, t_1^{\gamma_1}, z_1^{\gamma_1}, \ldots, u_r^{\beta_r}, v_r^{\beta_r}, t_r^{\gamma_r}, z_r^{\gamma_r}$. The remaining variables are denoted by $\zeta_1^{\mu_1} \ldots \zeta_{4r-4}^{\mu_{4r-4}}$ and $p_i\ p_j\ p_k\ p_s$ is a permutation of $p_1\ p_2\ p_3\ p_4$; $\sum_p$ denotes summations over all the permutations.
Then we replace in (A.9) the hafnian element $[w\ p]$ with its Taylor expansion for $p=\xi_1$

$$(A.11) \quad [w\ p]=[w\ \xi_1] + (p-\xi_1)^{\mu}\partial_{\mu}[w\ \xi_1] + \frac{1}{2}(p-\xi_1)^{\mu}(p-\xi_1)^{\nu}\partial_{\mu}\partial_{\nu}[w\ \xi_1] + \ldots$$

and note that, keeping in mind the property of the operation $\theta_{\lambda_1,\lambda_2..\lambda_r}$, because we are handling logarithmic divergences, only the first term of the expansion (A.10) gives contribute. This leads to expression (A.7) with R_4 given in (15). In the same way we prove equation (A.8). We first write

$$(A.12) \quad \left[\overset{o}{w}{}_1^{\nu_1}\ \overset{o}{w}{}_2^{\nu_2}\ u_1^{\beta_1}\ v_1^{\beta_1} \ldots z_r^{\gamma_r}\right] =$$

$$= \sum_c \left[\overset{o}{w}{}_1^{\nu_1}\ \overset{o}{w}{}_2^{\nu_2}\ p_1^{\delta_1}\ p_2^{\delta_2}\right]\left[\zeta_1^{\mu_1} \ldots \zeta_{4r-2}^{\mu_{4r-2}}\right] =$$

$$= \sum_p \sum_c \left[w_1^{\nu_1}\ p_i^{\nu_1}\right]\left[w_2^{\nu_2}\ p_j^{\nu_2}\right]\left[\zeta_1^{\mu_1} \ldots \zeta_{4r-2}^{\mu_{4r-2}}\right]),$$

$\sum_c$ denotes summations over all the possible ways of extracting the $p_1^{\delta_1}$, $p_2^{\delta_2}$ variables from the sequence $u_1^{\beta_1}, \ldots, z_r^{\gamma_r}$, the remaining variables are denoted by $\zeta_1^{\mu_1} \ldots \zeta_{4r-2}^{\mu_{4r-2}}$; p_i, p_j is a permutation of $p_1\ p_2$; $\sum_p$ is the sum over all permutations. By substituting (A.11) in the equation (A.11), retaining only effective contributions and making use of the covariance requirements [7] we find (A.8) with $R_2^{(1)}, R_2^{(2)}$ given in (15). Finally by keeping in mind (A.6)(A.7)(A.8) we can write equation (A.5) as follows

$$\delta_f \bar{K}_N = R_o \bar{K}_N + R_2^{(1)} Z^{\frac{N}{2}} \sum_{t=o}^{\infty} \sum_{p=o}^{\infty} \frac{(i\bar{g})^t}{(t)!} \frac{(i\bar{c})^p}{p!} \Theta_t(\underline{\lambda})$$

$$\int dS_t \int dQ_p \int d\xi_1 \Gamma_{\xi_1} \cdot\cdot \Gamma_{\xi_t} \left[x_1^{\alpha_1} \cdot\cdot x_N^{\alpha_N} \cdots \eta_p^4 \, \varrho_{\xi_1}^{\delta_1} \, \varrho_{\xi_1}^{\delta_1} \right] +$$

$$+ R_2^{(2)} Z^{\frac{N}{2}} \sum_{t=o}^{\infty} \sum_{p=o}^{\infty} \frac{(i\bar{g})^t}{t!} \frac{(i\bar{c})^p}{p!} \Theta_t(\underline{\lambda}) \left(\frac{N+4t+p}{2}\right) \int dS_t$$

$$\int dQ_p \, \Gamma_{\xi_1} \cdot\cdot \Gamma_{\xi_t} \left[x_1^{\alpha_1} \cdots \eta_p^4 \right] +$$

$$+ R_4 Z^{\frac{N}{2}} \sum_{t=o}^{\infty} \sum_{p=o}^{\infty} \frac{(i\bar{g})^t}{t!} \frac{(i\bar{c})^p}{p!} \Theta_t(\underline{\lambda}) \int dS_t \int dQ_p$$

$$\Gamma_{\xi_1} \cdot\cdot \Gamma_{\xi_t} \left[x_1^{\alpha_1} \cdots \eta_p^4 \, \varrho_{\xi_1}^{\delta_1} \, \varrho_{\xi_1}^{\delta_1} \, \varrho_{\xi_1}^{\rho_1} \, \varrho_{\xi_1}^{\rho_1} \right]$$

Where we have used the remarkable expansion:

$$\sum_{c_\rho} \left[h_1 \cdots h_{2\rho} \right] \left[\hat{k}_1 \cdots \hat{k}_\sigma \ \mu+1 \ \cdots 2n \right] =$$

$$= \sum_{c_r} \left[1 \cdots \mu \ \ell_1 \cdots \ell_\sigma \right] \left[m_1 \cdots m_{2r} \right]$$

where $\sum_{c_\rho}$ denote the sum over all the c_ρ combinations $h_1 < h_2 < \cdots < h_{2\rho}$; $k_1 < k_2 < \cdots < k_\sigma$ of the indices $12\cdots\mu$ $(\mu = 2\rho + \sigma)$, and with $\sum_{c_r}$ the sum over all c_r combinations $\ell_1 < \cdots < \ell_\sigma$; $m_1 < \cdots m_{2r}$ of the indices $\mu+1, \mu+2, \ldots 2n$ $(2n - \mu = 2r + \sigma)$

Then the relations

$$\frac{\partial}{\partial \bar{m}^2} \bar{K}_N = -\frac{i}{2} Z^{\frac{N}{2}} \sum_{t=o}^{\infty} \sum_{p=o}^{\infty} \frac{(i\bar{g})^t}{t!} \frac{(i\bar{c})^p}{p!} \Theta_t(\underline{\lambda})$$

$$\int dS_t \int dQ_r \int d\xi_1 \ \Gamma_{\xi_1} \cdots \Gamma_{\xi_t} \left[x_1^{\alpha_1} \cdots \varrho_\xi^\delta \ \varrho_\xi^\delta \right]$$

$$\frac{\partial}{\partial(i\bar{g})} \bar{K}_N = Z^{\frac{N}{2}} \sum_{t=o}^{\infty} \sum_{p=o}^{\infty} \frac{(i\bar{g})^t}{t!} \frac{(i\bar{c})^p}{p!} \Theta_t(\lambda)$$

$$\int dS_{t-1} \int dQ_p \; \Gamma_{\xi_1} \ldots \Gamma_{\xi_{t-1}} \int d\xi_t \, du_t \ldots dz_t \, \Gamma_{\xi_t}$$

$$\left[x_1^{\alpha_1} \ldots \eta_1^4 \; u_t^{\beta_t} \; v_t^{\beta_t} \; t_t^{\gamma_t} \; z_t^{\gamma_t} \right] ;$$

$$\bar{c} \frac{\partial}{\partial \bar{c}} \bar{K}_N = Z^{\frac{N}{2}} \sum_{t=0}^{\infty} \sum_{p=0}^{\infty} p \frac{(i\bar{g})^t}{t!} \frac{(i\bar{c})^p}{p!} \Theta_t(\lambda)$$

$$\int dS_t \int dQ_p \; \Gamma_{\xi_1} \ldots \Gamma_{\xi_t} \left[x_1^{\alpha_1} \ldots \eta_p^4 \right] ;$$

$$\bar{g} \frac{\partial}{\partial \bar{g}} \bar{K}_N = Z^{\frac{N}{2}} \sum_{t=0}^{\infty} \sum_{p=0}^{\infty} t \frac{(i\bar{g})^t}{t!} \frac{(i\bar{c})^p}{p!} \Theta_t(\lambda)$$

$$\int dS_t \int dQ_p \; \Gamma_{\xi_1} \ldots \Gamma_{\xi_t} \left[x_1^{\alpha_1} \ldots \eta_p^4 \right]$$

allow us to obtain equation (14). The operations $\frac{\partial}{\partial \bar{m}} \bar{K}, \frac{\partial}{\partial i\bar{g}}$ and $\bar{c} \frac{\partial}{\partial \bar{c}}, \bar{g} \frac{\partial}{\partial \bar{g}}$ correspond respectively to the differential vertices operators and counting operators introduced by Zimmermann.

APPENDIX B

In this appendix we shall prove by induction that the equation

$$(B.1)\quad \int dQ_{2p+1}\left[x_1^4\ \xi_1^{\delta_1}\ \zeta_1^{\delta_1}\cdots\xi_n^{\delta_n}\ \zeta_n^{\delta_n}\ \eta_1^4\cdots\eta_{2p+1}^4\right]=$$

$$(2p+1)\int dQ_{2p}\int dx_2\left[x_1^{\alpha_1}\ x_2^{\alpha_2}\ \xi_1^{\delta_1}\ \zeta_1^{\delta_1}\cdots\xi_n^{\delta_n}\ \zeta_n^{\delta_n}\ \eta_1^4\cdots\eta_{2p}^4\right]\qquad \alpha_1=\alpha_2=1,2,3$$

is satisfied for any p.
We first observe that

$$(B.2)\quad \int dQ_{2p+1}\left[x_1^4\ \eta_1^4\cdots\eta_{2p+1}^4\right]=$$

$$=(2p+1)\int dQ_{2p}\int dx_2\left[x_1^{\alpha_1}\ x_2^{\alpha_2}\ \eta_1^4\cdots\eta_{2p}^4\right]$$

$$(B.3)\quad \int dQ_{2p+1}\left[x_1^4\ \xi_1^{\delta_1}\ \zeta_1^{\delta_1}\ \eta_1^4\cdots\eta_{2p+1}^4\right]=$$

$$=(2p+1)\int dQ_{2p}\int dx_2\left[x_1^{\alpha_1}\ x_2^{\alpha_2}\ \xi_1^{\delta_1}\ \zeta_1^{\delta_1}\ \eta_1^4\cdots\eta_{2p}^4\right]$$

The equation (B.2) and (B.3) are easily proved keeping in mind equations (7) and (9). Then we suppose that (B.1) is satisfied for n=k and show that

$$(B.4)\quad \int dQ_{2p+1}\left[x_1^4\ \xi_1^{\delta_1}\ \zeta_1^{\delta_1}\cdots\xi_{k+1}^{\delta_{k+1}}\ \zeta_{k+1}^{\delta_{k+1}}\ \eta_1^4\cdots\eta_{2p+1}^4\right]$$

$$(2p+1)\int dQ_{2p}\int dx_2\left[x_1^{\alpha_1}\ x_2^{\alpha_2}\ \xi_1^{\delta_1}\ \zeta_1^{\delta_1}\cdots\xi_{k+1}^{\delta_{k+1}}\ \zeta_{k+1}^{\delta_{k+1}}\ \eta_1^4\cdots\eta_{2p}^4\right]$$

To verify (B.4) we develop the hafnians of the left and right handside according to equation (7)

$$(B.5)\quad \left[x_1^4\ \xi_1^{\delta_1}\ \zeta_1^{\delta_1}\cdots\xi_{k+1}^{\delta_{k+1}}\ \zeta_{k+1}^{\delta_{k+1}}\ \eta_1^4\cdots\eta_{2p+1}^4\right]\quad=$$

$$= \sum_{i=1}^{k+1} \Delta_c(x_1-\xi_i,\bar{m}) \left[\xi_1^{\delta_1} \zeta_1^{\delta_1} \ldots \not{\xi}_i^{\delta_i} \zeta_i^4 \ldots \xi_{k+1}^{\delta_{k+1}} \zeta_{k+1}^{\delta_{k+1}} \eta_1^4 \ldots \eta_{2p+1}^4 \right] + \sum_{i=1}^{k+1} \Delta_c(x_1-\zeta_i,\bar{m})$$

$$\left[\xi_1^{\delta_1} \zeta_1^{\delta_1} \ldots \xi_i^4 \not{\zeta}_i^{\delta_i} \ldots \xi_{k+1}^{\delta_{k+1}} \zeta_{k+1}^{\delta_{k+1}} \eta_1^4 \ldots \eta_{2p+1}^4 \right]$$

$$+ \sum_{i=1}^{2p+1} \Delta_c(x_1-\eta_i,\bar{m}) \left[\xi_1^{\delta_1} \zeta_1^{\delta_1} \ldots \eta_1^4 \ldots \not{\eta}_i^4 \ldots \eta_{2p+1}^4 \right];$$

(B.6) $$\left[x_1^{\alpha_1} x_2^{\alpha_2} \xi_1^{\delta_1} \zeta_1^{\delta_1} \ldots \xi_{k+1}^{\delta_{k+1}} \zeta_{k+1}^{\delta_{k+1}} \eta_1^4 \ldots \eta_{2p+1}^4 \right] =$$

$$= \sum_{i=1}^{k+1} \Delta_c(x_1-\xi_i;\bar{m}) \left[x_2^{\alpha_2} \xi_1^{\delta_1} \zeta_1^{\delta_1} \ldots \not{\xi}_i^{\delta_i} \zeta_i^{\alpha_1} \ldots \right.$$

$$\left. \ldots \xi_{k+1}^{\delta_{k+1}} \zeta_{k+1}^{\delta_{k+1}} \eta_1^4 \ldots \eta_{2p}^4 \right] +$$

$$+ \sum_{i=1}^{k+1} \Delta_c(x_1-\zeta_i,\bar{m}) \left[x_2^{\alpha_2} \xi_1^{\delta_1} \zeta_1^{\delta_1} \ldots \xi_i^{\alpha_1} \not{\zeta}_i^{\delta_i} \ldots \right.$$

$$\left. \ldots \xi_{k+1}^{\delta_{k+1}} \zeta_{k+1}^{\delta_{k+1}} \eta_1^4 \ldots \eta_{2p}^4 \right] +$$

$$+ \Delta_c(x_1-x_2,\bar{m}) \left[\xi_1^{\delta_1} \zeta_1^{\delta_1} \ldots \eta_1^4 \ldots \eta_{2p}^4 \right]$$

In order to obtain (B.5),(B.6) we have used equation(9). By substituting (B.5) and (B.6) in (B.4) and keeping in mind our assumption((B.1) is satisfied for n=k) the equation (B.4) follows. More in general we shall prove, by using the same method, the equation:

(B.7) $$p \int d^4x \int dQ_{p-1} \left[x^\ell x_1^{\alpha_1} \ldots x_N^{\alpha_N} \xi_1^{\delta_1} \zeta_1^{\delta_1} \ldots \right.$$

$$\left. \xi_n^{\delta_n} \zeta_n^{\delta_n} \eta_1^4 \ldots \eta_{p-1}^4 \right] =$$

$$= \int dQ_p \sum_{m=1}^{N} \delta_{\alpha_m,\ell}$$

$$\left[x_1^{\alpha_1} \ldots x_{m-1}^{\alpha_{m-1}} x_m^4 \; x_{m+1}^{\alpha_{m+1}} \ldots x_N^{\alpha_N} \; \xi_1^{\delta_1} \zeta_1^{\delta_1} \ldots \xi_n^{\delta_n} \zeta_n^{\delta_n} \eta_1^4 \ldots \eta_p^4 \right] +$$

$$- \int dQ_p \sum_{m=1}^{N} \delta_{\alpha_m,4}$$

$$\left[x_1^{\alpha_1} \ldots x_{m-1}^{\alpha_{m-1}} \; x_m^{\ell} \; x_{m+1}^{\alpha_{m+1}} \ldots x_N^{\alpha_N} \; \xi_1^{\delta_1} \zeta_1^{\delta_1} \ldots \xi_n^{\delta_n} \zeta_n^{\delta_n} \eta_1^4 \ldots \eta_p^4 \right]$$

$$\ell = 1,2,3 \qquad \alpha_i = 1,2,3,4$$

which allow us to verify the W-T identities of the σ-model for arbitrary number of external variables.

APPENDIX C

Let us consider the hafnian

$$\text{(C.1)} \quad \left[\xi_1^{\beta_1} \xi_1^{\beta_1} \; \xi_1^{\gamma_1} \; \xi_1^{\gamma_1} \ldots \xi_k^{\beta_k} \; \xi_k^{\beta_k} \xi_k^{\gamma_k} \xi_k^{\gamma_k} \eta_1^4 \; \eta_2^4 \ldots \eta_{2k+2}^4 \right]$$

We want determine the number N_k of connected graphs contained in (C.1).
By ispection one see that for k=1 and 2 we have $N_1 = 4!$ and $N_2 = 4^2 \cdot 6!$ respectively. We want to show that the following relation is true

$$\text{(C.2)} \quad N_k = 4^k \, (3k)!$$

Equation (C.2) can be written

$$\text{(C.3)} \quad N_k = M_k \, (2k+2)!$$

where

$$M_k = \frac{4^k (3k)!}{(2k+2)!}$$ is the number of ways in which the k ξ-vertices can be connected among them to give connected and loopless graphs.

As (C.3) is true for k=1 (and k=2) we can verify (C.3) by induction.
We assume that (C.3) is true for k=1,2..n and prove that it is valid for k=n+1, too. We note that the graphs of order (n+1) can be arranged in four groups corresponding to the configurations in which the (n+1) vertex is linked with one, two, three, four ξ-vertices respectively.
We have thus

$$M_{n+1} = 4\,(2n+2)\,M_n +$$

$$+ \frac{4\cdot 3}{2!} \sum_{h=1}^{n-1} \binom{n}{h} (2h+2)(2n-2h+2) M_h\, M_{n-h} +$$

(C.4)

$$+ \frac{4\cdot 3\cdot 2}{3!} \sum_{h=1}^{n-2} \sum_{k=1}^{n-h-1} \binom{n}{h} \binom{n-h}{k} (2h+2)\;(2k+2)$$

$$(2n-2h-2k+2)\; M_h M_k M_{n-h-k} +$$

$$+ \frac{4\cdot 3\cdot 2}{4!} \sum_{h=1}^{n-3} \sum_{k=1}^{n-h-2} \sum_{j=1}^{n-h-k-1} \binom{n}{h} \binom{n-h}{k} \binom{n-h-k}{j}$$

$$(2h+2)(2k+2)(2j+2)(2n-2h-2k-2j+2) M_h M_k M_j M_{n-h-j-k}$$

By using equation (C.3) and the identity

$$\sum_{k=o}^{n} \binom{\beta k+\alpha}{k} \binom{\beta n-\beta k+\gamma}{n-k} \frac{\alpha\gamma}{(\alpha+\beta k)(\gamma+\beta n-\beta k)} =$$

$$\frac{\alpha+\gamma}{\alpha+\gamma+\beta n} \binom{\beta n+\alpha+\gamma}{n}$$

we have

$$\text{(C.5)}\quad M_{n+1} = 4^{n+1} \frac{(3n+3)!}{(2n+4)!}$$

REFERENCES

1 N.N.Bogoliubov and O.S.Parasiuk,Acta Math.97,227 (1957)
O.S.Parasiuk,Ukrouski Math.J. 12,287 (1960)
N.N.Bogoliubov and D.W.Shirkov,Introduction to the theory of quantized fields, New York IntersciencеP.
2 K.Hepp, Comm.Math.Phys. 2,301 (1966)
3 J.F.Ashmore, Lettere al Nuovo Cimento 4, 289 (1972); G.t'Hooft, M.Veltman-Nuclear Physics B44, 189 (1972).
P.Butera, G.M.Cicuta, E.Montaldi-Renormalization and Space-time dimension Milano preprint 1973.
4 J.Schwinger, Ann.Phys. 2,407 (1957)
J.C.Polkinghorne, Il Nuovo Cimento, 8,179,781(1958)
M.Gell'Mann and M.Levy,Il Nuovo Cimento, 10,705 (1960)
B.W.Lee, Nucl.Phys.B9,649 (1969)
K.Symanzik, Commun.Math.Phys.,16,48 (1970)
S.Weinberg-Phys.Rev.Letters - 17-616(1966)
D.Bessis and G.Turchetti-The Cargèse Summer Sch.1970
5 F.Guerra and M.Marinaro, Il Nuovo Cimento, 60A, 756 (1969)
F.Guerra, Il Nuovo Cimento 1A, 523 (1971)
E.R.Caianiello "Combinatorics and renormalization in Quantum Field Theory" Benjamin
6 E.R.Caianiello, Il Nuovo Cimento 10 1634 (1953);11, 492 (1954)
7 E.R.Caianiello,F.Guerra,M.Marinaro, 60A,713 (1969)
M.Marinaro, Il Nuovo Cimento 9A 62 (1972)
8 F.Esposito,U.Esposito,F.Guerra-Il Nuovo Cimento 60A, 772 (1969)
9 W.Magnus F.Oberhettinger and R.P.Soni-Formulas and Theorems for the special functions of Math.Phys.- Springer Verlag
10 W.Zimmermann,1970 Brandeis Lectures, Lectures on elementary particles and quantum field theory vol.1 pp.397 Cambridge N.I.I.Press (1970)

RECENT PROGRESS ON COMPUTATIONAL METHODS IN FIELD THEORY

M. Pusterla[(+)] and G. Turchetti[(++)]

(+) Ist. Fisica Univ. Padova and I.N.F.N.

(++) Ist. Fisica Univ. Bologna and I.N.F.N.

The search of an adequate approximation method is important in theoretical physics, if people have to compare complicated formulas, derived from general physical theories, with reality. It becomes a necessity whenever the exact mathematical solution is missing and the only information available is fragmentary and consists, for instance, of a formal series expansion in a suitable parameter. This situation exists in field theory and the Padé approximants (P.A.) have been proposed long ago (1) as a convenient mathematical tool one can use to derive quantitative results from perturbation theory, particularly in the case of strong coupling. More specifically we are thinking of those realistic field theoretical models which are renormalizable and allow a power series in the renormalized coupling constant.

In this note we are not going to review the properties of the P.A. and their physical applications since this was the object of specialized conferences and is summarized in a very approachable collection of review papers and books (2). We rather want to focus the attention of the reader on some peculiarities that make the Padé method valid along the present line of research in field theory.

We recall the definition of P.A.: given a power series

$$\Sigma_n a_n z^n$$

the approximant $|N/M|(z)$ is defined as the ratio of two polynomials of degrees N and M such that

$$\Sigma_n a_n z^n - P_N(z)/Q_M(z) = O(z^{N+M+1})$$

The series may be formal, with zero radius of convergence and the set of P.A. $|N/M|(z)$ is shown to be unique by the previous definition.

The fundamental problem is the existence of sequences (in the Padé table) converging towards a unique analytic function f(z). This question is dealt quite extensively in the literature (3). We remind that under general conditions diagonal sequences are assumed (and in some cases proved) to converge to f(z), when this has an asymptotic expansion and also to provide the analytic continuation if f(z) is regular at the origin (consequently its Taylor series converges). The procedure has been extended to treat the case of functions which are matrices or operators in a Banach space.

How can a field theorist use the P.A. approach and when? In principle any time he deals with a function that is available as a power series in one parameter. In practice the expansion parameter is the coupling constant and the use of P.A. is only limited by the difficulties of computing high orders. In the case of trilinear local field interaction only the second order (one loop approximation) can be treated by semi-analytic methods which allow unlimited numerical accuracy. For scalar particles pure quadrilinear couplings are admitted and higher orders can be computed; the introduction of fermions requires trilinear interactions,for renormalizable theories, and limit to the $|1/1|$ the only significant P.A.. Since it is extremely hard to push further the perturbative expansion, a significant improvement can be achieved by working in a wider space,where one deals with the Green-functions rather than the S matrix.

In the Green-functions we may distinguish the discrete spin variables from the continuous momenta. If we fix the momenta to their on-shell values, we are left with a matrix, a submatrix of which is the physical amplitude. We call the P.A. of this matrix Green-function approximants and notice that they reduce to the usual S matrix approximants in the case of scalar particles. On the other hand the momenta can also be varied and a stationary solution looked for; we refer to this procedure as variational approximation.

For the variational approximation rigorous convergence theorems have been proved by Alabiso, Butera and Prosperi (4), both in potential scattering and for the Bethe-Salpeter equation (ladder approximation for scalar particles). It is well known that the bound state poles of the Green function (off shell T matrix) are independent of the momentum variables for fixed angular momentum.

$$\langle k'|T(E,g,J)|k\rangle \quad \text{poles at } g=g(E,J)$$

This property is lost when T is replaced by any of its P.A.; hence one can try to impose the independence from these parameters k,k', thus obtaining the best approximation in the search of bound states

$$|N/M|_{\langle k'|T(E,g,J)|k\rangle} \quad \text{poles at } g=g^{N,M}(E,J,k,k')$$

To this end one looks for the stationary value of g (the coupling) when varying k and k', usually with the constraint k=k'.

$$\delta g^{N,M}=0 \quad k=k(E,J) \qquad g_{staz.}=g^{N,M}(E,J)$$

The value of k at the stationary point is found to be very far from the mass shell. This variational approach was also applied to scattering states in potential theory and the Bethe-Salpeter equation. The extension to field theory has not yet been fully investigated.

The Green function approximation has been devised for field theoretical applications. Since the convergence of the P.A. critically depends on the dimensionality of the domain in which the function, that is going to be approximated, is defined, it becomes evident that an improvement has to be achieved, passing from an S matrix element to the whole Green-function.

The basic steps required by this method are the choice of a basis, the angular momentum expansion and the reduction due to the symmetry properties. Finally one is left with a set of matrices F^J which fulfill the extended unitarity condition:

$$F^J - F^{J+} = iF^J \Pi F^{J+}$$

where Π is the projector on the physical states. The $|N/M|$ $M \geq N$ P.A. fulfil the same equation and their physical amplitudes are exactly unitary. The analyticity properties in the energy variable are the perturbative ones except for additional poles which are good candidates to represent the spectrum. The relevant technical points are here briefly exposed for the πN and the NN systems (5 and 6).

The πN Green-function T(p,q,p',q') is a 4x4 matrix which decomposes on the set of Dirac matrices. Parity invariance reduces the number of independent amplitudes from 16 to 8, time reversal from 8 to 6. A suitable choice of basis vectors is the following

$$|\varepsilon,\lambda\rangle=\Gamma_\varepsilon u(p,\lambda) \qquad \Gamma_\varepsilon=\begin{cases} I & \varepsilon=+1 \\ \gamma_5 & \varepsilon=-1 \end{cases}$$

where $\varepsilon=\pm 1$ labels the parity of the state and $\lambda=\pm\frac{1}{2}$ the helicity, u being the usual Dirac spinor. On such a basis T splits into four 2x2 matrices,

$$\tilde{u}(p')\Gamma_{\varepsilon'}T\Gamma_{\varepsilon}u(p)$$

which can be treated as physical amplitudes of fictitious Green-functions $\Gamma_{\varepsilon'}T\Gamma_{\varepsilon}$. The kinematics is developed following this picture and the angular momentum expansion obtained following Jacob and Wick. We may formally introduce an angular momentum basis for T by defining $|\varepsilon,L;J\rangle=\Gamma_{\varepsilon}|L,J\rangle$ where ε appears as an additional quantum number (the nucleon parity). The space reflection invariance of T implies the parity of the initial and final state to be the same

$$\langle\varepsilon',L',J|T|\varepsilon,L,J\rangle\neq 0 \quad \text{only if } \varepsilon(-1)^{L}=\varepsilon'(-1)^{L'}$$

We are left with two 2x2 matrices $F^{J}_{\pm}$ (connecting the states $|+,J\pm\frac{1}{2}\rangle,|-,J\mp\frac{1}{2}\rangle$)which are symmetric if time reversal invariance holds and fulfill the extended unitarity equation.

For the NN case T is a 16x16 matrix for which beyond parity we must account for the exchange symmetry due to the presence of identical particles.The basis vectors are in this case

$$|\alpha\rangle\ \Gamma_{\alpha}|1,2\rangle \qquad \alpha=+,-,e,o \qquad |1,2\rangle=u(p_1)\otimes u(p_2)$$

where

$$\Gamma_{+}=I\otimes I \quad \Gamma_{-}=\gamma_5\otimes\gamma_5 \quad \Gamma_{e}=2^{-\frac{1}{2}}(I\otimes\gamma_5+\gamma_5\otimes I) \quad \Gamma_{o}=2^{-\frac{1}{2}}(I\otimes\gamma_5-\gamma_5\otimes I)$$

and again T splits into sixteen 4x4 matrices for which relevant kinematics is worked out. The exchange operator E (defined by $E|1,2\rangle=|2,1\rangle$) commutes with T and furthermore TE=ET=-T if the Pauli principle has to hold for the physical amplitude.Let us notice that if P is the Parity operator we have

$$E\Gamma_{\alpha}E=\omega_{\alpha}\Gamma_{\alpha} \qquad \omega_{\alpha}=\begin{cases}+1 & \alpha=+,-,e\\ -1 & \alpha=o\end{cases}$$

$$P\Gamma_{\alpha}P=\pi_{\alpha}\Gamma_{\alpha} \qquad \pi_{\alpha}=\begin{cases}+1 & \alpha=+,-\\ -1 & \alpha=e,o\end{cases}$$

The selection rules on the angular momentum states $|\alpha,L,S,J\rangle=\Gamma_{\alpha}|L,S,J\rangle$ (where S is the total spin and L the orbital momentum) issue from P and E invariance

$$P|\alpha,L,S,J\rangle=\pi_{\alpha}(-1)^{L}|\alpha,L,S,J\rangle$$
$$E|\alpha,L,S,J\rangle=\omega_{\alpha}(-1)^{L+S+I}|\alpha,L,S,J\rangle$$

where I is the isospin, a good quantum number as J.

The final result is that T splits into two 4x4 matrices and one 6x6 matrix having the singlet, the uncoupled triplet and the coupled triplet as physical amplitudes respectively.

To conclude we notice that for πN we go from 2 amplitudes for S to 6 for T while the ratio becomes 5 to 41 for NN. This increase of information should allow a better covergence, if any; this has indeed been observed on the ladder series for NN, obtained by iterating the Bethe-Salpeter equation,where a sequence of diagonal P.A. was computed(7).

One easily recognizes how the previous procedure can be extended to meson baryon and baryon baryon for the general mass case within the broken SU(3) symmetry.

REFERENCES

1) Gammel J. L. and Mc. Donald F. A. 1966 Phys. Rev.142, 1245
Bessis D. and Pusterla M. 1968 Nuovo Cimento 54a,243
Masson D. 1967 J. Math. Phys. 8,512

2) Baker G. A. and GammelJ. L. 1970 "The Padé approximants in theoretical physics" Academic presss
Graves-Morris "Padé approximants" The Institute of Physics (1972) and "Padé approximants and their applications" Academic Press (1972).

3) Akhiezer N. I. 1965 "The classical moment problem" Oliver and Boyd.
Wall H. S. 1948 "Analytic theory of continuous fractions" Princeton : van Nostrand.

4) Alabiso C. ,Butera P. and Prosperi G. M. 1971 Nucl. Phys. B'31,141.

5) Turchetti G. Nuovo Cimento Letters 6,497 (1973)

6) Bessis D., Mery P. and Turchetti G. to be published.

7) Gammel J. L. and Menzel M. T. "Bethe Salpeter equation for NN scattering:matrix Padé approximants" to be published on Phys. Rev.

SHORT REVIEW OF SMALL-DISTANCE-BEHAVIOUR ANALYSIS

K. Symanzik

Deutsches Elektronen-Synchrotron DESY

Hamburg, Federal Republic of Germany

INTRODUCTION

The objects of small-distance-behaviour analysis, so far as it is developped until now, are (the Fourier transforms of) the Euclidean Green's functions, or Schwinger functions (1). These are the restrictions of the analytic Wightman functions to Schwinger points (2): time components of all arguments imaginary, space components real, or, alternatively, the corresponding analytic continuations of vacuum expectation values (VEVs) of time-ordered products of field operators (Feynman amplitudes) to such points. The Fourier-transforms of the Schwinger functions are the analytic continuations (2) to imaginary energy, real space components, in all momenta arguments of the Fourier transforms of the Feynman amplitudes, or Minkowski Green's functions.

The problem is the large-momenta behaviour of the Fourier transforms of the Euclidean functions, in short the behaviour of the Feynman amplitudes for large Euclidean momenta. The reason for this restriction is that Weinberg's large-momenta theorem (3), a basic tool in this analysis, applies only to such momenta. Certain results can be expected to hold also for real (Minkowskian)-momenta Feynman amplitudes, as we will point out in these cases.

The tools are

a) Wilson (4) short-distance expansions, also in a slightly generalized sense, and suitable shuffling of subtractions in Bethe-Salpeter (BS) type equations;
b) mass vertex insertions and resulting partial differential equations (PDEs);
c) the renormalization group (5).

The phenomena observed are

a) the appearance of the Green's functions of a zero-mass theory as the asymptotic forms of the finite-mass-theory functions;
b) infrared (UR) singularities of these functions at exceptional momenta;
c) the necessity of assumptions if statements on the true large-momenta behaviour (in ϕ^4 theory) are to be made, and the apparent consistency of a certain set of such assumptions;
d) anomalous dimensions, and conformal invariance (in ϕ^4 theory) in the deep-Euclidean region, being implied by the set of assumptions mentioned under c);
e) the computability of the Euclidean large-momenta behaviour in certain theories, now called "asymptotic freedom".

All of the technical discussion will, for simplicity, be given at the example of the ϕ^4 theory, however, the techniques are applicable to all renormalisable theories. Throughout this review no proofs, but instead complete references, will be given.

1. MASSIVE AND MASSLESS ϕ^4 THEORY

1.1. Renormalization conditions for massive theory

The Green's functions are the VEVs

$$G(x_1 \ldots x_{2n}, y_1 \ldots y_\ell; m^2, g) =$$

$$= 2^{-\ell} \langle \left(\phi(x_1) \ldots \phi(x_{2n})\ N_2(\phi^2(y_1)) \ldots N_2(\phi^2(y_\ell))\right)_+ \rangle$$

where $N_2(\phi^2)$ is Zimmermann's (6) finite composite local operator. The Fourier transforms of the amputated one-particle-irreducible parts of these functions are the vertex functions (VFs)

$$\Gamma(p_1 \ldots p_{2n}, q_1 \ldots q_\ell; m^2, g) \qquad (\Sigma p + \Sigma q = 0).$$

The theory is (at least in perturbation theoretical expansions) specified (7) by the renormalization conditions (8,6), which we choose as (for any unexplained notation, here and later on, see Refs. 9, 10)

$$\Gamma(p(-p),;m^2,g)\big|_{p^2=m^2} = 0 \tag{1.1}$$

$$[\partial/\partial p^2]\ \Gamma(p(-p),;m^2,g)\big|_{p^2=m^2} = i \tag{1.2}$$

$$\Gamma(p_1 \ldots p_4,;m^2,g)_{\text{symm.pt.to } p_i{}^2=m^2} = -ig \tag{1.3}$$

$$\Gamma(00,0;m^2,g) = 1 \tag{1.4}$$

$$\Gamma(,00;m^2,g) = 0 \tag{1.5}$$

1.2. PDEs for vertex functions

As a consequence of renormalization theory, with

$$\mathcal{O}p_{2n,\ell} \equiv m^2[\partial/\partial m^2] + \beta(g)[\partial/\partial g] - 2n\gamma(g) + \qquad (1.6a)$$
$$+ \ell(2\gamma(g) + \eta(g))$$

the PDEs

$$\mathcal{O}p_{2n,\ell}\,\Gamma(p_1\ldots p_{2n},q_1\ldots q_\ell;m^2,g) + i\delta_{no}\,\delta_{\ell 2}\,\kappa(g) = \qquad (1.6b)$$
$$= -\,im^2\varphi(g)\,\Gamma(p_1\ldots p_{2n},q_1\ldots q_\ell 0;m^2,g)$$

hold (9,10). Herein the functions

$$\beta(g) = b_o\,g^2 + b_1\,g^3 + \ldots \qquad b_o = 3(32\pi^2)^{-1} \qquad (1.7a)$$
$$b_1 < 0 \qquad \text{(Ref. 11)}$$
$$\gamma(g) = c_o\,g^2 + c_1\,g^3 + \ldots \qquad c_o = (2^{11}3\pi^4)^{-1} \qquad (1.7b)$$
$$\eta(g) = h_o\,g + h_1\,g^2 + \ldots \qquad h_o = \frac{1}{3}\,b_o \qquad (1.7c)$$
$$\kappa(g) = k_o + k_1\,g + \ldots \qquad k_o = \frac{1}{3}\,b_o \qquad (1.7d)$$
$$\varphi(g) = f_o + f_2\,g^2 + \ldots \qquad f_o = 1 \qquad (1.7e)$$

are computed from consistency of (1.1-5) with (1.6). The coefficients listed on the right (including b_1) are renormalization convention independent (12,10).

(1.6) can be integrated to the form (10)

$$\Gamma(p_1\ldots p_{2n},q_1\ldots q_\ell;m^2,g) + \qquad (1.8)$$
$$+ im^2\int_{\lambda^2}^{1} d\lambda'^2\; a(g)^{n-\ell}\; a(g(\lambda'))^{-n+\ell}\; h(g)^{-\ell}\; h(g(\lambda'))^{\ell}\,\cdot$$
$$\cdot\,\varphi(g(\lambda'))\;\Gamma(p_1\ldots p_{2n},q_1\ldots q_\ell\;0;m^2\lambda'^2,g(\lambda')) =$$
$$= a(g)^{n-\ell}\; a(g(\lambda))^{-n+\ell}\; h(g)^{-\ell}\; h(g(\lambda))^{\ell}\,\cdot$$
$$\cdot\,\Gamma(p_1\ldots p_{2n},q\ \ldots q_1;m^2\lambda^2,g(\lambda)) -$$
$$-\,i\,\delta_{no}\,\delta_{\ell 2}\;a(g)^{-2}\;h(g)^{-2}\;[k(g) - k(g(\lambda))].$$

Here $g(\lambda)$ is defined by

$$\ln\lambda^2 = \int_g^{g(\lambda)} dg' \, \beta(g')^{-1}, \tag{1.9}$$

and

$$a(g) = \exp\left[2 \int_o^g dg' \, \beta(g')^{-1} \, \gamma(g')\right] = \tag{1.10a}$$

$$= 1 + 2 \, b_o^{-1} \, c_o \, g + \ldots$$

$$h(g) = \exp\left[\int^g dg' \, \beta(g')^{-1} \, \eta(g')\right] = \tag{1.10b}$$

$$= g^{1/3}(1 + ..g + \ldots)$$

$$k(g) = \int^g dg' \, \beta(g')^{-1} \, a(g')^2 \, h(g')^2 \, \kappa(g') = \tag{1.10c}$$

$$= - \, g^{-1/3}(1 + ..g + \ldots)$$

where, for the expansions in (1.10) to be correct, g is supposed to be chosen positive and smaller than the possible first positive zero g_∞ of $\beta(g)$, and in (1.10b-c) a convenient choice of integration constants is made.

Closely related to the derivation of (1.6) is that

$$\text{RHS of (1.8)} = \Gamma^{s(g,\lambda)}(p_1 \ldots p_{2n}, q_1 \ldots q_\ell; m^2, g) \tag{1.11a}$$

holds, where the superscript denotes the substitution (9) in the Lagrangean density

$$L \rightarrow L^{s(g,\lambda)} = L - s(g,\lambda) \, \frac{1}{2} \, m^2 \, \mathcal{S}(g) \, N_2(\phi^2) \tag{1.11b}$$

with (14)

$$s(g,\lambda) = \mathcal{S}(g)^{-1} \, a(g) \, h(g) \cdot \tag{1.11c}$$

$$\cdot \int_1^{\lambda^2} d\lambda'^2 \mathcal{S}(g(\lambda')) \, a(g(\lambda'))^{-1} \, h(g(\lambda'))^{-1}.$$

1.3. Asymptotic forms for nonexceptional momenta

Nonexceptional momenta (15) are defined by the condition that to all orders in renormalized perturbation theory,

$$\Gamma(p_1 \ldots p_{2n}, q_1 \ldots q_\ell 0; m^2\lambda^2, g) = \mathcal{O}((\ln \lambda)^c) \tag{1.12}$$

for $\lambda \to 0$, with c depending on the order considered. Weinberg's analysis (3) assures that (1.12) holds for generic (Euclidean, but actually also for generic Minkowskian) momenta. Then for the integral on the LHS of (1.8) the limit $\lambda \to 0$ exists (to all orders of perturbation theory at least), and

$$\Gamma^{s(g,0)}(p_1 \ldots p_{2n}, q_1 \ldots q_\ell; m^2, g) \equiv \tag{1.13}$$

$$\equiv \Gamma_{as}(p_1 \ldots p_{2n}, q_1 \ldots q_\ell; m^2, g)$$

exists (in the same sense) and satisfies (10)

$$\Gamma_{as}(p_1 \ldots p_{2n}, q_1 \ldots q_\ell; m^2, g) = \tag{1.14}$$

$$= a(g)^{n-\ell} a(g(\lambda))^{-n+\ell} h(g)^{-\ell} h(g(\lambda))^{\ell} \cdot$$

$$\cdot \Gamma_{as}(p_1 \ldots p_{2n}, q_1 \ldots q_\ell; m^2\lambda^2, g(\lambda)) -$$

$$- i \delta_{no} \delta_{\ell 2} a(g)^{-2} h(g)^{-2} [k(g) - k(g(\lambda))]$$

which is equivalent to the PDE (1.15)

$$\mathcal{O}p_{2n,\ell} \Gamma_{as}(p_1 \ldots p_{2n}, q_1 \ldots q_\ell; m^2, g) + i \delta_{no} \delta_{\ell 2} \kappa(g) = 0.$$

The functions $\Gamma_{as}(\ldots .)$ are, in view of (1.13), the VFs of a ϕ^4 theory with massless particles, the socalled praeasymptotic theory; the discreteness (16) of the particles follows from (1.14) and the consequence of (1.9) and (1.7a),

$$g(\lambda) = b_o^{-1}(\ln\lambda^{-2})^{-1} - b_o^{-3} b_1(\ln\lambda^{-2})^{-2} \ln \ln\lambda^{-2} + \tag{1.16}$$

$$+ \mathcal{O}((\ln\lambda)^{-2})$$

for $\lambda \to +0$, $0 < g < g_\infty$, which yield (15)

$$G_{as}(p(-p), ; m^2, g) = - \Gamma_{as}(p(-p), ; m^2, g)^{-1} = \tag{1.17}$$

$$= a(g)^{-1} i(p^2 + i\varepsilon)^{-1} \cdot$$

$$\cdot \{1 + 2 b_o^{-2} c_o(\ln[m^2(- p^2 - i\varepsilon)^{-1}])^{-1} + \mathcal{O}((\ln p^2)^{-2} \ln \ln p^2)\}$$

for $p^2 \to 0$, for Euclidean or Minkowskian momentum.

If for $\lambda \to 0$ in perturbation theory $\Gamma(p_1 \ldots p_{2n}, q_1 \ldots q_\ell; m^2\lambda^2, g)$ is expanded in the double power series in λ^2 and $\ln\lambda^2$, then $\Gamma_{as}(p_1 \ldots p_{2n}, q_1 \ldots q_\ell; m^2\lambda^2, g)$ is the (formal) sum of the terms proportional $(\ln\lambda^2)^k$, $k \geq 0$, and $\Gamma_{as}(p_1 \ldots p_{2n}, q_1 \ldots q_\ell; m^2, g)$ is the (formal) sum of the λ-independent terms. This construction (15) of Γ_{as} is traditional in renormalization group uses and here referred to as the "elementary recipe". The observation involved justifies the name "asymptotic form" for Γ_{as}.

1.4. Relation to directly constructed massless theory

Renormalization conditions appropriate for massless ϕ^4 theory with VFs Γ_o are

$$\Gamma_o(00,;m^2,V) = 0 \tag{1.17a}$$

$$\Gamma_o(p(-p),;m^2,V)\big|_{p^2=-m^2} = -i\, m^2 \tag{1.17b}$$

$$\Gamma_o(p_1 \ldots p_4,;m^2,V)_{\text{symm.pt.to } p_i{}^2=-m^2} = -i\, V \tag{1.17c}$$

$$\Gamma_o(\tfrac{1}{2}p\ \tfrac{1}{2}p,-p;m^2,V)\big|_{p^2=-m^2} = 1 \tag{1.17d}$$

$$\Gamma_o(,q(-q);m^2,V)\big|_{q^2=-m^2} = 0 \tag{1.17e}$$

The Γ_o satisfy the PDEs

$$\hat{\mathcal{O}}\!\rho_{2n,\ell}\, \Gamma_o(p_1 \ldots p_{2n}, q_1 \ldots q_\ell; m^2, V) + i\, \delta_{no}\, \delta_{\ell 2}\, \hat{\kappa}(V) = 0 \tag{1.18a}$$

with

$$\hat{\mathcal{O}}\!\rho_{2n,\ell} \equiv m^2[\partial/\partial m^2] + \hat{\beta}(V)[\partial/\partial V] - \tag{1.18b}$$

$$- 2n\, \hat{\gamma}(V) + \ell(\hat{\eta}(V) + 2\hat{\gamma}(V)),$$

the integrated form of which are the renormalization group relations of Gell-Mann and Low (5).

Use of (10)

$$\Gamma_{as}(p_1 \ldots p_{2n}, q_1 \ldots q_\ell; m^2, g) = \tag{1.19}$$

$$= Z_1(g)^{-n}\, Z_2(g)^{-\ell}\, \Gamma_o(p_1 \ldots p_{2n}, q_1 \ldots q_\ell; m^2, V(g)) +$$

$$+ i\, \delta_{no}\, \delta_{\ell 2}\, f(g)$$

in (1.6b) and (1.18a) yields (15)

$$dV(g)/dg \quad = \quad \hat{\beta}(V(g))/\beta(g) \tag{1.20}$$

consistent with (1.19) only if $\hat{b}_o = b_o$, $\hat{b}_1 = b_1$, and expressions (10) for $Z_1(g)$, $Z_2(g)$, and $f(g)$ not given here. It follows that for the computation of (the perturbation expansions of) the Γ_{as} from the directly constructed Γ_o only the functions (1.7 a - d) must be known.

2. EXCEPTIONAL MOMENTA IN ϕ^4 THEORY

2.1. Asymptotic forms at exceptional momenta

That the condition weaker than (1.12),

$$\Gamma(p_1 \ldots p_{2n}, q_1 \ldots q_\ell; m^2\lambda^2, g) \quad = \quad \mathcal{O}((\ln\lambda)^c) \tag{2.1}$$

for $\lambda \to 0$, does not hold for arbitrary (Euclidean) momenta follows already from Weinberg's analysis (3). (1.12) is violated if a nontrivial even partial sum of momenta (i. e. of an even number of p plus any number of q momenta) vanishes. Then the LHS of (1.12) is $\mathcal{O}(\lambda^{-2}(\ln\lambda)^c)$ and the integral in (1.8) does not allow $\lambda \to 0$. This corresponds to the nonexistence of $\Gamma_{as}(p_1 \ldots p_{2n}, q_1 \ldots q_\ell; m^2, g)$ at such momenta due to a UR singularity there, of the massless-theory function (2,15,16). The recipe described at the end of sect. 1.3. is then still usable (at least for Euclidean momenta) and yields functions $\Gamma_{\underline{as}}(p_1 \ldots p_{2n}, q_1 \ldots q_\ell; m^2\lambda^2, g)$ and $\Gamma_{\underline{as}}(p_1 \ldots p_{2n}, q_1 \ldots q_\ell; m^2, g)$, respectively, which are certain UR finite parts of the at these momenta UR singular Γ_{as} functions.

For the renormalization functions, the exceptional momenta sets are (p(-p)00,), (p(-p)p'(-p'),), and (p(-p),0). In sects. 2.1-3 we consider only these cases.

For the first momenta set, (1.6) becomes

$$\mathcal{O}p_{4,o}\ \Gamma(p(-p)00,;m^2,g) \quad = \tag{2.2}$$

$$= \ - \ i \ m^2 \ \wp(g) \ \Gamma(p(-p)00,0;m^2,g).$$

Short-distance expansion (SDE) of the RHS yields (6,15)

$$\Gamma(p(-p)00,0) \quad = \quad \Gamma(p(-p)00,)\ \Gamma(00,00) \ + \tag{2.3}$$

$$+ \ \Gamma_{rem}(p(-p)00,0)$$

with the remainder term having the property

$$\Gamma_{rem}(p(-p)00,0) = \mathcal{O}((p^2)^{-1} (\ln p^2)^c) \quad \text{for} \quad p^2 \to \infty. \tag{2.4}$$

(1.6) also gives

$$- i\, m^2 \mathcal{f}(g)\, \Gamma(00,00;m^2,g) = \mathcal{O}p_{2,1}\, \Gamma(00,0;m^2,g) = \eta(g) \tag{2.5}$$

due to (1.4). Thus (2.2) becomes

$$\mathcal{O}p_{2,-1}\, \Gamma(p(-p)00,;m^2,g) = \tag{2.6}$$
$$= - i\, m^2 \mathcal{f}(g)\, \Gamma_{rem}(p(-p)00,0;m^2,g)$$

which can be integrated to a form similar to (1.8). (2.4) now replaces (1.12) and leads to

$$\mathcal{O}p_{2,-1}\, \Gamma_{\underline{as}}(p(-p)00,;m^2,g) = 0 \tag{2.7}$$

or to the transformation formula (15)

$$\Gamma_{\underline{as}}(p(-p)00,;m^2,g) = \tag{2.8}$$
$$= a(g)^2\, a(g(\lambda))^{-2}\, h(g)\, h(g(\lambda))^{-1} \cdot$$
$$\cdot\, \Gamma_{\underline{as}}(p(-p)00,;m^2\lambda^2,g(\lambda)).$$

(1.13) is now replaced by (15)

$$\Gamma_{\underline{as}}(p(-p)00,;m^2,g) = \tag{2.9}$$
$$= \lim_{\lambda\to o} [h(g)\, h(g(\lambda))^{-1}\, \Gamma^{s(g,\lambda)}\, (p(-p)00,;m^2,g)]$$

which shows that

$$\Gamma_{as}(p(-p)00,;m^2,g) = 0 \tag{2.10}$$

in view of (1.13), (1.10b), and (1.16).

For the last of the momenta sets mentioned before, (1.6) becomes

$$\mathcal{O}p_{2,1}\, \Gamma(p(-p),0;m^2,g) = \tag{2.11}$$
$$= - i\, m^2 \mathcal{f}(g)\, \Gamma(p(-p),00;m^2,g).$$

SDE of the RHS yields (15)

$$\Gamma(p(-p),00) = \Gamma(p(-p)00,)\ \Gamma(,000) + \Gamma_{rem}(p(-p),00) \quad (2.12)$$

with

$$\Gamma_{rem}(p(-p),00) = \mathcal{O}((p^2)^{-1}\ (\ln p^2)^c) \quad \text{for} \quad p^2 \to \infty. \quad (2.13)$$

Since, from (1.6) and (1.5),

$$- i\ m^2 \mathcal{f}(g)\ \Gamma(,000;m^2,g) = \quad (2.14)$$

$$= \mathcal{O}p_{o,2}\ \Gamma(,00;m^2,g) + i\ \kappa(g) = i\ \kappa(g)$$

(2.11) becomes

$$\mathcal{O}p_{2,1}\ \Gamma(p(-p),0;m^2,g) - i\ \kappa(g)\ \Gamma(p(-p)00,;m^2,g) = \quad (2.15)$$

$$= - i\ m^2 \mathcal{f}(g)\ \Gamma_{rem}(p(-p),00;m^2,g).$$

Noticing that, due to (1.10c), $\mathcal{O}p_{o,2}[a(g)^{-2}\ h(g)^{-2}\ k(g)] = \kappa(g)$ and using (2.6), (2.15) can be written

$$\mathcal{O}p_{2,1}[\Gamma(p(-p),0;m^2,g) - \quad (2.16)$$

$$- i\ a(g)^{-2}\ h(g)^{-2}\ k(g)\ \Gamma(p(-p)00,;m^2,g)] =$$

$$= - i\ m^2 \mathcal{f}(g)[\Gamma_{rem}(p(-p),00;m^2,g) -$$

$$- i\ a(g)^{-2}\ h(g)^{-2}\ k(g)\ \Gamma_{rem}(p(-p)00,0;m^2,g)].$$

It follows

$$\mathcal{O}p_{2,1}[\underline{\Gamma_{as}}(p(-p),0;m^2,g) - \quad (2.17)$$

$$- i\ a(g)^{-2}\ h(g)^{-2}\ k(g)\ \underline{\Gamma_{as}}(p(-p)00,;m^2,g)] = 0$$

which, with (2.9), yields the transformation law (15)

$$\underline{\Gamma_{as}}(p(-p),0;m^2,g) = \quad (2.18)$$

$$= h(g)^{-1}\ h(g(\lambda))\ \underline{\Gamma_{as}}(p(-p),0;m^2\lambda^2,g(\lambda)) +$$

$$+ i\ a(g)^{-2}\ h(g)^{-2}\ [k(g) - k(g(\lambda))]\underline{\Gamma_{as}}(p(-p),0;m^2,g).$$

Also (15), from (2.16) and (2.9)

$$\underline{\Gamma_{as}}(p(-p),0;m^2,g) = \lim_{\lambda\to o} \{\Gamma^{s(g,\lambda)}(p(-p),0;m^2,g) + \quad (2.19)$$

$$+ i\ a(g)^{-2}h(g)^{-1}h(g(\lambda))^{-1}[k(g) - k(g(\lambda))]\Gamma^{s(g,\lambda)}(p(-p)00,;m^2,g)\}$$

such that, using (1.10c) and (1.16), for $\lambda \to 0$

$$\Gamma^{s(g,\lambda)}(p(-p),0;m^2,g) = -\frac{1}{3}(\ln\lambda^{-2})^{1/3}\, b_o^{1/3} \cdot \tag{2.20}$$
$$\cdot\ a(g)^{-2}\, h(g)^{-2}\, i\, \Gamma_{\underline{as}}(p(-p)00,;m^2,g) + \mathcal{O}(1)$$

showing that

$$\Gamma_{as}(p(-p),0;m^2,g) = \infty \tag{2.21}$$

in contrast to (2.10).

Finally

$$\mathcal{O}\rho_{4,o}\,\Gamma(p(-p)\ p'(-p'),;m^2,g) = \tag{2.22}$$
$$= -\,i\,m^2\,\varphi(g)\,\Gamma(p(-p)\ p'(-p'),0;m^2,g)\ .$$

The appropriate expansion as $p \to \infty$, $p' \to \infty$ simultaneously for the RHS is (15) (we assume $p \pm p' \neq 0$)

$$\Gamma(p(-p)\ p'(-p'),0) = \Gamma(p(-p)00,)\ \cdot \tag{2.23}$$
$$\cdot\ \Gamma(,000)\ \Gamma(p'(-p')00,) + \Gamma_{rem}(p(-p)\ p'(-p'),0)$$

such that with (2.14) and by similar steps as led to (2.17),

$$\mathcal{O}\rho_{4,o}\,[\Gamma_{\underline{as}}(p(-p)\ p'(-p');m^2,g) - i\,a(g)^{-2}\,h(g)^{-2}\ \cdot \tag{2.24}$$
$$\cdot\ k(g)\,\Gamma_{\underline{as}}(p(-p)00,;m^2,g)\,\Gamma_{\underline{as}}(p'(-p')00,;m^2,g)] = 0$$

with consequent transformation law (15). Similarly to (2.20), we have (15), for $\lambda \to 0$,

$$\Gamma^{s(g,\lambda)}(p(-p)\ p'(-p'),;m^2,g) = -\frac{1}{3}(\ln\lambda^{-2})^{1/3}\, b_o^{1/3}\ \cdot$$
$$\cdot\ a(g)^{-2}\, h(g)^{-2}\, i\,\Gamma_{\underline{as}}(p(-p)00,;m^2,g)\,\Gamma_{\underline{as}}(p'(-p')00,;m^2,g) + \mathcal{O}(1)$$

such that, similarly to (2.21)

$$\Gamma_{as}(p(-p)\ p'(-p'),;m^2,g) = \infty. \tag{2.25}$$

2.2. Asymptotic behaviour near exceptional momenta

By simple manipulations of the BS equation one derives (10) the formulae closely related to SDEs

$$\Gamma((\lambda p + r_1)(-\lambda p + r_2)\ r_3\ r_4,) = \tag{2.26}$$

$$= \Gamma(\lambda p(-\lambda p)00,)\ \Gamma(r_3 r_4,(r_1 + r_2)) + \mathcal{O}(\lambda^{-1}(\ln\lambda)^c),$$

$$\Gamma((\lambda p + r_1)(-\lambda p + r_2),r_3) = \Gamma(\lambda p(-\lambda p),0) + \tag{2.27}$$

$$+ \Gamma(,(r_1 + r_2)r_3)\ \Gamma(\lambda p(-\lambda p)00,) + \mathcal{O}(\lambda^{-1}(\ln\lambda)^c),$$

and

$$\Gamma((\lambda p + r_1)(-\lambda p + r_2)(\lambda p' + r_3)(-\lambda p' + r_4),) = \tag{2.28}$$

$$= \Gamma(\lambda p(-\lambda p)\lambda p'(-\lambda p'),) + \Gamma(\lambda p(-\lambda p)00,) \cdot$$

$$\cdot\ \Gamma(\lambda p'(-\lambda p')00,)\ \Gamma(,(r_1 + r_2)(r_3 + r_4)) + \mathcal{O}(\lambda^{-1}(\ln\lambda)^c)$$

which should be contrasted with (15) (2.29)

$$\Gamma((\lambda p_1 + r_1)\ldots(\lambda p_{2n} + r_{2n}),(\lambda q_1 + s_1)\ldots(\lambda q_\ell + s_\ell)) =$$

$$= \Gamma(\lambda p_1 \ldots \lambda p_{2n},\lambda q_1 \ldots \lambda q_\ell) + \mathcal{O}(\lambda^{3-2n-2\ell}(\ln\lambda)^c)$$

for nonexceptional momenta, all formulae meant for $\lambda \to \infty$. The asymptotic forms $\underline{\Gamma_{as}}$ of the functions containing λ on the RHSs of (2.26 - 28) have been analysed in sect. 2.1. and may be substituted for these Γ, obtaining thereby the asymptotic behaviour of the LHSs, with nonasymptotic functions $\Gamma(r_3 r_4,(r_1 + r_2))$ etc.

Consider replacing on the LHSs of (2.26-28) the functions Γ by $\underline{\Gamma_{as}}$. Then on the RHSs the functions $\underline{\Gamma_{as}}$ appear as before, and the functions not containing λ, $\Gamma(r_3 r_4,(r_1 + r_2))$ etc., are replaced by the corresponding Γ_{as}. Thus (10), replacing on the LHSs Γ by $\underline{\Gamma_{as}}$ makes a difference of order $\mathcal{O}((\ln\lambda)^c)$ i. e. the Γ_{as} are no approximation to the Γ not only at (where the Γ_{as} do not exist) but also near exceptional momenta.

2.3. Infrared singularities in the massless theory (10,17)

The formulae (2.26 - 28) have analoga in the directly constructed massless theory with VFs Γ_o of sect. 1.4. E.g., the one of (2.26) is

$$\Gamma_o((\lambda p + r_1)(-\lambda p + r_2)r_3 r_4,) = \underline{\Gamma_o}(\lambda p(-\lambda p)00,) \cdot \tag{2.30}$$

$$\cdot\ \Gamma_o(r_3 r_4,(r_1 + r_2)) + \Gamma_{orem}((\lambda p + r_1)(-\lambda p + r_2)r_3 r_4,).$$

The last term is, for $\lambda \to \infty$, $\mathcal{O}(\lambda^{-1}(\ln\lambda)^c)$, and for it a direct construction prescription that is manifestly UR finite can be given. It stays finite also for $r_1 + r_2 \to 0$ and even goes linearly to zero then. This allows two uses of (2.30): 1) Choose the r_i such that on the RHS (1.17d) can be used, whereupon (2.30) becomes a defining equation for the $\underline{\Gamma}_o$ function. From this function the corresponding Γ_{as} one can be gotten in a simple way, indicated e. g. by comparing (2.30) with the Γ_{as} form of (2.26) mentioned before. 2) Setting in (2.30) $r_1 + r_2 \to 0$ shows the singularity of the LHS near exceptional momenta since from the integrated form of (1.18) follows

$$\Gamma_o(\mu r_3 \mu r_4, \mu(r_1 + r_2); m^2, V) = \tag{2.31}$$
$$= \hat{h}(V)^{-1}\, b_o^{-1/3} (\ln \mu^{-2})^{-1/3} + \mathcal{O}((\ln\mu)^{-4/3} \ln\ln\mu)$$

for $\mu \to 0$, with $\hat{h}(V)$ a function analogous to (1.10b). The corresponding singularity of the Γ_{as} functions follows via (1.19). It is, as comparison of (2.9) with (2.30) upon use of (2.31) shows, quantitatively related to the singularity of the finite-mass-theory function at exceptional momenta for vanishing mass: $\ln \lambda^{-2}$ in (2.9) replaces $\ln[m^2(r_1 + r_2)^{-2}]$ in (2.30). Of course, (2.10) is recovered also in the limit of going to the exceptional momenta set within the Γ_{as} theory.

Analogous results hold for the Γ_o-analoga of (2.27) and (2.28).

2.4. Mass-switch-on in the massless theory (10)

(1.13) allows to interpret $\Gamma^{s(g,\lambda)}$ as the VF of the ϕ^4 theory with mass $m\lambda$, normalized such, however, that the limit $\lambda \to 0$ is continuous for nonexceptional momenta. (For exceptional momenta, it can then not exist, or is trivial, see (2.10), (2.21), (2.25).) One finds

$$\Gamma^{s(g,\lambda)}(p_1 \ldots p_{2n}, q_1 \ldots q_\ell; m^2, g) = \tag{2.32}$$
$$= \Gamma_{as}(p_1 \ldots p_{2n}, q_1 \ldots q_\ell; m^2, g) - \lambda^2 (\ln \lambda^{-2})^{2/3}\, m^2\, a(g)^{-1} \cdot$$
$$\cdot\, h(g)^{-1}\, b_o^{2/3}\, [\underline{\Gamma_{as}}(p_1 \ldots p_{2n} 00, q_1 \ldots q_\ell; m^2, g) +$$
$$+ \sum_{\text{partitions}} \Gamma_{as}(\ldots)\, G_{as}(\ldots)\, \Gamma_{as}(\ldots) \ldots G_{as}(\ldots)\, \Gamma_{as}(\ldots)] +$$
$$+ \mathcal{O}(\lambda^2 (\ln\lambda)^{-1/3} \ln\ln\lambda).$$

Here the last term in the squarebracket, present if $n + \ell \geq 2$, corresponds to all partitions of the arguments $(p_1 \ldots p_{2n} 00, q_1 \ldots q_\ell)$

into (at least two) groups, each group forming the arguments of a Γ_{as} function, and the G_{as} are propagators connecting the Γ_{as} to form a chain, with the two zero-momenta appearing in the arguments of the Γ_{as} functions forming the ends of the chain. It is for the squarebracket in (2.31) that the simple PDE

$$\mathcal{O}p_{2n,\ell-1}[\Gamma_{\underline{as}}(\ldots) + \ldots] \equiv \mathcal{O}p_{2n,\ell-1}\, \Gamma'_{\underline{as}}(\ldots) = 0$$

holds, with resulting transformation formulae analogous to (2.8).

Note that, in view of (1.11c), the mass-switch-on was effected by a change of the bare-mass term in the Lagrangian density of order $\lambda^2(\ln \lambda^{-2})^{1/3}$. I do not know whether a mass-switch-on smoother than the one in (2.32) can be obtained by changing also the interaction term in the Lagrangian density. To study this, precise estimates relating to Lowenstein's differential vertex operations (13) in an almost-massless theory would be required.

2.5. Infrared singularities in theories with scalar massless particles (10,17)

If in the theory considered besides scalar massless particles self-interacting like in ϕ^4 theory also massive particles appear already on the Langrangian level (i. e., not as bound states), then the UR singularities at small or near exceptional momenta can not be obtained directly from the renormalization group, as in sect. 2.3. One can, however, organize the contributions to VFs in such a way (which is essentially the "parquet approximation" of older renormalization-group applications (18), and which was first used for similar purpose as here by Larkin and Khmel'nitskii (19)) that one can prove, under a hypothesis verifyable in renormalized perturbation theory: either the UR singularities are (up to overall normalization factors) the same as in massless ϕ^4 theory, or there are no such singularities, the latter case corresponding to $V = 0$ in the construction of sect. 1.4. (Note e. g. the occurrence of V in (2.31) in the leading term on the RHS only in the overall factor.)

The first case would apply e. g. to the theory of massive and massless scalar particles, if the masslessness of one species of particles is imposed ad hoc. The second case, however, applies to the sigma-model in the Goldstone mode, where the conservation of the axial current(s) suppresses the self coupling of the pions to zero, such that e. g. the pion-pion amplitude is (in the renormalizable theory) nonzero only by virtue of the presence also of massive particles.

A straightforward generalization to nonscalar particles is the one to quantum electrodynamics (QED). The UR singularities in massless QED (16) (photon as well as electron massless) are unaffected (up to factors) by the presence of also massive (charged or neutral) particles. (Note that the UR singularities considered here are unrelated to the semiclassical ones of the ordinary QED Green's functions near the electron mass shell.) On the other hand, the self-coupling of the photons in ordinary (massive electron) QED is suppressed due to gauge invariance, i. e. current conservation, similarly as the ones of pions in the sigma-model.

3. EIGENVALUE CONSIDERATIONS IN ϕ^4 THEORY

3.1. The standard assumptions

For nonexceptional momenta, from sect. 1

$$\begin{aligned} &\Gamma(\lambda p_1 \ldots \lambda p_{2n}, \lambda q_1 \ldots \lambda q_\ell; m^2, g) = \qquad (3.1) \\ &= \lambda^{4-2n-2\ell}\, a(g)^{n-\ell}\, a(g(\lambda))^{-n+\ell}\, h(g)^{-\ell}\, h(g(\lambda))^{\ell} \cdot \\ &\cdot \Gamma_{as}(p_1 \ldots p_{2n}, q_1 \ldots q_\ell; m^2, g(\lambda)) - i\, \delta_{no}\, \delta_{\ell 2}\, a(g)^{-2} \cdot \\ &\cdot h(g)^{-2}\, [k(g) - k(g(\lambda))] - i\, m^2\, \lambda^{2-2n-2\ell} \cdot \\ &\cdot \int_0^1 d\mu^2\, a(g)^{n-\ell}\, a(g(\mu))^{-n+\ell}\, h(g)^{-\ell}\, h(g(\mu))^{\ell}\, \mathcal{J}(g(\mu)) \cdot \\ &\cdot \Gamma(p_1 \ldots p_{2n}, q_1 \ldots q_\ell 0; m^2 \lambda^{-2}\, \mu^2, g(\mu)) \end{aligned}$$

holds. (1.12) yields for the integral term the estimate

$$\mathcal{O}(\lambda^{2-2n-2}\, (\ln\lambda)^c) \quad \text{as} \quad \lambda \to \infty .$$

Thus, postpoing the estimate outside of perturbation theory, the behaviour as $\lambda \to \infty$ of the LHS is given by the first two terms on the RHS.

The standard assumptions are

1) There exists g_∞ such that

$$\lim_{g \uparrow g_\infty} \int^g dg'\, \beta(g')^{-1} = +\infty, \quad \text{and} \quad \beta(g) > 0 \quad \text{for} \quad 0 < g < g_\infty,$$

2) $\gamma(g) \to \gamma(g_\infty)$ continuously as $g \to g_\infty$ and likewise for $\eta(g)$, $\kappa(g)$, $\mathcal{J}(g)$,

3) $\Gamma_{as}(p(-p),;m^2,g)\big|_{g\to g_\infty}$ exists for some $p^2 < 0$

$\Gamma_{as}(p_1 \ldots p_4,;m^2,g)\big|_{g\to g_\infty}$ exists for e.g. some Euclidean set (p_i).

The limits in 3) can be taken as defining renormalization conditions for massless-theory $\Gamma_{as}(p_1 \ldots p_{2n},;m^2,g_\infty)$ functions. An obvious extension of 3) gives similarly all $\Gamma_{as}(p_1 \ldots p_{2n},q_1 \ldots q_\ell; m^2,g_\infty)$ functions.

Due to 2), in

$$a(g)\ a(g(\lambda))^{-1} = \lambda^{-4\gamma(g_\infty)}\ e^{-R(g,\lambda)} \tag{3.2a}$$

where

$$R(g,\lambda) = 2\int_1^{\lambda^2} \lambda'^{-2}\ d\lambda'^2\ [\gamma(g(\lambda') - \gamma(g_\infty)],$$

$$\lim_{\lambda\to\infty} \frac{R(g,\lambda)}{\ln\lambda} = 0$$

holds. Similarly for

$$h(g)\ h(g(\lambda))^{-1} = \lambda^{-2\eta(g_\infty)}\ e^{-R'(g,\lambda)}. \tag{3.2b}$$

Thus

$$\Gamma(\lambda p_1 \ldots \lambda p_{2n}, \lambda q_1 \ldots \lambda q_\ell; m^2, g) = \tag{3.3}$$

$$= \lambda^{4-2n-2\ell-4(n-\ell)\gamma(g_\infty)+2\ell\eta(g_\infty)} \cdot$$

$$\cdot\ e^{(-n+\ell)R(g,\lambda)+\ell R'(g,\lambda)}\ \Gamma_{as}(p_1 \ldots p_{2n}, q_1 \ldots q_\ell; m^2, g_\infty) +$$

$$+\ i\ \delta_{no}\ \delta_{\ell 2}\ a(g)^{-2}\ h(g)^{-2} \cdot$$

$$\cdot \begin{cases} [\ 4\gamma(g_\infty)+2\eta(g_\infty)]^{-1}\lambda^{8\gamma(g_\infty)+4\eta(g_\infty)}\ \kappa(g_\infty)\ e^{2R(g,\lambda)+2R'(g,\lambda)} & \text{if } 2\gamma(g_\infty) + \eta(g_\infty) > 0 \\ \ln\lambda^2\ \kappa(g_\infty)\ e^{2R(g,\lambda)+2R'(g,\lambda)} & \text{if } 2\gamma(g_\infty) + \eta(g_\infty) = 0 \\ [-\ k(g) + k(g_\infty)] & \text{if } 2\gamma(g_\infty) + \eta(g_\infty) < 0 \end{cases}$$

+ correction term from $g(\lambda) \neq g_\infty$ argument of Γ_{as}

+ correction term from massive $\neq$ massless theory

+ correction term for $\delta_{no}\ \delta_{\ell 2}$ part.

The first correction term requires for its estimate new assumptions such as left-differentiability of Γ_{as} at $g = g_\infty$, and similarly for the last correction term. The second correction term can be estimated essentially without any new assumption, cp. sect. 3.3. below.

3.2. Anomalous dimensions

Formula (3.3) expresses that

$$\dim \phi \;=\; 1 + 2\,\gamma(g_\infty) \qquad (3.4a)$$

$$\dim \hat{N}_2(\phi^2) \;=\; 2 + 2\,\eta(g_\infty) + 4\,\gamma(g_\infty) \qquad (3.4b)$$

are the dynamical dimensions in the sense of Wilson (4), if the standard assumptions are made. These are the dimensions ϕ and $\hat{N}_2(\phi^2)$ have in the scale invariant Gell-Mann-Low-limit theory (GL) $\Gamma_{as}(..,..;m^2,g_\infty)$. ($\hat{N}_2(\phi^2)$ differs by a UR divergent factor from $N_2(\phi^2)$ which is unsuitably normalized for a massless theory.)

Positivity of the metric requires

$$\gamma(g_\infty) \;\geq\; 0 \qquad (3.5a)$$

and

$$2\,\eta(g_\infty) + 4\,\gamma(g_\infty) \;\geq\; -1. \qquad (3.5b)$$

If $\gamma(g_\infty) = 0$, then the GL theory is the free one (20) and an argument due to Parisi (21) allows to conclude that in this case the zero g_∞ of $\beta(g)$ must be of higher than first order. In the next section also an upper bound for (3.5b) will be given.

A classical scale invariant field theory is also conformal invariant (22) under some conditions (23) on the Lagrange function, e. g. absence of derivative couplings is sufficient. A similar connection one expects in QFT. Indeed, Schroer has shown that the GL-limit theory, if it exists, is conformal invariant in the ϕ^4 model. On the other hand, Poincaré invariant gauges in QED do involve derivative couplings in such a way that asymptotic conformal invariance does not hold (25). Adler (26) has described a class of more complicated gauges for scale invariant QED that yield conformal invariance but which appear difficult to exploit.

3.3. Modification of a standard assumption

Assumption 3) of sect. 3.1. is less natural than 1) and 2); it should be made plausible that $\Gamma_{as}(\dots;m^2,g)$ does not behave like e. g. $a(g)$ which in general does not exist for $g = g_\infty$. From (3.1) with $\lambda = 1$, using assumption 2) of sect. 3.1., we have for $g = g_\infty$

$$\Gamma_{as}(p_1 \dots p_{2n}, q_1 \dots q_\ell; m^2, g_\infty) = \tag{3.6}$$

$$= \Gamma(p_1 \dots p_{2n}, q_1 \dots q_\ell; m^2, g) + i\, m^2 \int_0^1 d\mu^2 \,\cdot$$

$$\cdot\ \mu^{4(\ell-n)\gamma(g_\infty)+2\ell\eta(g_\infty)}\, \mathscr{f}(g_\infty)\ \Gamma(p_1 \dots p_{2n}, q_1 \dots q_\ell 0; m^2\mu^2, g_\infty)$$

where we have taken the limit under the integral sign and have introduced assumption

3') the massive-theory VFs exist for $g = g_\infty$.

The reasoning herefore is that $g = g_\infty$ is no particular value for the massive theory, while it is one for the preasymptotic theory in view of e. g. (1.14). If the integral in (3.6) converges at $\mu = 0$, the interchange of limits will have been permissible. The behaviour of the integrand at $\mu \to 0$ is a large-exceptional-momenta problem. The transformation formula generalizing (2.18) and used already in the derivation of (2.32) is

$$\underline{\Gamma_{as}}(p_1 \dots p_{2n}, q_1 \dots q_\ell 0; m^2\mu^2, g(\mu)) = \tag{3.7}$$

$$= a(g)^{-n+\ell+1}\ a(g(\mu))^{n-\ell-1}\ h(g)^{\ell+1}\ h(g(\mu))^{-\ell-1}\ \cdot$$

$$\cdot\ \underline{\Gamma_{as}}(p_1 \dots p_{2n}, q_1 \dots q_\ell 0; m^2, g) + i[k(g(\mu)) - k(g)]\ \cdot$$

$$\cdot\ a(g)^{-n-1+\ell}\ a(g(\mu))^{n-\ell-1}\ h(g)^{\ell-1}\ h(g(\mu))^{-\ell-1}\ \cdot$$

$$\cdot\ \underline{\Gamma'_{as}}(p_1 \dots p_\ell 00; q_1 \dots q_\ell; m^2, g).$$

Setting here $g \to g_\infty$, and using assumption 2) of sect. 3.1.yields

$$\underline{\Gamma_{as}}(p_1 \dots p_{2n}, q_1 \dots q_\ell 0; m^2\mu^2, g_\infty) = \tag{3.8}$$

$$= \mu^{4(n-\ell-1)\gamma(g_\infty)-2(\ell+1)\eta(g_\infty)}\ [\underline{\Gamma_{as}}(p_1 \dots p_{2n}, q_1 \dots q_\ell 0; m^2, g_\infty) +$$

$$+\ \underline{\Gamma'_{as}}(p_1 \dots p_{2n} 00, q_1 \dots q_\ell; m^2, g_\infty)\ i\,\kappa(g_\infty)$$

$$\cdot \left\{ \begin{array}{lll} c^{-1}\ \mu^{2c} & \text{if} & c < 0 \\ \ln \mu^2 & \text{if} & c = 0 \\ -\ c^{-1} & \text{if} & c > 0 \end{array} \right\} \ ]$$

where $c \equiv 4\ \gamma(g_\infty) + 2\ \eta(g_\infty)$. Inserting (3.8) into (3.6) one finds that the integral converges if $|c| < 2$ which, in view of (3.4b), is equivalent to $0 < \dim N_2(\phi^2) < 4$. This condition is also sufficient for the last correction term in (3.3) to be asymptotically negligible relative to the main term (cp. Appendix C of Ref. 27) as shown by introducing λ of (3.1) in the calculation just sketched.

In the case of exceptional momenta, it is assumption 3') that leads (28) to existence of the appropriate $\Gamma_{\underline{as}}(\ldots;m^2,g_\infty)$, the latter having been used already in (3.8).

3.4. Occurrence of logarithms in asymptotic forms

The scaling law for GL-limit theory VFs is obtained by setting $g = g_\infty$ in the relevant PDE. If the PDE is homogeneous, like (1.15) and (2.7), a pure power law results, from which the asymptotic behaviour of the $g \neq g_\infty$ VF differs by the factors $\exp R$ and $\exp R'$ as discussed in sect. 3.1. We here discuss the occurrence of logarithms only in the GL theory itself.

If the relevant PDE is inhomogeneous, special cases of which are (2.17) and (2.24), it depends on the scaling behaviour of the inhomogeneous part versus the one of the solution of the homogeneous PDE which one is the leading behaviour. If the two scaling laws are the same, a logarithm will result. An example of this is (3.8) (and an example showing the effect on the behaviour of $g < g_\infty$ VFs is (3.3)). $c = 0$ corresponds, in view of (3.4b), to $\dim N_2(\phi^2) = 2$, which is the canonical dimension. If the transformation law of the inhomogeneous part of the PDE involves already a logarithm, the one for the solution of the PDE may involve up to two, etc. Expressed more generally, functions $\Gamma_{\underline{as}}(\ldots;m^2,g_\infty)$ obey coupled PDEs and the relevant (reducible) representation of the dilatation group (29) may not be fully reducible, in the case of degeneracy of dimensions. (Here, the degeneracy resides in $\dim N_2(\phi^2) = 4 - \dim N_2(\phi^2)$.)

If $\dim N_2(\phi^2) = 2$, then

$$\Gamma_{as}(,q(-q);m^2,g_\infty) = \text{const} + \text{const}'\ \ln(-q^2)$$

due to the inhomogeneous term in (1.6b), as shown in (3.3). On the other hand, the dimension of a symmetry current (in $(\phi_i \phi_i)$ theory, for example) is canonical (4,23) as one proves best (9) requiring consistency of the relevant Ward identities with the PDEs, such that in the two-current correlation functions a logarithm is always generated by the inhomogeneous term brought about by the final subtractive renormalization that gave the $\delta_{n\rho}\, \delta_{\ell 2}$ term in (1.6). In this case, this logarithm dominates also the asymptotic behaviour of the massive-theory function (under assumptions 1), 3) of sect. 3.1.) due to the absence of γ-terms (9,23) in the PDE.

CONCLUSION

One has learnt to look at renormalizable field theories not so much term by term in the perturbation series, but as given by the formal sums of these series, which are the only yet available definition apart from seemingly untractable infinite systems of coupled integral equations, or via sequences of limiting procedures (30) (finite volume, UV cutoffs...) with so far uncontrollable effect.

Given this situation, reordering the perturbation series is what one should attempt. The systematic ways hereto found so far are the renormalization group, and operator product (and related) expansions. Combining these two, physical large-momenta problems can be treated insofar as they are reducible, by some expansion (e. g. by light cone expansion as in deep-inelastic e-p scattering (31)) or by highly plausible assumption (as in e^+e^- annihilation (32)) or these two suitably combined (33) to a large-Euclidean-momenta problem.

In any case, techniques have been developped that make the renormalization group applicable to cases it formerly wasn't, and the assumptions on which it rests and its present limitations have been clarified.

ACKNOWLEDGEMENT

The author thanks Professor R.E. Caianiello for his cordial invitation to lecture at this Summer Institute, and Frau R. Siemer for her careful typing of this manuscript.

REFERENCES

1) J. Schwinger, Proc. Natl. Acad. Sci. U. S. 44, 956 (1958); K. Symanzik, J. Math. Phys. 7, 510 (1966)

2) D. Ruelle, Thèse, Bruxelles, 1959; Nuovo Cimento 19, 356 (1961)

3) S. Weinberg, Phys. Rev. 118, 838 (1960)

4) K. G. Wilson, Phys. Rev. 179, 1499 (1969)

5) M. Gell-Mann, F.E. Low, Phys. Rev. 95, 1300 (1954)

6) W. Zimmermann, in "Elementary Particles and Quantum Field Theory"; Eds. S. Deser, M. Grisaru, H. Pendeton, Cambridge Mass., MIT Press 1971; Annals of Physics (N.A.) 77, 536, 570 (1973)

7) N.N. Bogoliubov, D.V. Shirkov, "Introduction to the Theory of Quantized Fields", New York, Interscience Publ. 1959; K. Hepp, Comm. meth. Phys. 2, 301 (1966)

8) K. Hepp, "Théorie de la Renormalisation", Berlin, Springer 1969

9) K. Symanzik, Comm. math. Phys. 18, 227 (1970)

10) K. Symanzik, DESY 73/6, to appear in Comm. math. Phys.

11) V.V. Belokurov, D.I. Kazakov, D.V. Shirkov, A.A. Slavnov, Dubna JINR-E-2-7320

12) K. Symanzik, in "Renormalization of Yang-Mills Fields and Applications to Particle Physics", C.N.R.S. Marseille, 72/P.470

13) Cf. J.H. Lowenstein, Comm. math. Phys. 24, 1 (1971)

14) The formulae of Appendix C of Ref. 15 contain an error.

15) K. Symanzik, Comm. math. Phys. 23, 49 (1971)

16) K. Symanzik, in: Springer Tracts in Modern Physics, 57 222 (1971)

17) Cf. also: K. Symanzik, Lectures at Grazer Universitätswochen 1973, Schladming, Austria (DESY T-73/3)

18) I.T. Diatlov, V.V. Sudakov, K.A. Ter-Martirosian, JETP 5, 631 (1957)

19) A.I. Larkin, D. E. Khmel'nitskiǐ, JETP 29, 1123 (1969)

20) K. Pohlmeyer, Comm. math. Phys. 12, 204 (1969)

21) G. Parisi, LNF 72/94, Frascati

22) G. Mack. Lectures at this Summer Institute

23) S. Coleman, R. Jackiw, Ann. Phys. (N.Y.) 67, 552 (1971)

24) B. Schroer, Lett. Nuovo Cimento 2, 867 (1971)

25) N.K. Nielsen, Aarhus University preprint, July 1973

26) S.L. Adler, Phys. Rev. 6, 3445 (1972)

27) G. Mack. K. Symanzik, Comm. math. Phys. 27, 247 (1972)

28) K. Symanzik (in preparation)

29) P. Otterson, W. Zimmermann, Comm. math. Phys. 24, 107 (1972); G.F. Dell'Antonio, Nuovo Cimento 12A, 756 (1972)

30) F. Guerra, Lectures at this Summer Institute

31) N. Christ, B. Hasslacher, A.H. Mueller, Phys. Rev. D6, 3543 (1972)

32) E.g., A. Zee, Phys. Rev. 8, 597 (1973)

33) A.H. Mueller, CO-2271-19, Columbia U. preprint

"DIAGRAMMAR" AND DIMENSIONAL RENORMALIZATION

G. 't Hooft*

CERN, Geneva, Switzerland

1. INTRODUCTION

There are several possible approaches to quantum field theory. One may start with a classical system of fields, interacting through non-linear equations of motion which are subsequently "quantized". Alternatively, one could take the physically observed particles as a starting point; then define a Hilbert space, local operator fields, and an interaction Hamiltonian. More ambitious, perhaps, is the functional integral approach, which has the advantage of being obviously Lorentz covariant.

All these approaches have one unpleasant and one pleasant feature in common. The unpleasant one is that in deriving the S-matrix for the theory, one encounters infinities of different types. In order to get rid of these one has to invoke a rather *ad hoc* "renormalization procedure", thus changing and undermining the theory halfway. The pleasant feature, on the other hand, is that one always ends up with a simple calculus for the S-matrix: the Feynman rules. Few physicists object nowadays to the idea that these Feynman diagrams contain more truth than the underlying formalism, and it seems only rational to abandon the aforementioned principles and use the diagrammatic rules as a starting point.

It is this diagrammatic approach to quantum field theory which we wish to advertise. The short-circuiting has several advantages. Besides the fact that it implies a considerable simplification, in particular in the case of gauge theories, one can simply superimpose

---

* On leave from the University of Utrecht, Netherlands.

the renormalization prescriptions on the Feynman rules. As for unitarity and causality, the situation has now been reversed: we shall have to investigate under which conditions these Feynman rules describe a unitary and causal theory. Within such a scheme many more or less doubtful or complicated theorems from the other approaches can be proven completely rigorously.

An extensive and pedagogical introduction to the diagrammatic approach is given in a CERN report, called "DIAGRAMMAR", to which we shall refer frequently[1]. These notes may be considered as an introduction to DIAGRAMMAR, but we shall treat the dimensional regularization and renormalization procedure in somewhat more detail. We shall repeat the arguments in DIAGRAMMAR in so far as we need them for our purposes.

The Feynman rules will always be <u>defined</u> by a Lagrangian $\mathcal{L}$,

$$\mathcal{L}(x) = \psi_i^*(x) V_{ij} \psi_j(x) + \tfrac{1}{2} \varphi_i(x) W_{ij} \varphi_j(x) + \mathcal{L}_I(\psi^*, \psi, \varphi) + J_\psi^* \psi + \psi^* J_\psi + \varphi J_\varphi \ , \tag{1.1}$$

where ψ_i and ϕ_i are complex and real fields respectively, V_{ij} and W_{ij} are space-time operators that must have an inverse. $\mathcal{L}_I$ is a possibly non-local, polynomial function of the fields ψ^*, ψ, and ϕ. The last three terms are source terms, which may be of convenience in defining off-mass-shell Green's functions.

The propagators will then always be <u>defined</u> to be $(-V - i\varepsilon)^{-1}_{ij}$ and $(-W - i\varepsilon)^{-1}_{ij}$, and the vertices will be <u>generated</u> by the interaction terms in $\mathcal{L}_I$. We assume that the reader is more or less familiar with the use of Feynman rules so that we do not need to repeat here the prescriptions for obtaining an S-matrix, starting with the rules for vertices and propagators.

The infinities of quantum field theory can now be removed in different ways. In simple cases one may replace the propagators V^{-1} and W^{-1} by others that converge more rapidly at $k \to \infty$, for instance,

$$-V^{-1} \to -V^{-1} - (-V + M)^{-1} \ , \tag{1.2}$$

where M is a large number. In gauge theories, however, a replacement such as (1.2) would break gauge invariance, and it becomes extremely difficult to see how gauge invariance can be restored in the limit $M \to \infty$.

For gauge theories one can get rid of the infinities in a better way: the dimensional regularization method. In fact, dimensional regularization can <u>only</u> be formulated for diagrams, <u>not</u> for canonical systems or functional integrals. It is here that the diagrammatic approach is particularly useful.

One of the most important items in DIAGRAMMAR is what we call the canonical transformation. This is the most general field transformation in the Lagrangian (1.1), of the form

$$\begin{aligned}
\psi(x) &\to \psi(x) + F(\psi, \psi^*, \varphi, x), \\
\psi^*(x) &\to \psi^*(x) + F^*(\psi, \psi^*, \varphi, x), \\
\varphi(x) &\to \varphi(x) + G(\psi, \psi^*, \varphi, x),
\end{aligned} \tag{1.3}$$

where F and G are in general non-local functions of the fields ψ and ϕ. If we decompose the fields into real fields ψ_i, then (1.3) can be written as

$$\psi_i(x) \to \psi_i(x) + F_i(\psi, x). \tag{1.4}$$

The new Lagrangian will again have bilinear and interaction parts. The transformation is also performed in the source terms of (1.1).

Now it is not difficult to verify that the tree diagrams (i.e. diagrams with no closed loops) from the old and the new Lagrangian are exactly equivalent. The reason is that the substitutions in the vertex and source parts of (1.1) exactly cancel the substitutions in the propagator (= bilinear) parts. But the same statement for diagrams with loops does not hold without an additional change in the Lagrangian. As is shown in DIAGRAMMAR, the necessary addition can conveniently be described in terms of a new ghost particle having a complex field $\alpha_i(x)$. Its interaction is defined by the Lagrangian

$$\mathcal{L}^{ghost} = \alpha_i^*(x)\alpha_i(x) + \int d_4x' \, \alpha_i^*(x) \frac{\delta F_i(\psi, x)}{\delta \psi_j(x')} \alpha_j(x'), \tag{1.5}$$

and by the additional prescription that there should be one more minus sign for each loop of α's (there are only closed loops of α's). So the substitution (1.4) together with the addition of (1.5) leads to a set of rules which is equivalent to the old one, in the sense that for instance vacuum-vacuum transition amplitudes in the presence of sources $\mathcal{J}$ remain the same.

This result justifies the functional integral method, in which field transformations are always accompanied by "Jacobian factors". The addition (1.5) corresponds exactly to the Jacobian for the most general type of transformation. Thus we close the gap between the diagrammatic approach and the functional integral approach for finite theories: a functional integral "proof" of Ward-Takahashi identities[2)] can always be substituted by a diagrammatic one or vice versa. But again, if infinities are to be studied, we think the direct diagrammatic description is more accurate.

2. THE CUTTING RULES

The dimensional renormalization procedure is, like most of the others, formulated via an induction process. Counter terms are added order by order in the perturbation expansion, increasing the possible number of closed loops by one at each step*). It is essential to have a profound insight into the nature of the possible infinities arising at each order. To this end we shall derive a relation satisfied by any set of Feynman diagrams. It is called causality relation in DIAGRAMMAR.

The starting point is the decomposition of the propagator into positive and negative energy parts:

$$\Delta_{ij}(x) = \theta(x_0)\,\Delta^{+}_{ij}(x) + \theta(-x_0)\,\Delta^{-}_{ij}(x)\,, \tag{2.1}$$

$$\Delta^{\pm}_{ij}(x) = \frac{1}{(2\pi)^3}\int d_4k\; e^{ikx}\,\vartheta(\pm k_0)\,\rho(k^2)\,, \tag{2.2}$$

with $x = x_i - x_j$. Here we used the notation

$$\Delta_{ij}(x) = \frac{1}{(2\pi)^4 i}\,(-V-i\varepsilon)^{-1}_{ij}\,(x_i, x_j)\,.$$

Although we shall not really use it here, we take the spectral functions ρ to be real, hence

$$\Delta^{\pm}_{ij} = (\Delta^{\mp}_{ij})^{*}\,.$$

Also

$$\Delta^{\pm}_{ij} = \Delta^{\mp}_{ji}\,. \tag{2.3}$$

Consequently

$$\Delta^{*}_{ij}(x) = \vartheta(x_0)\,\Delta^{-}_{ij}(x) + \vartheta(-x_0)\,\Delta^{+}_{ij}(x)\,. \tag{2.4}$$

---

*) Although perfectly correct, this loop expansion method is often redundant, in the sense that more counter terms are added than are strictly necessary. More economic is, for instance, the BPH procedure[3], but we prefer the loop expansion because in gauge theories diagrams are often arranged according to their number of loops.

As usual,

$$\vartheta(x_0) = \frac{1}{2\pi i}\int_{-\infty}^{\infty} d\tau \, \frac{e^{i\tau x_0}}{\tau - i\varepsilon} = \begin{cases} 1 & \text{if } x_0 > 0 , \\ 0 & \text{if } x_0 < 0 \end{cases}$$

$$\vartheta(x_0) + \vartheta(-x_0) = 1 . \tag{2.5}$$

Now consider a diagram with n vertices. Such a diagram represents in coordinate space an expression containing many propagators depending on arguments $x_1, \ldots, x_n$. In DIAGRAMMAR, such an expression is denoted by

$$F(x_1, x_2, \ldots, x_n) .$$

The diagram has a particular topology. For instance we can take the diagram of Fig. 1

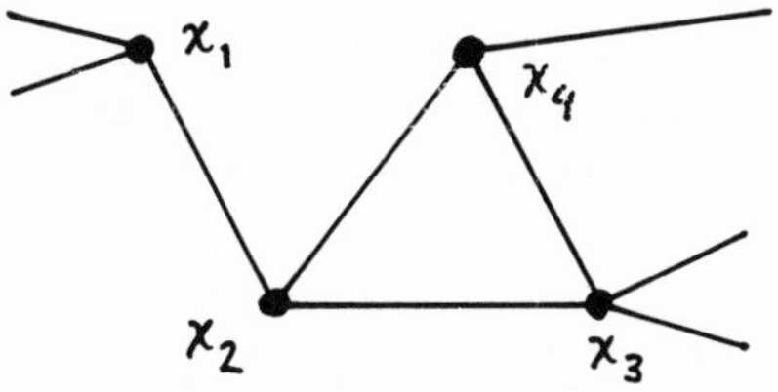

Fig. 1 Diagram for the function $F(x_1, x_2, x_3, x_4)$.

to which may correspond the function

$$F(x_1, x_2, x_3, x_4) = (ig)^4 \Delta_F(x_1 - x_2)\Delta_F(x_2 - x_3)\Delta_F(x_3 - x_4)\Delta_F(x_4 - x_2). \tag{2.6}$$

Belonging to the same topological system, different functions F are now defined, with some of the arguments x_i underlined. They are obtained from the original function F the following way:

i) A propagator Δ_{ki} is unchanged if neither x_k nor x_i are underlined.

ii) A propagator Δ_{ki} is replaced by Δ_{ki}^{+} if x_k but not x_i is underlined.

iii) A propagator Δ_{ki} is replaced by Δ_{ki}^{-} if x_i but not x_k is underlined.

iv) A propagator Δ_{ki} is replaced by Δ_{ki}^{*} if x_k and x_i are underlined.

v) For any underlined x replace one factor i by -i.

Apart from that, the rules for the vertices are unchanged.

Consider F in configuration space. One of the points x in Fig. 1, say x_i, must have the largest time component, $x_{i0} > x_{j0}$ for all j. From (2.1) and (2.4) we then derive

$$F(\ldots, x_i, \ldots) + F(\ldots, \underline{x}_i, \ldots) = 0. \tag{2.7}$$

where the underlining of the other arguments is the same in both terms. The sign is a consequence of def. (v). From Eq. (2.7) another equation follows which is <u>independent</u> of the order of the time components of the points x:

$$\sum_{\text{underlinings}} F(x_1, \underline{x}_2, \ldots, x_i, \ldots, \underline{x}_n) = 0, \tag{2.8}$$

where we sum over all possible ways in which the arguments can be underlined (including no underlinings and all x underlined). To prove (2.8) from (2.7) one merely selects the x with the largest time and arranges the terms in (2.8) in pairs, which are then all zero according to (2.7). Equation (2.8) is the crucial equation to determine whether a certain set of Feynman rules leads to a unitary S-matrix (see DIAGRAMMAR).

Now again consider $F(x_1, \ldots, x_n)$ and take any two variables, say x_i and x_j. Let us suppose that the time component of x_j is larger than that of x_i. Then the following equation holds independently of the order of the other time components:

$$\sum_{\substack{\text{underlinings}\\ \text{except } x_i}} F(x_1, \ldots, x_i, \ldots x_n) = 0 \qquad \text{if } x_{i0} < x_{j0}. \tag{2.9}$$

A similar equation holds if $x_{j0} < x_{i0}$, and so we have

$$F(x_1, x_2, \ldots, x_n) + \vartheta(x_{j0} - x_{i0}) \sum_i{}' F(x_1, \ldots x_i, \ldots, \underline{x}_n)$$
$$+ \vartheta(x_{i0} - x_{j0}) \sum_j{}' F(x_1, \ldots, x_j, \ldots, \underline{x}_n) = 0. \tag{2.10}$$

The prime indicates absence of the term without underlined variables. The index i implies absence of terms with x_i underlined.

Now we can also take together those terms with neither x_i nor x_j underlined:

$$
\begin{aligned}
&F(x_1, x_2, \ldots, x_n) + \sum_{ij}{}' F(x_1, \ldots, x_n) \\
&+ \vartheta(x_{j0} - x_{i0}) \sum_{\substack{j \text{ underlined,} \\ i \text{ not}}} F(x_1, \ldots, x_i, \underline{x}_j, \ldots, x_n) \\
&+ \vartheta(x_{i0} - x_{j0}) \sum_{\substack{i \text{ underlined,} \\ j \text{ not}}} F(x_1, \ldots, \underline{x}_i, x_j, \ldots, x_n) = 0.
\end{aligned}
\tag{2.11}
$$

This equation can be visualized in a figure if we use a shaded line to divide the diagrams into two regions: the underlined vertices are in the shaded region at the right, the ones not underlined are at the left (Fig. 2).

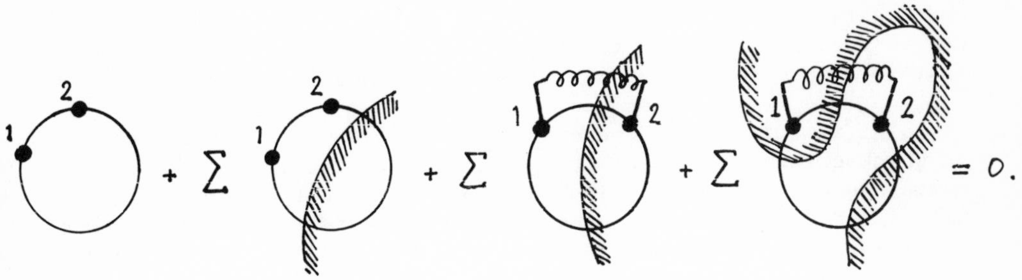

Fig. 2 Equation (2.11) in diagrams.

The blob stands for any diagram or collection of diagrams. The points 1 and 2 indicate two arbitrarily selected vertices. The "self-inductance" denotes the θ function (Fig. 3).

$$\frac{1}{2\pi i} \; \frac{1}{-k_0 - i\varepsilon} \; \delta_3(\vec{k})$$

Fig. 3 The θ function in momentum configuration.

The summation is over all possible ways in which the diagrams can be cut with the points 1 and 2 in the given position. Note that lines going through the cutting line are on-mass-shell, and integrations through the cutting line are phase-space integrations.

The reason why Eq. (2.11) is called causality relation is that the same relation can again be applied to that part of the second diagram in Fig. 2 which is in the unshaded region. Repeating the procedure a number of times we find that the original diagram can be expressed completely as a sum of two terms, one proportional to $\theta(x_{20} - x_{10})$ and having only positive energy flow from 1 to 2, and one vice versa.

The causality equation is particularly useful for renormalization if the regularization procedure does not violate the decomposition properties (2.1)-(2.4) of the propagator.

Suppose we have an irreducible diagram containing ℓ loops. Suppose further that all infinities of subgraphs with $\ell - 1$ or less loops have been removed by means of local counter terms (there is no subgraph with ℓ loops). Then we shall prove that the infinity of our diagram can be removed by a local counter term, whose dimension can be obtained by power counting. A local counter term is a vertex that is a pure polynomial in terms of the momenta.

Let $x_1, \ldots, x_r$ be the external vertices, that are the vertices to which external lines of the graph are attached. The external momenta are $p_1, \ldots, p_{r-1}, -(p_1 + \ldots + p_{r-1})$. Let us apply Fig. 2, with 1 and 2 replaced by 1 and r. The first term is the diagram of interest. The second term has at least one loop cut through because the original diagram was irreducible. The second term therefore consists of a phase-space integral of diagrams with less loops and is hence finite because of the induction assumption. The third and similarly the fourth term in Fig. 2 contain one additional integration, which we denote explicitly:

$$\int dk_0 \frac{1}{2\pi i} \frac{1}{k_0 - i\varepsilon} G(\vec{p}_1, p_{10} + k_0, p_2, \cdots, p_{r-1}), \tag{2.12}$$

where G stands for the third blob in Fig. 2 but without the θ function self-inductance. For the momentum p_1 we denoted space and time components explicitly. Note that G is again a finite function.

This equation is an ordinary dispersion relation that might diverge in the limit $M \to \infty$, or $n \to 4$, or whatever our regularization procedure was. But of course, before taking such a limit, one can make subtractions such as:

$$\begin{aligned}\int dk_0 \frac{1}{2\pi i} \frac{1}{k_0 - i\varepsilon} \Big\{ & G(\vec{p}_1, p_{10} + k_0, p_2, \cdots) \\ & - G(\vec{p}_1, k_0, p_2, \cdots) - p_{10} G'(\vec{p}_1, k_0, p_2, \cdots) - \cdots \Big\} \\ & + C_0(\vec{p}_1, p_2, \cdots p_{r-1}) + p_{10} C_1(\vec{p}_1, p_2, \cdots) + \cdots,\end{aligned} \tag{2.13}$$

until the integral converges. The number of necessary subtractions can be deduced from power counting. Then after the limit $M \to \infty$ or

$n \to 4$ the only place where an infinity can develop is in the unknown terms C_0, $p_{10}C_1$, ..., which are polynomials in p_{10}. By extending the argument to the other external momenta, and by Lorentz invariance, we see that the infinity can only be a polynomial in all external momenta, which was the thing to prove.

From this theorem we may conclude the following. If the regularized theory also respects Eqs. (2.1)-(2.4) for finite values of the regularization parameter, then the perturbation series for the S-matrix can be made finite by adding, order by order in the perturbation series, local counter terms.

3. DIMENSIONAL REGULARIZATION, INTUITIVELY

Consider the following integral,

$$I = \int d_n p \, \frac{1}{(p^2 + 2kp + m^2 - i\varepsilon)^\alpha} \quad . \tag{3.1}$$

If n is a positive integer and $2\alpha > n$ it is well defined, and not difficult to calculate:

$$I = \frac{i\pi^{n/2}}{(m^2 - k^2)^{\alpha - n/2}} \quad \frac{\Gamma(\alpha - n/2)}{\Gamma(\alpha)} \quad . \tag{3.2}$$

The dimensional regularization procedure consists of attributing to the integral (3.1) a formal meaning also if

i) n is not a positive integer

ii) $2\alpha < \mathrm{Re}\, n$.

Formula (3.2) will then hold for general n and α, as we shall see more explicitly in the next section. At first sight this may not seem to be much of an improvement, because if we let $\alpha < n/2$ and $n \to$ integer, then $\Gamma(\alpha - n/2)$ in Eq. (3.2) will exhibit a pole and also the "formal" integral will be infinite. But now we have a convenient regularization parameter n.

Take as an example $\alpha = 1$. The infinite part as $n \to 4$ is

$$I \to i\pi^2 (m^2 - k^2) \, \frac{-1}{2 - n/2} \quad + \text{finite as } n \to 4. \tag{3.3}$$

In connection with the theorem in the previous section it is not very astonishing that this is a polynomial in terms of k and m^2. We shall add to the amplitude the effect of a local counter term,

$$i\pi^2 (m^2 - k^2) \frac{2}{4-n} . \tag{3.4}$$

We then get

$$I^{renormalized} = \lim_{n\to 4} \left(\frac{i\pi^{n/2}}{(m^2-k^2)^{1-n/2}} \frac{\Gamma(1-n/2)}{\Gamma(1)} + \frac{2i\pi^2(m^2-k^2)}{4-n} \right)$$

$$= i\pi^2 (m^2-k^2)(\log(m^2-k^2) + C) , \tag{3.5}$$

with

$$C = \lim_{n\to 4} \left(\pi^{n/2-2} \Gamma(1-n/2) + \frac{2}{4-n} \right) .$$

The constant C can be removed by finite renormalizations. In fact, C depends on our initial choice of the unit of mass, simply because the dimension of the counter term (3.4) differs from that of the amplitude (3.2) at $n \neq 4$, and therefore should also contain a factor

$$\mu^{n-4} ,$$

with μ some arbitrary unit of mass.

The result (3.5) is exactly the same as what one would get from dispersion relations or more conventional regularization techniques. The dimensional renormalization procedure now consists of taking the "bare" masses and coupling constants to be dependent on n (in the perturbation expansion they will have poles at $n = 4$), in such a way that the limit $n \to 4$ is finite and makes sense. The program, and some of its features, is pictured schematically in Table 1.

4. RIGOROUS DEFINITIONS FOR THE DIMENSIONAL REGULARIZATION PROCEDURE

In this section we first consider regularization with which we denote the procedure to make the theory finite in terms of some cut-off parameter, here $n - 4$. The renormalization program concerns the limit towards the physical situation, here $n \to 4$ and shall be considered later (step 3 in Table 1).

First of all, according to Table 1, we must define what we mean by taking a non-integer number of dimensions. This is simply done by requiring the formula

$$\int d^n k \, f(k^2) = \frac{\pi^{n/2}}{\Gamma(\frac{n}{2})} \int_0^\infty f(r^2) \, r^{n-2} \, dr^2 , \tag{4.1}$$

Table 1 The dimensional renormalization procedure

$n = 4 - \varepsilon$, ε non-rational	$n = 4$
1. Can only be well defined for Feynman diagrams	
Except for the nature of the infinities there is a continuous limit towards $n = 4$ →	
No logarithmic divergencies, therefore we can make	Also logarithmically divergent integrals present. Therefore no such thing as analytic subtractions or Parti Fini can be well defined. We have to use, for instance:
2. "Analytic" subtractions. (Parti Fini of integrals)	2′. Pauli-Villars regularization
← These theories are different. Limit $n \rightarrow 4$ does not exist because of singularities at $n = 4$. →	
Gauge invariance maintained, no anomalies; scaling behaviour as expected (but of course ε-dependent)	Gauge invariance broken; Bell-Jackiw-Adler anomalies to be expected. Anomalous scaling behaviour.
3. Now we make the bare masses and coupling constants n-dependent (poles at $n = 4$), such that the limit $n \rightarrow 4$ exists →	Dimensionally renormalized theory. No Bell-Jackiw-Adler anomalies. Gauge invariance maintained (but renormalized fields might transform differently). Anomalous scaling behaviour, due to ε dependence at the left.

to be valid also for non-integer (even negative or complex) n. This formula can be applied if we have a finite number of vector integration variables. First we perform all integrations over the angles, and then the remaining ones using the formula (4.1). So our definition is good for Feynman diagrams with a finite number of loops, but it seems to be impossible to extend it to Hilbert spaces or path integrals. Note that there is not yet any discontinuity in formula (4.1) at positive integer n.

Next the ultraviolet divergent integrals are considered. We propose a subtraction method which does not work for logarithmically divergent integrals, so we are forced to go first to non-integer, or even non-rational, n. Our subtraction method which we call "analytic" subtraction, is probably equivalent to taking the "Parti Fini" of an integral[4] and can be formulated in various ways. One way is to decompose the integral into parts which are all convergent in a certain region of n, and then continuate analytically; or one may redefine the integral by making partial integrations and subsequently neglecting the (infinite or zero) boundary terms. This has been defined very precisely in DIAGRAMMAR and elsewhere[5]. Perhaps the most convincing and unambiguous way is to decompose integrands in k-space into finite terms (when integrated), and terms of the form

$$f(k_1 - a_1, k_2 - a_2, \ldots), \tag{4.2}$$

where f has the property

$$f(\lambda p_1, \lambda p_2, \ldots) = \lambda^\alpha f(p_1, p_2, \ldots), \tag{4.3}$$

and a_1, a_2, ... are arbitrary. The integrals over the functions f are then <u>defined</u> to be zero, so that we are left with the finite integrations ("Parti Fini"). This definition is unambiguous because the above decomposition is unambiguous. It only does not work for logarithmically divergent integrals because a logarithmically divergent function f satisfying (4.3) is divergent both at the origin <u>and</u> at infinity, so we cannot compensate logarithmic infinities at infinity only, without creating new ones at some point a.

The formula in the previous section can be deduced from these definitions, and indeed we notice the singularity at n → 4 develop, where we would have to deal with a logarithmically divergent part of the integral.

The importance of the analytic subtraction procedure is that summations, differentiations, and integrations commute, and that the result is independent of the choice of the integration variables. [Remember that Bell-Jackiw-Adler anomalies[6] are a consequence of the ambiguity of the integration variable. They evidently cannot occur in $4-\varepsilon$ dimensions!] This implies that all formal manipulations with Feynman diagrams in Sections 1 and 2 are also strictly legal for analytically subtracted diagrams.

As for the cutting rules (Section 2), it is only with respect to the time direction that Fourier integration and decomposition of the propagator are necessary. Dimensional regularization is therefore applied with respect to space-dimension. The cutting rules can exhibit infra-red divergent threshold singularities if n is taken to be too small (see DIAGRAMMAR) but this is of no influence to our argument here because we only need to consider the vicinity of $n = 4$.

From the theorem of Section 2 we therefore conclude that order-by-order renormalization by means of local counter terms (polynomials in k, but singular in $n = 4$) leads to a completely finite perturbation theory.

5. SCALE TRANSFORMATIONS

Quantum field theories behave quite remarkably under scale transformations. The clue needed to calculate this behaviour is the Callan-Symanzik equation[7)]. In first approximation the scaling behaviour is governed by the one-loop corrections to the amplitudes. The leading terms of these integrals are rather easy to calculate, but the next to leading parts seem to be rather ugly.

An alternative way to calculate the scaling behaviour is to use the dimensional renormalization procedure. The advantage of that method, which has been published elsewhere[8)], is perhaps its elegance, and moreover the fact that for the leading terms one never needs to calculate real integrals, but only their "infinite" part [compare Eq. (3.3)]. The central idea is that in $4-\varepsilon$ dimensions the analytically subtracted amplitudes obey the "naïve" scaling rules, which can easily be obtained by dimension-counting. So we merely need to examine closely the $\varepsilon \to 0$ limit, keeping in mind that the original Lagrangian has been altered by counter terms such as (3.4) in order to obtain a physically interesting limit, and the observation in Section 3 that these counter terms must be given the correct dimension by means of some unit of mass, μ.

In perturbation theory one can choose freely in which of the perturbation terms to put the finite part of such counter terms. Exploiting this freedom we derive equations for the scaling behaviour on the one hand, and for the higher-order singularities at $n = 4$ on the other. In fact, the coefficients in front of the kth order poles can be expressed in terms of the coefficients in front of the single poles. In other words, the parameters in the analytically subtracted theory, λ_B^i, must have an n-dependence of the form

$$\lambda_B^i = \mu^{D^i(n)}\left[\lambda_R^i + \frac{a_1^i(\lambda)}{n-4} + \frac{a_2^i(\lambda)}{(n-4)^2} + \cdots\right] , \tag{5.1}$$

where μ is a unit of mass and $D^i(n)$ is the dimension of λ^i as a function of n; we find how to express the Callan-Symanzik coefficient β in terms of $a_1^i(\lambda)$, and we find relations that express the coefficients a_2^i, a_3^i, ... in terms of a_1^i. These relations are sufficiently simple to enable us to sum the series (5.1). A more thorough discussion of the resulting expressions for coupling constants is given by Symanzik[9].

One can take full advantage of the fact that only the infinities (i.e. poles at n = 4) of the integrals enter in our expressions for β. These infinities can be obtained by means of a simple algebra, deirectly from the original Lagrangian of the theory[10].

A surprise emerges if one applies the above-sketched techniques to field theories of different kinds. A renormalizable field theory always turns into itself by a scale transformation, but with different coefficients. All conventional theories, like ϕ^4 or QED, have the property that the coupling constants thus become large if we scale towards the far ultraviolet region, which implies that the short-distance behaviour is described by a badly divergent perturbation expansion and is therefore uncomputable; on the other hand, the infra-red divergencies can often be summed to give something finite. The opposite, however, is the case in pure non-Abelian gauge theories, or non-Abelian gauge theories with a limited number of fermions. These theories have a well-computable short-distance behaviour but have an unpredictable infra-red behaviour.

The scaling behaviour of the theory of quantum electrodynamics can be visualized the following way: if we consider an electric charge and its electric field as a function of the distance, then we observe that the charge is "screened" by the virtual pairs created by the electric field. So at a large distance the effective charge is smaller than at a small distance: the effective coupling constant e increases if we decrease the length scale.

The question now rises why the above heuristic argument does not apply to the non-Abelian gauge bosons themselves*). An argument -- but a not very convincing one -- might be the following. From Ref. 10 it can be deduced that the opposite effect stems mainly from the same terms that determine the (quite large) magnetic moment of the gauge bosons. Indeed, if we take a magnetic dipole and consider its field at a certain distance, then it is conceivable that virtual boson pairs tend to align their spins (anti-) parallel to the magnetic field (ferromagnetism) thus amplifying instead of screening

*) We thank L. Susskind for discussions on this point.

the field. By Lorentz invariance, moving virtual pairs should also amplify electric fields, thus giving rise to a negative screening effect. Because the magnetic moment is relatively large, this effect is much stronger than the positive screening due to their electric charge. We can assign to each particle a number denoting their contribution to the screening. For SU(2), the relative numbers are:

gauge bosons (including possible ghosts)			$-11/6$
spin $1/2$ fermions,	isospin	$1/2$	$+1/6$
	"	1	$2/3$
	"	$3/2$	$5/3$
	"	2	$10/3$
spin 0 bosons,	isospin	$1/2$	$1/24$
	"	1 (real fields)	$1/12$
	"	$3/2$	$5/12$
	"	2 (real fields)	$5/12$

To find the total screening, one simply adds these numbers for all occurring particles. If the number accidently adds up to zero then the sign of the screening will be determined by higher-order effects that have not been calculated.

REFERENCES

1) G. 't Hooft and M. Veltman, "Diagrammar", CERN 73-9 (1973).

2) B.W. Lee, lectures at this Summer School.

3) K. Hepp, Comm. Math. Phys. 2, 301 (1966).

4) E.R. Caianiello, lectures at this Summer School.

5) G. 't Hooft and M. Veltman, Nuclear Phys. B44, 189 (1972).

6) J.S. Bell and R. Jackiw, Nuovo Cimento 51, 47 (1969).
S.L. Adler, Phys. Rev. 177, 2426 (1969).

7) C.G. Callan, Phys. Rev. D2, 1541 (1970).
K. Symanzik, Comm. Math. Phys. 18, 48 (1970).

8) G. 't Hooft, CERN preprint TH 1666, to be published in Nuclear Phys. B.

9) K. Symanzik, lectures at this Summer School; see also his Cargèse and Erice lectures, 1973.

10) G. 't Hooft, CERN preprint TH 1692, to be published in Nuclear Phys. B.

QUANTUM GRAVITY AS A GAUGE FIELD THEORY

G. 't Hooft*

CERN, Geneva, Switzerland

1. INTRODUCTION

The recent advances in the understanding of gauge theories make possible a fresh approach to the quantum theory of gravitation. First, we now know precisely how to obtain Feynman rules for a gauge theory; secondly, the dimensional regularization scheme provides a powerful tool for handling divergencies[1]. Indeed, several authors have already published work using these methods[2].

One may ask why one should be interested in quantum gravity. Certainly, a theory for quantized gravity will not be experimentally verifiable or deniable for many centuries to come. Even if particles with energies corresponding to the fundamental gravitational length (10^{28} eV) will eventually be obtained, then still it is very unlikely that any gravitation effect will be observable. Quantum gravity effects might be dominant in the first 10^{-43} sec of the universe or in the last 10^{-43} sec of the formation process of a "black hole", but we are more interested in a question of logic. Firstly, how can a theory of gravity be fitted into the scheme of the other basic interactions? It seems to be unavoidable to "quantize" gravity in order to make the link in a self-consistent way. Secondly, interesting paradoxical situations might arise if we consider those curious stable gravitational objects called "black holes". How do they fit into our spectrum of elementary particles? What is the smallest possible black hole? Can they be created in pairs? etc. Thirdly, it may well be true that an eventual "supertheory" has a built-in cut-off for the other interactions. In view of the fact that it

---

* On leave from the University of Utrecht, Netherlands.

seems possible to formulate all known interactions in terms of field theoretical models that show only logarithmic divergencies, the smallness of the gravitational coupling constant need not be an obstacle. For the time being no reasonable or convincing analysis of this type of possibility has been presented, and we have no ambitions in that direction.

Much work on quantum gravity has already been reported in the literature[3]; in particular we mention the work of DeWitt[4]. And now, with our knowledge of gauge theories, we can solve some problems which were left open by these authors (this must have been our main motive for this investigation). Our contribution (we think) is first the resolution of the problem, Which of the fields $g_{\mu\nu}$, $g^{\mu\nu}$, $\sqrt{g}\, g^{\mu\nu}$, or perhaps a_i^μ (the Vierbein field), is to be chosen as our "elementary" field? (Section 5). Secondly we make an important remark concerning the counter terms for the theory of pure gravity (Section 6).

A much more detailed version of our calculations will be published elsewhere[5]. As I was asked to give an elementary talk, the larger fraction of this contribution can be found in any textbook on general relativity. We only arrange things in a pedagogical way, presenting gravity as a gauge field theory.

2. THE GAUGE TRANSFORMATION

The underlying principle of the theory of general relativity is invariance under general coordinate transformations,

$$x'^\mu = f^\mu(x) . \tag{2.1}$$

It is sufficient to consider infinitesimal transformations,

$$x'^\mu = x^\mu + \eta^\mu(x) , \quad \eta \text{ infinitesimal} . \tag{2.2}$$

Or, in other words, a function A(x) is transformed into

$$A'(x) = A(x + \eta(x)) = A(x) + \eta^\lambda(x)\,\partial_\lambda A(x) . \tag{2.3}$$

If A does not undergo any other change, then it is called a <u>scalar</u>. We call the transformation (2.3) simply a <u>gauge transformation</u>, generated by the (infinitesimal) gauge function $\eta^\lambda(x)$, to be compared with Yang-Mills isospin transformations, generated by gauge functions $\Lambda^a(x)$.

For the derivative of A(x) we have

$$\partial_\mu A'(x) = \partial_\mu A(x) + \eta^\lambda{}_{,\mu}\,\partial_\lambda A(x) + \eta^\lambda \partial_\lambda \partial_\mu A(x), \tag{2.4}$$

where $\eta^\lambda_{,\mu}$ stands for $\partial_\mu \eta^\lambda$, the usual convention. Any object A_μ transforming the same way, i.e.

$$A'_\mu(x) = A_\mu(x) + \eta^\lambda_{,\mu} A_\lambda(x) + \eta^\lambda \partial_\lambda A_\mu(x) , \tag{2.5}$$

will be called a covector. We shall also have contravectors $B^\mu(x)$ (note that the distinction is made by putting the index upstairs), which transform like

$$B^{\mu\prime}(x) = B^\mu(x) - \eta^\mu_{,\lambda} B^\lambda(x) + \eta^\lambda \partial_\lambda B^\mu(x) , \tag{2.6}$$

by construction such that

$$A_\mu(x) B^\mu(x)$$

transforms as a scalar. Similarly, one may have tensors with an arbitrary number of upper and lower indices.

Finally, there will be density functions $\omega(x)$ that transform like

$$\omega'(x) = \omega(x) + \partial_\lambda[\eta^\lambda(x) \omega(x)] . \tag{2.7}$$

They enable us to write integrals of scalars

$$\int \omega(x) A(x) d_4(x) ,$$

which are completely invariant under local gauge transformations (under certain boundary conditions).

For the construction of a complete gauge theory it is of importance that the gauge transformations form a group. Of course they do, and hence we have a Jacobi identity. Let $u(i)$ be the gauge transformations generated by $\eta^\mu(i,x)$. Then if

$$[u(1), u(2)] = u(3) ,$$

then

$$\eta^\mu(3,x) = \eta^\lambda(2,x) \partial_\lambda \eta^\mu(1,x) - \eta^\lambda(1,x) \partial_\lambda \eta^\mu(2,x) . \tag{2.8}$$

3. THE METRIC TENSOR

In much the same way as in a gauge field theory[6], we ask for a dynamical field that fixes the gauge of the vacuum by having a non-vanishing vacuum expectation value. (Contrary to the Yang-Mills case it seems to be impossible to construct a reasonable "symmetric" theory.) To this end we choose a two-index field, $g_{\mu\nu}(x)$, which is symmetric in its indices,

$$g_{\mu\nu} = g_{\nu\mu} \ , \tag{3.1}$$

and its vacuum expectation value is

$$\langle g_{\mu\nu}(x)\rangle_0 = \delta_{\mu\nu} \ , \tag{3.2}$$

(our metric corresponds to a purely imaginary time coordinate).

We wish to argue that this is probably the only possible way to break the gauge symmetry, which implies that the "graviton", which is usually described by the field $g_{\mu\nu}$, cannot be decomposed into more elementary constituents (just like the photon): we must have a tensor [scalars and vectors do not break the symmetry sufficiently*)], and it may not have more than two indices, otherwise there will be too many time components with the wrong sign for metric or energy density. The time components of $g_{\mu\nu}$ can be transformed away by gauge transformations.

With $g_{\mu\nu}$, or its inverse, $g^{\mu\nu}$, we can now define lengths and time-intervals at each point in space-time:

$$|\ell|^2 = g_{\mu\nu}\,\ell^\mu \ell^\nu \ ,$$

and $g_{\mu\nu}$ can be used to raise or lower indices:

$$A^\mu = g^{\mu\nu} A_\nu \ , \quad A_\nu = g_{\nu\mu} A^\mu \ , \ \text{etc.} \tag{3.3}$$

Just as in the Yang-Mills case, we can now define covariant derivatives:

$D_\mu A = \partial_\mu A$ (the derivative of a scalar transforms as a vector),

---

*) One might object that the so-called "Vierbein field", necessary for introducing fermions, is a vector field. But the Vierbein field theory is physically equivalent to the theory described in these notes.

$$D_\mu A_\nu = \partial_\mu A_\nu - \Gamma^\alpha{}_{\mu\nu} A_\alpha ,$$
$$D_\mu B^\nu = \partial_\mu B^\nu + \Gamma^\nu{}_{\mu\alpha} B^\alpha . \tag{3.4}$$

The field $\Gamma^\alpha_{\mu\nu}$ is called the Christoffel symbol and is yet to be defined. First we write down how it should transform under a gauge transformation, such that the above-defined covariant derivatives be real tensors:

$$\Gamma^{\lambda\prime}_{\mu\nu} = \Gamma^\lambda{}_{\mu\nu} + (\text{ordinary terms for 3-index tensor}) + \partial_\mu \partial_\nu \eta^\lambda . \tag{3.5}$$

We see that no harm is done by making the restriction that

$$\Gamma^\lambda{}_{\mu\nu} = \Gamma^\lambda{}_{\nu\mu} , \tag{3.6}$$

because the symmetric part of Γ only is enough to make (3.4) covariant.

We now define the field Γ by requiring

$$D_\lambda g_{\mu\nu} = 0 , \tag{3.7}$$

(from which follows: $D_\lambda g^{\mu\nu} = 0$). We see that we have exactly the right number of equations. By writing Eq. (3.7) in full we find that it is easy to solve

$$\Gamma^\lambda{}_{\mu\nu} = \tfrac{1}{2} g^{\lambda\alpha} (\partial_\mu g_{\alpha\nu} + \partial_\nu g_{\alpha\mu} - \partial_\alpha g_{\mu\nu}) . \tag{3.8}$$

Note that $\Gamma^\lambda_{\mu\nu}$ is not a covariant tensor.

Allowing for one more derivative one can construct a covariant tensor, called Riemann or curvature tensor (see the standard textbooks):

$$R^\alpha{}_{\beta\gamma\delta} = \Gamma^\alpha{}_{\beta\delta,\gamma} - \Gamma^\alpha{}_{\beta\gamma,\delta} + \Gamma^\alpha{}_{\tau\delta} \Gamma^\tau{}_{\beta\gamma} - \Gamma^\alpha{}_{\tau\gamma} \Gamma^\tau{}_{\beta\delta} . \tag{3.9}$$

The comma denotes ordinary differentiation. Indices can be raised or lowered following (3.3). Without putting in any further dynamical equation, one finds the following identities,

$$R_{\alpha\beta\gamma\delta} = R_{\gamma\delta\alpha\beta} = -R_{\alpha\beta\delta\gamma} ,$$
$$R_{\alpha\beta\gamma\delta} + R_{\alpha\gamma\delta\beta} + R_{\alpha\delta\beta\gamma} = 0 , \tag{3.10}$$
$$R_{\alpha\beta\gamma\delta;\mu} + R_{\alpha\beta\delta\mu;\gamma} + R_{\alpha\beta\mu\gamma;\delta} = 0 .$$

The semicolon denotes covariant differentiation. Further, we define

$$R_{\mu\nu} = R^{\alpha}{}_{\mu\nu\alpha} \,, \qquad R = R_{\mu\nu} g^{\mu\nu} \,,$$
$$G_{\mu\nu} = R_{\mu\nu} - \tfrac{1}{2} R g_{\mu\nu} \,, \tag{3.11}$$

which satisfy, according to (3.10):

$$R_{\mu\nu} = R_{\nu\mu} \,, \qquad D_{\mu} G_{\mu\nu} = 0 \,. \tag{3.12}$$

The metric tensor also enables us to define a density function [see (2.7)],

$$\omega(x) = \sqrt{\det(g_{\mu\nu}(x))} \,. \tag{3.13}$$

In the quantum theory we shall encounter a fundamental problem: instead of $g_{\mu\nu}$ we could go over to a new metric $g'_{\mu\nu}$ with, for instance,

$$g'_{\mu\nu} = f_1(R) g_{\mu\nu} + f_2(R) R_{\mu\nu} \,. \tag{3.14}$$

So in a curved space there is some arbitrariness in the choice of metric (Section 6). We bypass this problem here

4. DYNAMICS

The question now is whether we can make the fields $g_{\mu\nu}$ propagate. Indeed we can, because we can construct a gauge invariant action integral

$$S = -\frac{c^2}{16\pi K} \int \omega R \, d_4 x \,, \tag{4.1}$$

where K is to be identified with the usual gravitational constant:

$$K = 6{,}67 \cdot 10^{-11} \ m^3 \, kg^{-1} \, sec^{-2} \,. \tag{4.2}$$

For simplicity we shall take the units in which

$$\frac{c^2}{16\pi K} = 1 \tag{4.3}$$

At a later stage one could put K back in the expressions to find that the expansion in numbers of closed loops will correspond to an expansion with respect to K.

One can also add other fields in the Lagrangian, for instance

$$\mathcal{L} = \omega\{-R - \tfrac{1}{2}g^{\mu\nu}\partial_\mu\varphi\,\partial_\nu\varphi - \tfrac{1}{2}m^2\varphi^2\} , \tag{4.4}$$

where ϕ is a scalar field. We shall not repeat here the usual arguments to show that variation of the Lagrangian (4.4) really leads to the familiar gravitational interactions between masses, and to unfamiliar interactions between objects with a great velocity ("gravitational magnetism"). The equation for the gravitational field will be

$$G_{\mu\nu} = -\tfrac{1}{2} T_{\mu\nu} , \tag{4.5}$$

where $T_{\mu\nu}$ is the usual energy-momentum tensor (Einstein's equation). The action (4.1) has much in sommon with the action

$$-\tfrac{1}{4} G^a_{\mu\nu} G^a_{\mu\nu}$$

in Yang-Mills theories. As we shall indicate in the next section, a massless graviton with helicity ±2 will propagate. Notice that we have been led to Einstein's theory of gravity almost automatically. It seems to be the simplest choice if we ask for a theory with invariance under general coordinate transformations.

5. QUANTIZATION

The first thing we must do is make a shift

$$g_{\mu\nu} = \delta_{\mu\nu} + A_{\mu\nu} , \tag{5.1}$$

and consider $A_{\mu\nu}$ as the quantum fields. Here the problem mentioned in the introduction presents itself: what if we start with

$$g^{\mu\nu} = \delta^{\mu\nu} + B^{\mu\nu} \quad ? \tag{5.2}$$

The answer is that as long as we take as our elementary field any local function of the $g_{\mu\nu}$, the obtained physical amplitudes will be the same. The transformation from one function (for example, $g_{\mu\nu}$) to the other (for example, $g^{\mu\nu}$) will be accompanied by a Jacobian, or closed loops of fictitious particles (see our lectures on gauge theories). But the propagators of these particles are constants or pure polynomials in k, because the transformation is local. If we now turn on the dimensional regularization procedure, which has to be used in order to get gauge invariant results, then the integrals over polynomials,

$$\int d^n k \quad Pol(k) \, ,$$

vanish[7]. This is the reason why it makes no difference whether we start from Eq. (5.1) or Eq. (5.2). Non-local functions of $g_{\mu\nu}$ are not allowed. These non-local transformations would give rise to fictitious particles that do contribute in the cutting rules[1], and they are outlawed once unitarity has been established for the choice (5.1) or (5.2). By choosing a convenient gauge, comparable with the Coulomb gauge in QED, it is indeed not difficult to establish that the theory is unitary, in a Hilbert space with massless particles with helicity ±2.

We can work out the bilinear part of the Lagrangian in Eq. (4.1) in terms of the fields $A_{\mu\nu}$:

$$\mathcal{L} = -\tfrac{1}{4}(\partial_\mu A_{\alpha\beta})^2 + \tfrac{1}{8}(\partial_\mu A_{\alpha\alpha})^2 + \tfrac{1}{2}L_\mu^2 + \text{higher orders,} \tag{5.3}$$

with

$$L_\mu = \tfrac{1}{2}\partial_\mu A_{\alpha\alpha} - \partial_\alpha A_{\mu\alpha} \, . \tag{5.4}$$

For practical calculations it seems to be convenient to choose the gauge

$$\mathcal{L}^C = -\tfrac{1}{2}L_\mu^2 \tag{5.5}$$

Just in order to show the divergent character of the complications involved, we show here the Faddeev-Popov ghost for this gauge, obtained by subjecting L_μ to an infinitesimal gauge transformation:

$$\mathcal{L}^{F.-P} = -\partial_\alpha\varphi^*_\mu \partial_\alpha\varphi^\mu + \varphi^*_\mu\big[A_{\lambda\alpha,\alpha}\varphi^\lambda{}_{,\mu} + A_{\mu\lambda}\varphi^\lambda{}_{,\alpha\alpha} + A_{\mu\lambda,\alpha}\varphi^\lambda{}_{,\alpha} + A_{\mu\alpha,\alpha\lambda}\varphi^\lambda + A_{\mu\alpha,\lambda}\varphi^\lambda{}_{,\alpha} - \tfrac{1}{2}A_{\alpha\alpha,\lambda}\varphi^\lambda{}_{,\mu} - \tfrac{1}{2}A_{\alpha\alpha,\mu\lambda}\varphi^\lambda - A_{\lambda\alpha,\mu}\varphi^\lambda{}_{,\alpha}\big]. \tag{5.6}$$

The Lagrangian (4.1), expanded in powers of $A_{\mu\nu}$, with the gauge-fixing term (5.5) and the ghost term (5.6), form a perfect quantum theory. It is, however, more complicated than necessary, because gauge invariance is given up right in the beginning by adding the bad terms (5.5) and (5.6). The background field method[4,5,8] is much more elegant because gauge invariance is exploited in all stages of the calculations, thus simplifying things a lot. By using this method one can also prove the following. All infinities of the theory can be split up into gauge-invariant terms and gauge-dependent quantities, such that the gauge-dependent parts merely describe field renormalizations. Renormalization of physical quantities, such as observable couplings and masses, can always be written in a gauge-invariant way. It is especially these infinities in which we are interested. In the dimensional regularization procedure the infinities present themselves as poles at $n = 4$. We now have calculated the one-loop infinities in a gravity model.

6. THE ONE-LOOP INFINITIES

The problems consisted in calculating the residues of the (single) poles at $n = 4$ for one-loop diagrams. The dimension of these residues can be obtained by simple power counting. It is four, independent of the number of external lines, which means that it is to be cancelled by counter terms with four derivatives in the Lagrangian. In a theory of pure gravity, the only invariant terms with four derivatives are

$$\begin{aligned} &\omega R_{\alpha\beta\gamma\delta} R^{\alpha\beta\gamma\delta} , \\ &\omega R_{\mu\nu} R^{\mu\nu} , \\ \text{and} \quad &\omega R^2 . \end{aligned} \tag{6.1}$$

Now, owing to a very remarkable accident, first shown by DeWitt[4)], the combination

$$\int \omega \, d_4 x \left\{ R_{\alpha\beta\gamma\delta} R^{\alpha\beta\gamma\delta} + R^2 - 4 R_{\mu\nu} R^{\mu\nu} \right\} , \tag{6.2}$$

only depends on the values of $g_{\mu\nu}$ at the boundary, and thus does not contribute as a Lagrangian. So we can remove the first term from formula (6.1).

We found that for pure gravity in our gauge, the necessary counter-Lagrangian has the form

$$\frac{1}{8\pi^2 (n-4)} \omega \left[\frac{1}{120} R^2 + \frac{7}{20} R_{\mu\nu} R^{\mu\nu} \right] . \tag{6.3}$$

In conventional field theory one now would argue that such R^2 and $R_{\mu\nu}R_{\mu\nu}$ terms do not occur in the original Lagrangian, so the counter term (6.3) cannot be obtained by renormalizing the original Lagrangian. Our theory is infinite or depends on an infinite number of parameters. We wish to stress here that this argument is false. There are two equivalent ways of seeing this.

i) Any infinitesimal addition of terms such as (6.3) to the Lagrangian does not change the physical content of our theory, up to the order considered, simply because the first-order Lagrange equations read

$$\begin{aligned} R_{\mu\nu} &= 0 \\ R &= 0 , \end{aligned} \tag{6.4}$$

so that formula (6.3) vanishes.

ii) The counter term (6.3) can be absorbed (or obtained) by a field renormalization of an unusual type:

$$g_{\mu\nu} \to g_{\mu\nu} + \alpha R_{\mu\nu} + \beta R g_{\mu\nu} \tag{6.5}$$

The Lagrangian in terms of the new $g_{\mu\nu}$ reads

$$\mathcal{L} \to \mathcal{L} + \frac{\delta \mathcal{L}}{\delta g_{\mu\nu}} (\alpha R_{\mu\nu} + \beta R g_{\mu\nu}) , \tag{6.6}$$

with

$$\frac{\delta \mathcal{L}}{\delta g_{\mu\nu}} = -\omega G^{\mu\nu} = \omega(-R^{\mu\nu} + \tfrac{1}{2} R g^{\mu\nu}) \tag{6.7}$$

α and β can be fixed to yield formula (6.3).

Note, however, that the external source terms in $\mathcal{L}$ might be sensitive to the replacement (6.5). [In the presence of sources (6.4) also does not hold.] This is the origin of our remark at the end of Section 3. It is impossible to tell in advance whether the sources couple to $g_{\mu\nu}$ or to $g'_{\mu\nu}$ in Eq. (3.14). But this corresponds to the unobservability of external line renormalizations in the S-matrix.

Things change drastically if we add matter fields in our Lagrangian, as in Section 4. As an exercise we tried

$$\mathcal{L}^{inv} = \omega \left\{ -R - \tfrac{1}{2} g^{\mu\nu} \partial_\mu \varphi \, \partial_\nu \varphi \right\} , \tag{6.8}$$

where ϕ is a massless Klein-Gordon field interacting with gravity. Now we must replace the equations of motion (6.4) by (4.5), here

$$R_{\mu\nu} = -\frac{1}{2}\partial_\mu\varphi\,\partial_\nu\varphi$$

and (6.9)

$$D^2\varphi = 0 .$$

Because of the new interactions, formula (6.3) is now replaced by a much more lengthy expression. After insertion of the equations of motion (6.9), we now find that one term survives. The required counter term is

$$\Delta\mathcal{L} = \frac{1}{8\pi^2(n-4)}\,\frac{203}{80}\,\omega R^2 . \tag{6.10}$$

In this case the above arguments apply to show that this model is really unrenormalizable; Eq. (6.10) does not vanish on-mass-shell and cannot be absorbed by a field renormalization.

7. OUTLOOK AND CONCLUSION

We conclude that we understand how to calculate the one-loop quantum corrections in pure gravity, and that we have shown that all infinities for physical processes cancel there. An interesting result is that the metric tensor $g_{\mu\nu}$ is no longer unique; it is contaminated with admixtures of $R_{\mu\nu}$ and $Rg_{\mu\nu}$. The conjecture that the two- and many-loop corrections are of a similar nature is plausible, but we have no proof. An extremely preliminary investigation of the two-loop infinities, which would have dimension six, yields as a result that even after insertion of the classical equations of motion one term could survive:

$$\omega\, R_{\alpha\beta\gamma\delta}\, R_{\gamma\delta\mu\nu}\, R_{\mu\nu\alpha\beta} . \tag{7.1}$$

As yet, we have not been able to establish an equation like (6.2) which would be necessary in order to eliminate this term. We have not been sufficiently motivated to tackle this problem further.

The addition of matter will in general make the theory definitely unrenormalizable, even at the one-loop level. We tried to add several other interaction terms, such as

$$a\,\omega R\varphi^2 , \tag{7.2}$$

to the Lagrangian (6.8), but instead of one infinity, (6.10), we then obtain a huge series of infinite terms, all of the unrenormalizable type. Only a miraculous cancellation, or a miraculous symmetry, could restore renormalizability.

Even if we could manage to obtain eventually a renormalizable theory (pure gravity gives a good chance), we are still far from getting answers to the questions mentioned in the introduction. The expansion with respect to the number of loops corresponds to a long-distance approximation, as one can easily show. It is a challenge to build a small-distance theory (small compared to the fundamental length!), without the wildest speculations.

* * *

REFERENCES

1) G. 't Hooft and M. Veltman, "DIAGRAMMAR", CERN 73-9 (1973).

2) D.H. Capper and G. Leibbrandt, Nuovo Cimento Letters 6, 117 (1973).
D.H. Capper, G. Leibbrandt and M. Ramón Medrano, ICTP Trieste preprint 73/26, April 1973.

3) R.P. Feynman, Acta Phys. Polon. 24, 697 (1963).
S. Mandelstam, Phys. Rev. 175, 1580, 1604 (1968).
E.S. Fradkin and I.V. Tyutin, Phys. Rev. D2, 2841 (1970).
C.J. Isham, A. Salam and J. Strathdee, Phys. Rev. D3, 867 (1971), Nuovo Cimento Letters 5, 969 (1972).

4) B.S. DeWitt, *in* Relativity, Groups and Topology, Summerschool of Theoretical Physics, Les Houches, France, 1963 (Gordon and Breach, New York, London).
B.S. DeWitt, Phys. Rev. 162, 1195 and 1239 (1967).

5) G. 't Hooft and M. Veltman, CERN preprint TH 1723 (1973).

6) G. 't Hooft, Nuclear Phys. B35, 167 (1971).

7) See Ref. 2 or Appendix B of Ref. 1.

8) J. Honerkamp, Nuclear Phys. B48, 269 (1972); Proc. Conf. on "Renormalization of Yang-Mills Fields and Applications to Particle Physics", Marseille 1972 (CNRS, Marseille, 1972).

SELF-CONSISTENT QUANTUM FIELD THEORY AND SYMMETRY BREAKING

H. Umezawa

Department of Physics
University of Wisconsin-Milwaukee
Milwaukee, Wisconsin 53201

This lecture consists of two parts. In part I, we present a study of spontaneously broken symmetries in the terminology of quantum field theory. Though many of our results in this part help our physical and intuitive understanding of subjects and are also useful for practical computations, some of them are derived by means of certain approximations. A rigorous derivation of these results are presented in part II in which use was made of the path-integral method. The latter method is very useful for formal analysis of problems of broken symmetries.

1.1 The Spontaneous Breakdown of Symmetries

The aim of this lecture is to summarize results of an application of quantum field theory to systems which do not phenomenologically manifest the symmetries possessed by their basic equations. Many such phenomena are widely known in solid state physics. For example, the crystals do not possess translational invariance though the Hamiltonian of molecular gas is translationally invariant. In ferromagnetism we observe a similar situation concerned with the rotational invariance, and in superconductivity and superfluidity we are concerned with the phenomenological disappearance of phase invariance, and so on.

A remarkable feature of these phenomena is that the original symmetry is not simply lost in observation but is replaced by very systematic and stable "new symmetries": periodic nature in crystals, full polarization in ferromagnets, etc. It is natural to expect that appearance of new symmetries is a manifestation of the original invariance of microscopic theory: due to the original invariance the original symmetry cannot simply be destroyed, but is replaced by a symmetry of a different kind.

Another interesting feature of the systems under consideration is the appearance of certain Bose quanta which originate various long range correlations: phonon in crystals, magnon in ferromagnets, etc. For example, in the case of crystals, the phonon is the one which correlates the motion of a large number of molecules in such a way that the molecules are prevented from going out of their stationary positions (i.e. the lattice points). These bosons have a primary importance in observable phenomena, because external stimuli of small energy can only excite these boson levels: for example, small external stimuli on a crystal simply excite molecular vibrations (i.e. excitation of phonons). This lecture aims at showing what kind of role these bosons play in the course of replacing the original symmetries by phenomenological ones.

Since the crystal molecules in observation are under the control of long range correlations (induced by the above mentioned bosons), their behavior differs very much from that of molecules which are not influenced by this correlation. The energy spectrum of these molecules can, in principle, be determined by means of certain external stimulus of a relatively large energy. To excite several molecules and bosons mentioned above, we need to supply the system with the energy which is equal to the sum of excitation energies of these molecules and bosons. This situation is similar to the one in the particle physics: in experiments of particle reactions the energy of the object is given by the sum of energies of incoming particles which are measured separately: the energy thus defined is the one which conserves through particle reactions. In this sense we say that the observed particles are free particles. They are called the quasi-particles in the solid state physics, while we may use incoming particles as observed particles in particle physics. Although these observed particles are free in the sense that they do not have any interaction energy, they can perform various reactions in collisions.

The boson introduced previously is a member of quasi particles. In the case of crystal, phonon is not a modified form of original molecule: the original molecule is changed into the quasimolecule. According to the terminology of quantum field theory which will be introduced in the next section, the crystal phonon is a bound state of many molecules.

1.2 The Dynamical Rearrangement of Symmetries

We have seen in the previous section that we have two languages to describe the systems under consideration: one is to describe phenomena in terms of "free" physical particles such as the quasi-particles (i.e. phonons, quasi-molecules, quasi-electrons, etc.) or the incoming particles, and another is to express the basic equations in terms of original basic fields. Let ϕ stand for physical fields associated with the physical particles and ψ for the basic fields. The latter is usually called the Heisenberg field.

We can then understand the phenomenological disappearance of original symmetry in the following way. Note first that the original symmetry is expressed in terms of Heisenberg fields ψ while the phenomenological symmetry is associated with the physical fields ϕ. In one word, the disappearance of original symmetry means that the symmetry associated with ψ differs from that associated with ϕ. Let us now examine this statement in detail. We first ask: What do we mean by the symmetry associated with ψ? To answer this we recall the existence of a set of basic equations for ψ. Let us call these equations the Heisenberg equations and denote it by

$$F(\psi) = 0 . \tag{1}$$

These equations can be a set of differential equations (as in usual field theory) or can be a set of algebraic relations (as in the field theory of currents), or something else. Suppose a mapping

$$\psi \rightarrow \psi' = Q(\psi) \tag{2}$$

under which the Heisenberg equations (1) are invariant:

$$F(\psi') = 0 . \tag{3}$$

We then state that the Heisenberg equations possess the symmetry associated with the mapping (2). We shall now turn our attention to the physical fields ϕ. Since the physical particles associated with ϕ are "free" in the sense that their energies are determined as certain functions (say E_k) of their momenta, ϕ must satisfy a set of linear and homogeneous equations. These equations will be denoted by

$$\lambda(\partial)\phi = 0 . \tag{4}$$

Let us now recall that the observed excitation energies are the energies of the physical particles. Adapting the statement in the quantum mechanics that the states which appear in observations belong to certain Hilbert space, <u>we require that the Hilbert space is the Fock space associated with the physical particles</u>. Let us denote this Fock space by $\mathcal{H}$. To construct $\mathcal{H}$, we shall introduce the annihilation operators of physical particles (and their anti-particles): these operators will be respectively denoted by $\alpha_{\vec{k}}^{(r)}$ and $\beta_{\vec{k}}^{(r)}$ where $\vec{k}$ signifies the particle or antiparticle-momentum and r the helicity states. For the sake of simplicity, we simply denote these operators by $\alpha_{\vec{k}}$ and $\beta_{\vec{k}}$, ignoring the helicity superscript r.

$$[\alpha_{\vec{k}}, \alpha^{\dagger}_{\vec{\ell}}]_{\pm} = \delta(\vec{k}-\vec{\ell}), \text{ etc.} \tag{5}$$

We shall introduce an orthonormalized complete set of square-integrable functions $\{f_i(\vec{k})\}$:

$$\int d^3k\ f_i^*(\vec{k}) f_j(\vec{k}) = \delta_{ij} . \tag{6}$$

As is well known, this set is a countable one. We then introduce the annihilation operators for particles in wave-packet states:

$$\alpha_i = \int d^3\vec{k}\ f_i(\vec{k})\alpha_{\vec{k}} , \tag{7}$$

$$\beta_i = \int d^3\vec{k}\ f_i(\vec{k})\beta_{\vec{k}} . \tag{8}$$

These satisfy

$$[\alpha_i, \alpha_j^{\dagger}]_{\pm} = \delta_{ij}, \text{ etc. } . \tag{9}$$

Using α_i and β_i and following the well known steps, we can build the Fock space $\mathcal{H}$. The vacuum state in $\mathcal{H}$ will be denoted by $|0\rangle$:

$$\alpha_i\ |0\rangle = 0, \quad \beta_i\ |0\rangle = 0 \tag{10}$$

The physical field ϕ is then defined by

$$\phi(x) = \int \frac{d^3k}{(2\pi)^{3/2}} [u(\vec{k})\alpha_{\vec{k}}\ e^{i\vec{k}\vec{x}-iE_k t} + v(\vec{k})\beta_{\vec{k}}\ e^{-i\vec{k}\vec{x}+iE_k t}] , \tag{11}$$

where $u(\vec{k})$ and $v(\vec{k})$ are the wave functions satisfying (4):

$$\lambda(iK)u(\vec{k}) = 0 \tag{12}$$

$$\lambda(-iK)v(\vec{k}) = 0 \tag{13}$$

where $K_\mu = (\vec{k}, iE_k)$. Then, any operator which appears in our consideration must be made of ϕ and $\phi^{\dagger}$. In particular, ψ must be an operator which consists of ϕ and $\phi^{\dagger}$:

$$\psi = \psi(\phi) \tag{14}$$

We shall later present a more detailed study of (14). Let us now suppose that a mapping

$$\phi \to \phi' = q(\phi) \tag{15}$$

induces the mapping in (2):

$$\psi(\phi') = \psi' . \tag{16}$$

The phenomenological symmetry is the one associated with the mapping (15). When the mapping q in (15) has a form different from Q in (2), the original symmetry associated with Q is lost in observation and is replaced by the phenomenological symmetry associated with q. We then say that the Q-symmetry is dynamically rearranged into the q-symmetry.[1,2]

Eq. (14) conditions the Heisenberg operators ψ in two ways: firstly, ψ should consist of the physical fields ϕ, and secondly ψ must satisfy the Heisenberg equations (1). It is natural to expect that the invariance of the Heisenberg equations under the Q-transformation in (2) restricts the choices for possible forms of the dynamical rearrangement of symmetry (i.e. the choices for the q-transformation). This will be illustrated by several examples in section (1.4).

Let us now study (14) more carefully. To do this we shall introduce a basis of $\mathcal{H}$ by the set

$\{|\alpha_1 \ldots\rangle\}$ defined by

$$|\alpha_i \ldots\rangle = c\, \alpha_i^\dagger \ldots |0\rangle . \tag{17}$$

Here c is the normalization constant, and α_i stand for α_i or β_i. We then require that all the matrix elements $\langle a|\psi|b\rangle$ are well-defined when $|a\rangle$ and $|b\rangle$ belong to the above set. This requirement can be expressed as[1]

$$\psi(x) = \chi + c\phi(x) + \int d^4\xi \int d^4\zeta\, F(x-\xi, x-\zeta) :\phi(\xi)\phi(\zeta): + \ldots \tag{18}$$

Here ϕ stands for both ϕ and $\phi^{\dagger}$. The dots denote those terms which contain the higher order normal products of ϕ. χ is zero unless ψ is a boson field. The symbol $c\phi$ means $\sum_i c_i\phi_i$ when there are many kinds of physical fields. Since we require the existence only of the matrix elements $\langle a|\psi|b\rangle$, equality in (18) means only the equality of matrix elements of both sides of (18) (i.e. the weak equality). In the following arguments in part I, equality always means the weak equality unless otherwise stated. It is of essential importance that the states $|a\rangle$ and $|b\rangle$ in $\langle a|\psi|b\rangle$ are wave-packet states: otherwise, the matrix elements of left-hand side of (18) might not be defined when the domains for the time-integration extend to ∞ (or $-\infty$). The expansion (18) is called the dynamical map of ψ.

Since ψ is defined by the weak relation (18), the field equation (1) must be read as a weak relation, too. In other words, the equality in the Heisenberg equation (1) is the weak equality. To know how to read the Heisenberg equation, we must know also how to calculate the products of ψ. We shall adopt the following rule for the product:

$$\begin{aligned}
&[\int d^4\xi \int d^4\zeta\ F_1(x-\xi,\ x-\zeta)\ :\phi(\xi)\phi(\zeta):] \\
&\times[\int d^4\xi' \int d^4\zeta'\ F_2(y-\xi',\ y-\zeta')\ :\phi(\xi')\phi(\zeta')] \\
&= \int d^4\xi \int d^4\zeta \int d^4\xi' \int d^4\zeta'\ F_1(x-\xi,\ x-\zeta)F_2(y-\xi',\ y-\zeta') \\
&\quad :\phi(\xi)\phi(\zeta):\ :\phi(\xi')\phi(\zeta'): \qquad (19)
\end{aligned}$$

When this definition cannot define the product of $\psi(x)$ and $\psi(y)$ at the same point (x=y), we may need certain limiting process of the form

$$\lim_{\varepsilon\to 0} \psi(x)\psi(x+\varepsilon)\ .$$

In this lecture we are not going to elaborate this point any further.

We shall now present several comments on the dynamical map (18). In writing down (18), we assumed that the translational invariance is not phenomenologically broken. We shall show later how to break phenomenologically the translational invariance.

It is obvious that $\hat{\phi}(x) = u^{-1}\phi(x)u$ can also be used as physical fields when u is a unitary operator. Among an infinite number of choices for the physical fields, we shall choose the incoming fields as the physical fields. This choice is established by requiring that the expansion coefficients in (18) are retarded functions:[1,3]

$$F(x-\xi,\ x-\zeta) = \theta(t_x-t_\xi)\theta(t_x-t_\zeta)\overline{F}(x-\xi,\ x-\zeta),\ \text{etc.} \tag{20}$$

Another comment on (18) is that there does not necessarily exist "one-to-one" correspondence between the members of the set, $\{\psi\}$, of Heisenberg fields and those of the set, $\{\phi\}$, of physical fields. To define a correspondence between the Heisenberg fields and physical fields, we shall follow the LSZ-method[4] to define the asymptotic limit of the Heisenberg fields. To illustrate the asymptotic limit, we shall simplify the situation by assuming ϕ to be a scalar field. Then, defining

$$a_i(t) = i\int d^3x\ u_i^*(x)\ \overleftrightarrow{\frac{\partial}{\partial t}}\ (\psi(x)-\chi) \tag{21}$$

with

$$u_i(x) = \int \frac{d^3k}{(2\pi)^{3/2}}\,\frac{1}{\sqrt{2E_k}}\,f_i(\vec{k})\ e^{i\vec{k}\vec{x}-iE_k t}\ , \tag{22}$$

the asymptotic limits are defined by $\lim(t\to\pm\infty)a_i(t)$. Here it should be recalled that the equality in (18) is the weak equality. (18) gives

$$\lim_{t\to-\infty} a_i(t) = c\alpha_i \tag{23}$$

We shall write this as

$$\text{asymptotic limit } (t\to-\infty) \text{ of } \psi = c\phi \ . \tag{24}$$

Suppose that there are physical fields which do not correspond to the asymptotic limit ($t\to-\infty$) of any member of the set, $\{\psi\}$. We then state that these physical particles are composite.[1,5] We then look for a product of $\psi(x)$, the asymptotic limit ($t\to-\infty$) of which is a composite physical field. When we need at least n number of ψ to form such a product, we call the composite particle a n-body bound state.

This concept of compositeness has been successfully applied to the deuteron.[6,7] The experimental data for low energy neutron-proton scattering and the experimental value of deuteron binding energy have led to the conclusion that the deuteron is composite according to the above definition of compositeness. In this consideration use was made of the approximation in which the off-energy-shell transition matrix is approximately replaced by the on-shell transition matrix: this approximation seems reasonable in view of the fact that the deuteron is a loosely bound system.

We shall close this section by an important comment on the Hamiltonian. Since the observable values of energies are given by summing the energies of physical particles, the energy operator is the Hamiltonian of the physical fields:

$$H_o(\phi) = \int d^3k \; E_k(\alpha^\dagger_{\vec{k}}\alpha_{\vec{k}} + \beta^\dagger_{\vec{k}}\beta_{\vec{k}}) \ . \tag{25}$$

Making use of the relation

$$\frac{\partial}{\partial t}\,\phi(x) = i[H_o,\ \phi(x)] \tag{26}$$

in which equality is the strong equality, we can derive from (18) the relation[1]

$$i\langle a|\,[H_o,\ \psi(x)]\,|b\rangle = \langle a|\ \frac{\partial}{\partial t}\,\psi(x)\,|b\rangle \ . \tag{27}$$

Let us now denote the Hamiltonian of Heisenberg fields by H. Then (27) shows

$$[H, \psi(x)] = [H_o, \psi(x)] \tag{28}$$

where equality is the weak one. This relation can be extended as

$$[H, \text{product of } \psi(x)] = [H_o, \text{product of } \psi(x)] \ . \tag{29}$$

In derivation of (27) we used once the integration in part, the use of which is justified when we recall that $|a\rangle$ and $|b\rangle$ are wave-packet states.

Another way of deriving (27) is to calculate the Fourier form of matrix elements of right hand side of (18) by considering (20). For example we have

$$\langle\beta_i|\psi(x)|\alpha_j\rangle = \int\frac{d\omega_1}{2\pi}\int\frac{d\omega_2}{2\pi}\int\frac{d^3k^{(1)}}{(2\pi)^{3/2}}\int\frac{d^3k^{(2)}}{(2\pi)^{3/2}}\,\overline{F}(-\vec{k}^{(1)},\omega_1;\,\vec{k}^{(2)},\omega_2)$$

$$v(\vec{k}^{(1)})f_i^*(\vec{k}^{(1)})u(\vec{k}^{(2)})f_j(\vec{k}^{(2)})\ e^{i(-\vec{k}^{(1)}+\vec{k}^{(2)})\vec{x}}$$

$$\frac{e^{i(E_1-E_2)t_x}}{(E_1+\omega_1-i\varepsilon)(E_2-\omega_2+i\varepsilon)}$$

$$+\ (v\leftrightarrow u,\ f_i^*\leftrightarrow f_j,\ \omega_1\to-\omega_1,\ \omega_2\to-\omega_2)$$

where $\overline{F}(-\vec{k}^{(1)},E_1,\vec{k}^{(2)},E_2)$ is the Fourier component of $F(x-\xi_1,x-\xi_2)$ in (20), and E_1 and E_2 mean respectively $E_{k^{(1)}}$ and $E_{k^{(2)}}$. The time dependence of the above expression explicitly shows that (27) is satisfied, because E_1 and E_2 are the eigenvalues of H_o in the bra- and ket-states respectively.

1.3 The Self-Consistent Method

As soon as we are given a set of Heisenberg equations, we find ourselves in a dilemma: we do not really know how to read the Heisenberg equations, because we do not know anything about the Hilbert space $\mathcal{H}$. According to our consideration in the previous section the Hilbert space must be a Fock space of the physical fields, which can be identified only after we solve the Heisenberg equations. We can recognize a similar dilemma in the Lehmann-Symanzik-Zimmermann formalism[4] too: there, the incoming fields are established as an asymptotic (weak) limit of the Heisenberg fields, while to perform the weak limit requires a knowledge of the Hilbert space which is the Fock space associated with the incoming fields. A root of this dilemma is the existence of infinite number of irreducible representations (of an infinite set of canonical variables) which are unitarily inequivalent to each other:[8] we do not have such a dilemma in quantum mechanics with finite degrees of freedom.

The above dilemma is a kind of problem which is usually resolved by a self-consistent consideration. In the case under consideration, we are concerned with the self-consistency between the Heisenberg equations and the choice of the Fock space of physical fields.[9] We catch a glimpse of the self-consistent method already in the renormalization theory, in which mass parameters which are to be determined are introduced by the requirement that the physical particles are stable.

The self-consistent approach proceeds as follows:[1] we first prepare, as a candidate for the set of physical fields, a set of free fields by appealing to various physical considerations. Regarding these free fields as physical incoming fields and leaving their energy spectra unknown, we write the Heisenberg fields ψ in the expansion form (18). (1) then leads to a set of equations for the expansion coefficients. By solving these equations we determine the expansion (18) together with the energy spectra of the physical particles. When these equations do not admit any solution, we modify the initial set of free physical fields and repeat the computations. Such a modification of the initial set of physical fields is frequently made by introducing more free fields. This is the way many composite particles are successively brought into the set of physical particles.[1] As an example let us assume a field equation for the nucleon Heisenberg field. We then choose, as the initial set of quasiparticle fields, an isodoublet free Dirac field which is regarded as the physical nucleon field. Writing the nucleon Heisenberg field in the form (18) in terms of normal products of physical nucleons and then inserting this expression into the Heisenberg field equation (and the canonical equal-time commutator) for the nucleon Heisenberg field, we obtain a set of equations for the expansion coefficients. We may then find that these equations do not have a solution unless they are modified by introducing a new member in the initial set of physical fields. This new member may turn out to be the physical deuteron, which then is regarded as a composite particle. As was pointed out in the previous section, this approach has led to a very reasonable result for the low-energy neutron-proton scattering when the experimental value of deuteron mass was used.[6,7]

As was pointed out previously, the need for the self-consistency condition lies in the fact that, in quantum field theory, there exist infinitely many Fock spaces which are unitarily inequivalent to each other. By the same reason, there is nothing which guarantees that application of the self-consistent method to a given Heisenberg equation leads to a unique choice of the physical fields.

It is quite instructive to demonstrate the full course of the self-consistent method by means of a solvable model such as the Nθ-model.[10] The Nθ-model is a model which consists of the N- and θ-Heisenberg fields with an interaction of the form $N^\dagger\theta^\dagger\theta N$. The Heisenberg equations in the Nθ-model are invariant under the transformations $N\to e^{i\alpha_1}N$ and $\theta\to e^{i\alpha_2}\theta$. We first assume two physical fields, N^{in} and θ^{in}. For simplicity we restrict our attention to a particular solution of the following property: the transformations, $N^{in}\to e^{i\alpha_1}N^{in}$ and $\theta^{in}\to e^{i\alpha_2}\theta^{in}$ induce $N\to e^{i\alpha_1}N$ and

$\theta \to e^{i\alpha_2}\theta$ (i.e. the phase symmetry is not dynamically rearranged.) Then it can be shown that the self-consistent method cannot be completed unless we introduce another physical field V^{in} which is a ($N\theta$)-composite particle.[5,10,11] It can be shown also that $N^{in} \to e^{i\alpha_1}N^{in}$, $\theta^{in} \to e^{i\alpha_2}\theta^{in}$, $V^{in} \to e^{i(\alpha_1+\alpha_2)}V^{in}$ induce $N \to e^{i\alpha_1}N$ and $\theta \to e^{i\alpha_2}\theta$.

A remarkable result given by the $N\theta$-model is the relation[10]

$$H = H_o + W_o \tag{30}$$

where W_o is a c-number, H_o is the free Hamiltonian for N^{in}, θ^{in} and V^{in} and H is the Hamiltonian for the Heisenberg field N and θ. Briefly speaking, (30) is derived as follows: Since use of the self-consistent method determines the expression (i.e. the dynamical map (18)) of N and θ in terms of N^{in}, θ^{in} and V^{in}, we can express H (which consists of N and θ) in terms of N^{in}, θ^{in} and V^{in}. The result turns out to be the relation* (30). It is very instructive to see an explicit computation how the free Hamiltonian of the composite particle V^{in} emerges from the Hamiltonian of the Heisenberg fields N and θ. Since the equality in (18) is the weak one, so is (30).

The relation (30) can be derived quite generally from (29) when $\mathcal{H}$ is an irreducible representation for ψ satisfying the basic equation (1).

As is well-known, the perturbation theory with the adiabatic trick gives another way of deriving (30). A defect of this is that the composite particles cannot be created at any finite order term in perturbative expansion. The states of free physical (non-composite) particles are used as the unperturbed states and the adiabatic factor $\exp(-\varepsilon|t|)$ is introduced in the interaction. We shall further assume that the volume, V, of the system is finite and make use of the periodicity condition. Then, the perturbative

*This relation has been derived in many other cases[12,13,14,15,16] by explicit calculations which employed certain approximations.

calculation gives[9] (in the limit $\varepsilon\to 0$)

$$H = H_o + Q + W_o \tag{31}$$

where Q is a matrix of the following nature:

$$\langle i|Q|j\rangle = Q_{ij}\ \delta_{W_i,W_j}\ \delta(\vec{p}_i-\vec{p}_j)\ . \tag{32}$$

Here $|i\rangle$ and $|j\rangle$ are the eigenstates of H_o:

$$H_o|i\rangle = W_i|i\rangle\ .$$

The symbol $\vec{p}_i$ in (32) denotes the total momentum in the state $|i\rangle$. Unless the condition $\vec{p}_i = \vec{p}_j$ naturally leads to $W_i = W_j$, the condition $W_j = W_i$ is a restriction in addition to the condition $\vec{p}_i = \vec{p}_j$ and makes the matrix element $\langle a|Q|b\rangle$ as small as $(1/V)^{1/3}$ when $|a\rangle$ and $|b\rangle$ are normalized states of particles. On the other hand, $\vec{p}_i = \vec{p}_j$ naturally leads to $W_i = W_j$ only in the case of one-particle transitions, which are induced by $\alpha^\dagger_{\vec{p}}\ \alpha_{\vec{p}}$ or by $\beta^\dagger_{\vec{p}}\ \beta_{\vec{p}}$. However, since E_k in H_o is the physical energy of a single-particle, Q should not create any self-energy, implying that Q contains no terms of the form $\alpha^\dagger\alpha$ or $\beta^\dagger\beta$. We thus find[9] that

$$\langle a|Q|b\rangle = O(1/L) \qquad (L = V^{1/3})\ , \tag{33}$$

where O(1/L) stands for quantities, magnitude of which is of the order of 1/L. In the limit $V\to\infty$, (31) <u>becomes</u> (30). Let us now calculate the S-matrix by using Q as the interaction. In the first order approximation we find

$$S = 1 + \frac{1}{ih}\int dt\ Q \tag{34}$$

Since the reaction occurs in the domain of volume V, the range of time-integration in (34) is of the order L. (34) shows that (S-1) can remain unvanishing in the limit $L\to\infty$, while Q itself vanishes according to (33). This consideration supplies us with an intuitive reason as to why the scattering of "free" physical particles are possible, even though the Hamiltonian, H, acts as the free

Hamiltonian H_o.

Calculation of the S-matrix in the self-consistent method is the same as that in the L-S-Z formalism.[4] The outgoing field can be obtained by studying the asymptotic limit ($t\to+\infty$) of $\psi-\chi$, in which ψ is expressed in the form of dynamical map (18).

We can use the relation (30) as a condition for the Fock space of physical fields.[9] In other words, one way of selecting the Hilbert space is to require that H becomes a free Hamiltonian.

As an important comment, we shall prove[13] that W_o in (30) is an extreme value of vacuum expectation value of H. This statement means the following: Suppose that we have a Fock space of physical particles, for which (30) is satisfied:

$$H = \int d^3k\ E_k\ \alpha_k^\dagger \alpha_k + W_o \tag{35}$$

Let us now move to a nearby Fock space in which the annihilation operator is α_k':

$$\alpha_k' = \alpha_k + \delta\alpha_k \ .$$

The change, $\delta\alpha_k$, is infinitesimal. H can be rewritten as

$$H = \int d^3k\ E_k(\alpha_k'^\dagger \alpha_k' - \alpha_k'^\dagger \delta\alpha_k - \delta\alpha_k^\dagger \alpha_k') + W_o \ .$$

Let us denote the vacuum in the new Fock space by $|0'\rangle$:

$$\alpha_k'|0'\rangle = 0, \quad \langle 0'|\alpha_k'^\dagger = 0 \ .$$

We then find

$$\langle 0'|H|0'\rangle = W_o = \langle 0|H|0\rangle \tag{36}$$

where $|0\rangle$ is the vacuum associated with α_k. (36) proves that W_o is an extreme value. This shows a similarity between the self-consistent quantum field theory and the variational method. However, the condition (30) may not exhaust all possible extreme values of vacuum expection value of H.

Closing this section we shall introduce a theorem which has been useful in many-body problems. To introduce the theorem, let us suppose a Bose infield ϕ_B^{in} satisfying a free field equation

$$\lambda(\partial)\phi_B^{in} = 0 .$$

The expression of the Heisenberg field ψ in terms of infields is denoted by $\psi(\phi^{in}, \phi_B^{in})$ in which ϕ^{in} stands for all the infields except ϕ_B^{in}. Then a transformation $\phi_B^{in} \to \phi_B^{in} + G(x)$ with

$$\lambda(\partial)G(x) = 0$$

induces a transformation

$$\psi(\phi^{in}, \phi_B^{in}) \to \psi(\phi^{in}, \phi_B^{in} + G(x)) .$$

The statement of the theorem is that $\psi(\phi^{in}, \phi_B^{in} + G(x))$ satisfies the Heisenberg equation for $\psi(\phi^{in}, \phi_B^{in})$. We shall present a proof for this in Part II by using the path-integral method.

1.4 Examples of Dynamical Rearrangement of Symmetry

In this section we illustrate the dynamical rearrangement of symmetries by means of several models. In these considerations the expansion coefficients in the dynamical map (18) are calculated by means of the Bethe-Salpeter equations which can be derived from the Heisenberg field equations when the canonical equal-time commutation relations for the Heisenberg fields are assumed. The pair approximation (or the random phase approximation in the many-body problem) is employed in the course of the self-consistent calculation. A formal derivation of dynamical rearrangement of symmetries without use of any approximation will be presented in Part II.

Example 1. the Goldstone model[17]

The Heisenberg equation is

$$(\Box - m_o^2)\psi = \lambda(\psi^\dagger\psi)\psi \tag{37}$$

where ψ is a complex scalar field. This equation is invariant under the phase transformation

$$\psi \to e^{i\theta}\,\psi. \tag{38}$$

The current defined by

$$j_\mu = \psi^\dagger \overleftrightarrow{\partial}_\mu \psi \tag{39}$$

conserves

$$\partial_\mu j_\mu = 0\ . \tag{40}$$

We apply the self-consistent method to this model by requiring

$$\langle 0|\psi|0\rangle = c \neq 0 \tag{41}$$

where $|0\rangle$ is the vacuum of the Fock space, $\mathcal{H}$, of physical incoming fields. The result[1,18] shows that there exist at least two kinds of Hermitian physical fields, B^{in} and ϕ^{in}, among which B^{in} is massless:

$$\left.\begin{aligned} (\Box - m^2)\phi^{in} &= 0 \\ \Box\, B^{in} &= 0 \end{aligned}\right\} \tag{42}$$

The dynamical map (18) has the form

$$\psi = e^{i\frac{1}{\eta}B^{in}}\, F(\phi^{in},\ \partial B^{in})\ , \tag{43}$$

where η is a c-number constant which is proportional to c in (41), and $F(\phi^{in}, \partial B^{in})$ is a quantity which consists of ϕ^{in} and ∂B^{in}. The symbol ∂B^{in} stands for B^{in} carrying any power of space- or time-derivatives.

(43) shows that the transformation

$$\left.\begin{aligned} B^{in} &\to B^{in} + \eta\theta \\ \phi^{in} &\to \phi^{in} \end{aligned}\right\} \tag{44}$$

induces the transformation, (38), of the Heisenberg field ψ. This is the dynamical rearrangement of the phase symmetry. Note that ϕ^{in} does not undergo any transformation (44): ϕ^{in} is frozen. On the other hand, B^{in} shifts itself by the c-number $\eta\theta$. In other words B^{in} changes the density of the Bose condensation. (44) implies that the phase transformation of ψ is taken care of by a regulation of Bose condensation of the massless physical particles B^{in}.

Due to the massless nature of B^{in}, the free field equations, (42), are invariant under the transformation (44). This is a manifestation of the phase invariance of the Heisenberg equation. This can be understood as follows. The free field equations (42) are the results of the requirement that ψ satisfies the Heisenberg equation (37). Now suppose the transformation $\phi^{in} \to \phi^{in'}$ and $B^{in} \to B^{in'}$ which changes ψ into $\psi' = e^{i\theta}\psi$. Since ψ' also satisfies (37), we see that $\phi^{in'}$ and $B^{in'}$ must satisfy (42): in other words, a transformation of physical fields which induces an invariant transformation of Heisenberg fields keeps the free field equations of physical fields unchanged. This statement is true quite generally.

Making use of the expression of ψ in terms of ϕ^{in} and B^{in} (i.e. the dynamical map), we can express the current j_μ in terms of the physical fields. The result(1,18) is

$$j_\mu = j_\mu^F + j_\mu^B \tag{45}$$

with

$$j_\mu^F = (\delta_{\mu\nu} - \frac{\partial_\mu \partial_\nu}{\Box}) J_\nu(\phi^{in}, \partial B^{in}) \tag{46}$$

and

$$j_\mu^B = \eta\, \partial_\mu\, B^{in} \tag{47}$$

Here $J_\nu(\phi^{in}, \partial B^{in})$ is a certain quantity which consists of ϕ^{in} and

∂B^{in}. Note that the existence of the factor $(\delta_{\mu\nu} - \partial_\mu \partial_\nu/\Box)$ in j^F guarantees that j_μ^F conserves:

$$\partial_\mu j_\mu^F = 0 \ . \tag{48}$$

The boson current j_μ^B conserves because B^{in} is massless:

$$\partial_\mu j_\mu^B = 0 \ . \tag{49}$$

(48) and (49) shows that j_μ conserves as it should do.

Taking advantage of the existence of the factor $(\delta_{\mu\nu} - \partial_\mu \partial_\nu/\Box)$ in j_μ^F, we can prove that

$$\int d^3x \ j_4^F(x) = 0 \ . \tag{50}$$

Let us now note the relation

$$[D, \psi(x)] = \psi(x) \tag{51}$$

which is a result of the canonical commutation relation. Here

$$D = i\int d^3x \ j_4(x) \ . \tag{52}$$

(51) and (41) lead us to conclude that the vacuum is not an eigenstate of D. The same conclusion is obtained also from the relation

$$D = \eta \int d^3x \ \frac{\partial}{\partial t} \ B^{in} \ , \tag{53}$$

which is obtained from (45), (47) and (50). It is well known that it does not make any sense to talk about such D because all the matrix elements, $\langle a|\exp[i\theta D]|b\rangle$, vanish when the volume of the system is infinite. In other words, $\exp[i\theta D]$ does not map $\mathcal{H}$ on $\mathcal{H}$ itself. This is the reason why in our considerations we have not talked about the generator D. However it is instructive to study D_f defined by

$$D_f = i\int d^3x \ f(\vec{x}) j_4(x) \tag{54}$$

in which $f(\vec{x})$ is a square-integrable function because $\exp[i\theta D_f]$ is a sensible object. We then find

$$\lim_{f\to 1} e^{i\theta D_f} \psi e^{-i\theta D_f} = e^{i\theta} \psi \tag{55}$$

and also

$$\lim_{f\to 1} e^{i\theta D_f} \phi^{in} e^{-i\theta D_f} = \phi^{in} \tag{56}$$

$$\lim_{f\to 1} e^{i\theta D_f} B^{in} e^{-i\theta D_f} = B^{in} + \eta\theta \ . \tag{57}$$

These results coincide with the relations in (44).

Using the expressions of ψ written in terms of ϕ^{in} and B^{in}, we can express the Hamiltonian, H, in terms of the physical fields. The result shows that H is the free Hamiltonian of the physical fields, confirming explicitly the general statement in the previous section:

$$H = H_o + W \ . \tag{58}$$

Let us close this section by a comment on the asymptotic limit of ψ. Since $F(\phi^{in}, \partial B^{in})$ in (43) has the form

$$F(\phi^{in}, \partial B^{in}) = C + Z^{\frac{1}{2}} \phi^{in} + \ldots \tag{59}$$

with a real positive constant $Z^{\frac{1}{2}}$, (43) gives

$$\left.\begin{aligned} \psi_1 &= C + Z^{\frac{1}{2}} \phi^{in} + \ldots \\ \psi_2 &= \frac{C}{\eta} B^{in} + \ldots \end{aligned}\right\} \tag{60}$$

where dots stand for normal products of higher orders. ψ_1 and ψ_2 are respectively the real and imaginary part of ψ:

$$\psi = \psi_1 + i\psi_2 \ . \tag{61}$$

(60) shows that $(\psi_1 - C)$ and ψ_2 respectively approach $Z^{\frac{1}{2}}\, \phi^{in}$ and $\frac{C}{\eta}\, B^{in}$ in the limit $t \to -\infty$.

In the above consideration for the Goldstone model, we meet some divergent integrals. We treated these integrals by means of a Lorentz-invariant cut-off.[18,19]

Example 2. The Nambu model for a Dirac field ψ

The Heisenberg equation[20] is

$$\gamma\partial\psi = -\lambda\{(\bar{\psi}\psi)\psi - (\bar{\psi}\gamma_5\psi)\gamma_5\psi\} \tag{62}$$

This is invariant under the chiral-phase transformation:

$$\psi \to e^{i\theta\gamma_5}\,\psi\ . \tag{63}$$

The conserving current which is associated with the invariance is

$$j_\mu = i\bar{\psi}\gamma_\mu\gamma_5\psi\ . \tag{64}$$

We apply the self-consistent method to this model by requiring

$$\langle 0|\bar{\psi}\psi|0\rangle = C \neq 0\ . \tag{65}$$

The result[1,12] is very similar to the one in the case of the Goldstone model. We find two physical fields ϕ^{in} and B^{in} satisfying

$$\left.\begin{aligned} (\gamma\partial+m)\phi^{in} &= 0 \\ \Box B^{in} &= 0 \end{aligned}\right\} \tag{66}$$

The dynamical map has the form

$$\psi = e^{i\frac{1}{\eta}B^{in}\gamma_5}\, F(\phi^{in},\ \partial B^{in}) \tag{67}$$

which should be compared with (43). All the arguments from (44) through (58) hold in the case of the Nambu model too. Thus the chiral phase symmetry for ψ is dynamically rearranged into (44) for the physical fields. The chiral phase invariance is preserved in the sense that the free equations (66) for the physical fields are invariant under the transformation (44), which shows that the chiral phase transformation is taken care of solely by a regulation of Bose condensation of the massless field B^{in}. It is remarkable that ϕ^{in} becomes unchanged under the chiral phase transformation so as to preserve the invariance even after ϕ^{in} acquires a mass. The current j_μ splits into two terms, j_μ^F and j_μ^B, which are respectively of the forms (46) and (47). The Hamiltonian becomes the free Hamiltonian for ϕ^{in} and B^{in}, when expressed in terms of physical fields. The divergent integrals which appear in the calculation of the dynamical map were treated by means of a Lorentz-invariant cut-off.[12]

Example 3. A scale-invariant model for a real scalar field ψ

The Heisenberg equation is

$$\Box \psi(x) = \lambda \psi^3(x) \tag{68}$$

which is invariant under the scale transformation:

$$\psi(x) \to \psi'(x) = e^{\theta} \psi(e^{\theta} x) \tag{69}$$

Here θ is the transformation parameter. Our task is to solve (68) under the condition

$$\langle 0|\psi^2(x)|0\rangle = C \neq 0 . \tag{70}$$

We do not use the condition $\langle 0|\psi(x)|0\rangle \neq 0$ in order to avoid breaking of the invariance of (68) under $\phi \to -\phi$.

Calculations[21] show that there appear two kinds of physical fields, one of which is massless:

$$\left.\begin{aligned} (\Box - m^2)\phi^{in}(x,m) &= 0 \\ \Box B^{in}(x) &= 0 \end{aligned}\right\} \tag{71}$$

The dynamical map[21] has the form

$$\psi(x) = F[m(1+\frac{1}{\eta}B^{in}), \exp(\frac{1}{\eta}B^{in} m\frac{\partial}{\partial m})\phi^{in}, \text{ their derivatives}] \tag{72}$$

where F is a functional with the dimension of inverse of length. Note that

$$\exp(\frac{1}{\eta} B^{in}(x)\ m \frac{\partial}{\partial m})\phi^{in}(x,m) = \phi^{in}(x,m(1+\frac{1}{\eta} B^{in})) \tag{73}$$

In (72) and (73), the mass derivative, $\partial/\partial m$, is so defined that it does not operate on creation- and annihilation operators (α, $\alpha^{\dagger}$, β, $\beta^{\dagger}$). Let us now consider the transformation

$$\left.\begin{aligned} &B^{in}(x) \rightarrow e^{\theta}\ B^{in}(e^{\theta}x) + \eta(e^{\theta}-1) \\ &\phi^{in}(x,m) \rightarrow e^{\theta}\ \phi^{in}(e^{\theta}x,\ e^{-\theta}m) \end{aligned}\right\} \tag{74}$$

This induces the following changes:

$$m(1+\frac{1}{\eta} B^{in}(x)) \rightarrow e^{\theta}m\ (1+\frac{1}{\eta} B^{in}(e^{\theta}x)) \tag{75}$$

$$\exp(\frac{1}{\eta} B^{in}(x) m \frac{\partial}{\partial m})\phi^{in}(x,m) \rightarrow e^{\theta}\exp(\frac{1}{\eta} B^{in}(e^{\theta}x) m \frac{\partial}{\partial m})\phi^{in}(e^{\theta}x,m) \tag{76}$$

Since the dimensions of $m(1+\frac{1}{\eta} B^{in})$ and $\exp(\frac{1}{\eta} B^{in}\ m \frac{\partial}{\partial m})\phi^{in}$ are both the inverse of length, the changes in (75) and (76) transform F in (72) as

$$F \rightarrow [e^{\theta}F]_{x\rightarrow e^{\theta}x}\ .$$

We thus see[(21)] that the transformation, (74), of the physical fields induce the scale transformation, (69), of the Heisenberg field. This is the dynamical rearrangement of the scale symmetry.

Since, in the transformation for ϕ^{in} in (74), both x and m change according to their dimensions, the free field equation for ϕ^{in} in (71) is invariant under the transformation (74). Further, the free field equation for B^{in} in (71) is also invariant under the transformation (74). As was pointed out previously, these invariances are a trace of scale invariance of the Heisenberg equations.

When a transformation changes all the dimensional quantities according to their dimensions, it is called the dimensional transformation.[22] Note that any equation is invariant under the dimensional transformation due to the dimensional consistency. The transformation of ϕ^{in} in (74) is the dimensional transformation. This is the reason why the equation for ϕ^{in} is invariant even after ϕ^{in} acquired a mass.

Let us note also that the transformation (74) is regulating the Bose condensation of B^{in}.

For an infinitesimal θ, (69) and (74) take the following form:

$$\delta\psi = \theta(1+(x\cdot\partial))\psi \tag{77}$$

$$\left.\begin{aligned} \delta\phi^{in} &= \theta(1+(x\cdot\partial)-m\frac{\partial}{\partial m})\phi^{in} \\ \delta B^{in} &= \theta(1+(x\cdot\partial))B^{in}+\eta \end{aligned}\right\} \tag{78}$$

where $\delta\psi$, $\delta\phi^{in}$ and δB^{in} are the changes of ψ, ϕ^{in} and B^{in} respectively.

The conserving current, j_μ, associated with the scale invariance can be constructed by means of the Callen-Coleman-Jackiw method:[23]

$$j_\mu = \theta_{\mu\nu}\, x_\nu \; . \tag{79}$$

Here $\theta_{\mu\nu}$ is the energy-stress tensor satisfying

$$\partial_\mu\theta_{\mu\nu} = 0 \;, \qquad \theta_{\mu\mu} = 0 \tag{80}$$

and has the form

$$\theta_{\mu\nu} = T_{\mu\nu} - \frac{1}{6}(\Box\delta_{\mu\nu} - \partial_\mu\partial_\nu)\,\psi^2$$

with

$$T_{\mu\nu} = -\partial_\mu\psi\,\partial_\nu\psi + \frac{1}{2}\delta_{\mu\nu}(\partial_\lambda\psi\,\partial_\lambda\psi + \frac{1}{2}\psi\Box\psi) \; .$$

When we define

$$D_f = i\int d^3x\, f(\vec{x})j_4(x) \;, \tag{81}$$

this gives

$$\lim_{f\to 1} [D_f, \psi] = i(1+(x\cdot\partial))\psi \ . \tag{82}$$

On the other hand, by expressing D_f in terms of physical fields, we can derive[21]

$$\left.\begin{aligned} \lim_{f\to 1} [D_f, \phi^{in}] &= i(1+(x\cdot\partial) - m\frac{\partial}{\partial m})\phi^{in} \\ \lim_{f\to 1} [D_f, B^{in}] &= i(1+(x\cdot\partial)B^{in} + i\eta \end{aligned}\right\} \tag{83}$$

These relations correspond to the relations in (78).

Calculations show that j_μ has a term which is linear in $\partial_\mu B^{in}$ and which induces the c-number term, η, in (83). This term in the current has the form $(1/3)\eta x_\nu \partial_\mu \partial_\nu B^{in}$ which contributes to D_f by the term

$$D_f^B = \eta \int d^3x\ f(\vec{x})\ \frac{\partial}{\partial t}\ B^{in} + \ldots$$

Here the dots denote those terms which do not contribute to (83) in the limit $f\to 1$.

Let us comment on the commutator

$$\lim_{f\to 1} [H, D_f] = iH \tag{84}$$

where H is the Hamiltonian for the Heisenberg operators. Denoting the Hamiltonian for the physical fields by H_o we are led[21] to the relation (30):

$$H = H_o + W_o \tag{85}$$

However, the relations in (83) lead to

$$\lim_{f\to 1} [H_o, D_f] = i(1-m\frac{\partial}{\partial m})H_o \tag{86}$$

the form of which is different from (84). This is not a contradiction, because the relation (85), in which the equality is the weak equality, cannot always be trusted in commutation relations $[H, A]$ or $[H_o, A]$ when A is a space integration of a local quantity. This is because the matrix elements of $H-H_o$ are of the order $(1/L)$ where L is the linear size of the world considered. In ref. (24), we have shown how (84) can be derived from (86) by calculating the different $H-H_o$. (86) shows why the ϕ^{in} can acquire a nonvanishing mass even when the Heisenberg equations are scale invariant. At this point, it may be worthwhile to recall a theorem which states that a scale invariant theory permits only vanishing mass or a continuous mass-spectrum. Although such a continuous mass spectrum can be recognized in a Hilbert space which is much larger than each Fock space, our selection in the Fock space selects one mass-value out of the continuous spectrum: the mass value can change when we choose a different Fock space. This explains why the result of our calculation tells only that $m \neq 0$: our result cannot specify the value of m. This is not a difficulty, because whatever may be the value of m, we can use it as the unit of mass. The difference in forms (84) and (86) is an example in which the algebra is changed in the course of dynamical rearrangement of symmetries.

In the calculation of scale-invariant model, we meet certain divergent integrals. Unlike in the previous models, we cannot eliminate these divergences by introducing a cut-off energy, which is not dimensionless. Therefore, we have eliminated the divergence in the following way.[21] Noticing that the divergent integrals always carry the bare coupling constant λ, we performed the limit $\lambda \to 0$ in such a way that the product of λ and divergent integrals remain finite. It should be noted that such a limiting process does not eliminate particle reactions: in other words, the S-matrix is not unity even in the limit $\lambda \to 0$. This is because this limit eliminates only the vertices which are not connected with any line of the composite particle (B^{in}). The particles reactions can be mediated by the composite particles.

Example 4. The superconductivity [13,25,26,27]

In this case we are concerned with the dynamical rearrangement of the phase symmetry, as we were in Example 1. However, the situation here is different from that in Example 1, because the electrons are influenced by Coulomb interactions.

In the case under consideration ψ denotes the Heisenberg field for electrons:

$$\psi = \begin{pmatrix} \psi_{\uparrow} \\ \psi_{\downarrow} \end{pmatrix} \tag{87}$$

The Heisenberg equation is phase-invariant and gauge-invariant. Here the phase transformation is defined by

$$\psi \to e^{i\theta} \psi , \tag{88}$$

while the gauge transformation is

$$\left.\begin{aligned} \psi &\to e^{i\lambda(x)} \psi \\ A_{\mu} &\to A_{\mu} + i\partial_{\mu} \lambda(x) \end{aligned}\right\} \tag{89}$$

where the gauge $\lambda(x)$ stands for any real function. We look for a solution of the Heisenberg equation by requiring

$$<0|\psi_{\uparrow}(x)\psi_{\downarrow}(x)|0> \neq 0 \tag{90}$$

Calculations show that, besides the quasi electron ϕ^{in}, there exists a Bose field B^{in} and a gauge field χ. The expression of ψ in terms of physical fields takes the form[13,25]

$$\psi = e^{i\frac{1}{\eta}(B^{in}+\chi)} F(\phi^{in}, u_{\mu}) \tag{91}$$

Here u_{μ} is

$$\left.\begin{aligned} \vec{u} &= \vec{A}^{T} - \frac{1}{e\eta} \vec{\nabla} B^{in} \\ u_{o} &= \frac{e\eta}{\nu^{2}} \pi^{in} \end{aligned}\right\} \tag{92}$$

where π^{in} is the canonical conjugate of B^{in}, $\vec{A}^T$ is the transverse part of the vector potential and ν is the plasma frequency.

The free field equations for ϕ^{in}, B^{in} and π^{in} are

$$\left.\begin{aligned} &(\frac{\partial}{\partial t} + iE(\vec{\nabla}^2))\phi^{in} = 0 \\ &(\frac{\partial^2}{\partial t^2} - \nu^2)\vec{\nabla}^2 B^{in} = 0 \\ &\frac{\partial}{\partial t}\pi^{in} = \frac{\nu^2}{e^2\eta^2}\vec{\nabla}^2 B^{in} \end{aligned}\right\} \qquad (93)$$

where $E(\vec{\nabla}^2)$ is a derivative operator made of $\vec{\nabla}^2$. There is no equation for χ: this manifests the gauge-invariance.

The second equation in (93) admits not only oscillating solutions with the plasma frequency, but also any constant as its solution.(13,25,26) It is due to this reason that (93) is invariant under the transformation(13,25)

$$\left.\begin{aligned} &\phi^{in} \to \phi^{in}, \ \chi \to \chi, \ \pi^{in} \to \pi^{in} \\ &B^{in} \to B^{in} + \eta\theta \end{aligned}\right\} \qquad (94)$$

(91) shows that (94) is the transformation which induces the phase transformation $\psi \to e^{i\theta}\psi$ (the dynamical rearrangement of phase symmetry). The gauge transformation $\psi \to e^{i\lambda(x)}\psi$ is taken care of by

$$\left.\begin{aligned} &\phi^{in} \to \phi^{in}, \ \pi^{in} \to \pi^{in}, \ B^{in} \to B^{in} \\ &\chi(x) \to \chi(x) + e\eta\lambda(x) \end{aligned}\right\} \qquad (95)$$

When we express the electric current j_μ in terms of physical fields, it takes the form

$$j_\mu = j^F_\mu(\phi^{in}, u_\mu) + j^B_\mu \qquad (96)$$

$$j^B_\mu(x) = -\nu^2 C(\vec{\nabla}^2)\, u_\mu(x) \qquad (97)$$

where $C(\vec{\nabla}^2)$ is a derivative operator satisfying $C(0) = 1$, and in (96), j_μ^F is a certain conserving current which is not a linear term of physical fields and which satisfies

$$\partial_\mu j_\mu^F = 0, \qquad \int d^3x\, j_o^F = 0, \qquad <0|j_\mu^F|0> = 0\ . \tag{98}$$

On the other hand, use of the second equation in (93) leads to

$$\partial_\mu u_\mu = 0 \tag{99}$$

which shows that j_μ^B is also conserved:

$$\partial_\mu j_\mu^B = 0 \tag{100}$$

Making use of (92), (97) and (98), we obtain

$$eD = -\int d^3x\, j_o = -\, e\eta \int d^3x\, \pi^{in} \tag{101}$$

which is indeed the generator both for the transformation (94) for physical fields and for the phase transformation of ψ.

When $|0>$ is the ground state, we can put the vector potential in the form

$$\vec{A} = \vec{A}^T + \vec{\nabla}\chi, \qquad A_o = \frac{\partial\chi}{\partial t}\ .$$

Then the Maxwell equation gives (the Meissner equation)

$$(\Box - \nu^2\, C(\vec{\nabla}^2))\vec{A}^T = 0 \tag{102}$$

where uses were made of the relations $<0|j_\mu|0> = 0$, $<0|B^{in}|0> = 0$ and $<0|\pi^{in}|0> = 0$. (92), (102) and (93) show that u_μ-quanta of low momentum has the plasma energy ν. It is remarkable that there is no quanta of gapless energy in this case and still the equation for B^{in} is invariant under the transformation $B \to B + \eta\theta$ due to the existence of $\vec{\nabla}^2$ in the second equation (i.e. the equation for B^{in}).*

Let us close our consideration of superconductivity by a comment on the "boson transformation" by which we can create "electric currents" in a superconducting state. The idea of boson transformation is based on the theorem presented at the end of the previous section: when a physical boson field satisfies a free field equation

$$\lambda(\partial)\ B^{in} = 0\ , \tag{103}$$

a transformation

$$B^{in}(x) \rightarrow B^{in} + G(x) \tag{104}$$

with

$$\lambda(\partial)G(x) = 0 \tag{105}$$

induces a transformation

$$\psi \rightarrow \psi' \tag{106}$$

under which the Heisenberg equation is invariant. The transformation (104) is called the boson transformation.

Though a superconductor can show various phenomenological behaviors according to different boundary conditions, these phenomenological situations differ from each other by different ways of condensation of B^{in}-quanta. Furthermore, all the phenomenological situations must result from a same Heisenberg equation for ψ. Since the boson transformations with all possible solutions of (105) cover all possible ways of condensation of B^{in}-quanta which do not modify the Heisenberg equation, we are led to the conclusion that all the phenomenological situations can be covered by

*The fact that there exists no quantum of gapless energy is well-known (e.g. 26, 13, 25, 27, 28, 29). However, it is less known that the equation for B^{in} admits any constant as its solution. This was explicitly shown in articles in ref. (13, 25, 29). In the article in ref. 26, it was pointed out that all the low levels except the zero energy level moves up to the plasma level.

the boson transformations.

In the case of superconductivity, the second equation in (93) shows that (105) reads as the Laplace equation, $\vec{\nabla}^2 G = 0$, when we confine our attention to phenomena which do not change as fast as plasma oscillation (cf. (93)). When we write as $G(x) = \eta\bar{f}(x)$ (with $\bar{f}(x)$ satisfying $\vec{\nabla}^2\bar{f} = 0$), (91) shows that $\bar{f}(x)$ is a space-dependent phase. (92) together with (97) shows that the boson transformation (104) induces the current.

$$<0|\vec{j}^B(x)|0> = \frac{\nu^2}{e} C(\vec{\nabla}^2)\vec{\nabla}\bar{f}(x) \tag{107}$$

which modifies the Maxwell equation (102) as

$$(\Box - \nu^2 C(\vec{\nabla}^2))\vec{A}^T = -\frac{\nu^2}{e} C(\vec{\nabla}^2)\vec{\nabla}\bar{f}(x) \ . \tag{108}$$

This is the basic phenomenological equation for superconductivity.[13]

This illustrates how we can use the boson transformation in the study of phenomena which are not invariant under space-translation. It is worth noting that measurements of neutron scattering by vortex currents in superconductivity has determined the function $C(\vec{\nabla}^2)$ in (108): the result agreed remarkably well with the theoretical expression of $C(\vec{\nabla}^2)$.

Example 5. The Anderson-Higgs-Kibble phenomenon

We shall now study a modified Nambu model in which a pseudo-vector field A_μ is introduced in (62) for the Nambu model. The Heisenberg equations are

$$\left.\begin{aligned} &(\gamma_\mu, \partial_\mu - ieA_\mu\gamma_5)\psi = -\lambda\{(\bar{\psi}\psi)\psi - (\bar{\psi}\gamma_5\psi)\gamma_5\psi\} \\ &[\Box\delta_{\mu\nu} - \partial_\mu\partial_\nu]A_\nu = ej_\mu \end{aligned}\right\} \tag{109}$$

with

$$j_\mu = i\,\bar{\psi}\gamma_\mu\gamma_5\psi \tag{110}$$

Here the vector potential A_μ is not a classical c-number field, but a Heisenberg operator.

The Heisenberg equations in (109) are invariant under the chiral phase transformation

$$\psi \to e^{i\theta\gamma_5}\psi \tag{111}$$

and also under the gauge transformation

$$\left.\begin{aligned} &\psi \to e^{ie\lambda(x)\gamma_5}\psi \\ &A_\mu \to A_\mu + \partial_\mu\lambda \end{aligned}\right\} \tag{112}$$

The situation in this example is very similar to the one in the previous example of superconductivity. The physical fields are a physical fermion ϕ^{in}, a gauge field χ, a massive Proca field ϕ_μ^{in} and a pseudoscalar field B^{in} which is <u>independent of space and time</u>. There is no equation for χ: this is a manifestation of gauge invariance.

The dynamical map takes the form[29]

$$\psi = e^{i\frac{1}{\eta}(B^{in}+\chi)} F(\phi^{in}, \phi_\mu^{in}, \pi^{in}) \tag{113}$$

where π^{in} is the canonical conjugate of B^{in}. The phase transformation is taken care of by $B^{in} \to B^{in} + \eta\theta$, while the gauge transformation by $\chi \to \chi + e\eta\lambda$. We leave details to the articles in ref. (29) and (30).

1.5 Physical Interpretation and Concluding Remarks

In several examples we learned how the symmetries of Heisenberg fields ψ are rearranged into symmetries of the infields. We observed that the symmetry transformation of infields are those under which the free field equations for infields are invariant and that there always appears a boson which regulates its condensation state at its energyless level as $B^{in} \to B^{in} + \eta\theta$. This can be intuitively understood as follows.[31] To create a ground state, $|0\rangle$, which manifests certain systematic patterns (e.g. in ferromagnets), one needs certain quanta condensed in the ground state in such a way that these quanta "print" their quantum numbers (e.g. spin of magnon) on the ground state and thus create the patterns. These quanta need to obey the Bose statistics in order to be able to condense in the ground state. In other words, in the course of dynamical rearrangement there is created a zero energy level of a Bose quanta, and the boson condensation in this level creates a systematic pattern in the ground state. Such a boson is called the Goldstone boson. When there is no Heisenberg field of the Bose type (e.g. in the Nambu model), the dynamics naturally creates the Goldstone boson as a composite particle. When there appears a Bose condensation in which the condensation density is dependent of space or time, the result is a space- or time-dependent phenomena. This is the intuitive meaning of the boson transformation illustrated in the example of superconductivity: for example, an x-dependent current density in superconducting metals is created by the boson transformation $B^{in}(x) \rightarrow B^{in}(x) + \eta f(x)$ in which $f(x)$ is a c-number function satisfying the equation for the Goldstone field B^{in}.

In all the examples of dynamical rearrangement of symmetries in the previous section the infields performed certain linear transformations with the exception of the Goldstone fields which also carried certain c-number constants in their transformations. This is due to the fact that the infields satisfy free field equations which are linear and homogeneous: the linear transformations are the simplest choices for transformations under which the linear and homogeneous equations are invariant. This consideration leads us to the following question:[31] what is the most general set of transformations under which linear and homogeneous equations of free fields are invariant. This question is important because the transformations of free fields as a result of dynamical rearrangement of symmetries must belong to the set.

Let us consider a case in which the above mentioned invariant transformations ($\phi^{in} \to \phi^{in'} = T(\phi^{in})$ with a certain functional T) of free fields ϕ^{in} contains nonlinear terms of higher order than the first. In order to let a product of n ϕ^{in} satisfy the free field equation of ϕ^{in}, this product must carry a projection operator which selects only the energies which are on the single-particle energy shell. To have such a projection operator, which is an δ-function of energy, $T(\phi^{in})$ must contain a time-integration over the entire domain of time (from $-\infty$ to $+\infty$). [An example of such invariant transformation is given by the mapping $\phi^{in} \to \phi^{out}$.] In this article, we do not go into any study of this possibility.

Let us confine our attention to cases in which the invariant transformations of infields are of the form

$$\phi_i^{in} = C_{ij}\, \phi_j^{in} + C_i \tag{114}$$

Here ϕ_i^{in} stands for infields, and C_{ij} and C_i are c-number constants. Let us first assume that $C_i = 0$. Then, the infields must form a set of irreducible representations of a symmetry group associated with the infields, in such a way that there is no mass differences among those which belong to a same representation.(31) As an example, let us consider a mulitiplet of Heisenberg fields ψ_i (i=1, ..., n) which belongs to an irreducible representation of a Lie group symmetry (say G). Let ϕ_i^{in} denote the infield which is the asymptotic limit of ψ_i. There is no reason that these ϕ_i^{in} (i=1, ..., n) cover the whole set of infields: there can appear many composite particles, as we have seen in the previous section. Further, there is no reason(31) that the set, $\{\phi_i^{in}\}$, must be an irreducible representation of G, because the dynamical rearrangement of symmetries permits infields to transform differently from the Heisenberg operators. As was pointed out, when ϕ_i^{in} and ϕ_j^{in} have different masses they belong to different representations. In the past there have been several perturbative calculations of mass differences in SU(3)-invariant models. In these calculations an octet representation, $\{\psi_i\}$, of Heisenberg hadron fields was assumed. The calculations showed that the eight infields, which correspond to the eight Heisenberg fields in the octet representation, do not have a same mass. This result does not imply an appearance of mass differences in an octet representation of the infields: they imply that the eight infields do not belong to a same irreducible representation of the symmetry group (say G^{in}) associated with the infields.

When all the infields transform as in (114) with $C_j = 0$, the generators of the transformation are sums of terms of form $\alpha_i^\dagger \alpha_j$ so that the vacuum is an eigenstate of these generators. This is impossible when a symmetry is spontaneously broken. Therefore, there should exist a set of fields[31] (say B_i^{in}, i=1, ..., k), for which $C_i \neq 0$: they transform as

$$B_i^{in} = C_{ij}\ B_j^{in} + C_i\ , \tag{115}$$

in which not all C_i are zero. This is an invariant transformation when and only when the minimum energy of B_i^{in} is zero. <u>This result is just the Goldstone theorem</u> and the fields B_i^{in} are the Goldstone fields. The above consideration shows[31] also that the Goldstone fields must form an irreducible representation of the form $\{B_i^{in} + b_i\}$ where the quantities b_i are certain c-number constants. In other words, the Goldstone fields transform as

$$B_i^{in} + b_i \rightarrow C_{ij}\ (B_j^{in} + b_j)$$

where C_{ij} denotes the transformation coefficients. This means that B_i^{in} transform as

$$B_i^{in} \rightarrow C_{ij}\ B_j^{in} + (C_{ij}\ b_j - b_i)\ ,$$

in which $(C_{ij}\ b_j - b_i)$ manifests the regulation of Bose condensation. The above argument shows that the set $\{b_i\}$ acts as a spurion: the conservation law of observable quantum number associated with the symmetry under consideration holds when and only when the quantum number of this spurion is taken into account: this is the reason[31] for observable breakdown of conservation law.

Our important question now is to ask how the two groups, G and G^{in}, are related to each other. The situation is very simple when G and G^{in} are equal to each other. When this is the situation in the case of spontaneous breakdown of the chiral SU(2) × SU(2) symmetry, then not only the pions but also the σ-particle must be Goldstone particles. We have seen, however, in the example of the scale invariance that the algebraic relations can be modified by the dynamical rearrangement of symmetries: G^{in} can be different from G. When G^{in} is different from G in the case of dynamical rearrangement of the chiral SU(2) × SU(2) symmetry σ does not need to be a Goldstone particle. As we have seen in the example of the scale invariant model, the difference between the two groups, G and G^{in}, can be induced by infrared effects of the Goldstone quanta. Another way of studying the possibility for $G \neq G^{in}$ is given by the path-integral method (cf. Part II). A careful derivation[33,34] of dynamical rearrangement of symmetries in the path-integral formalism contains certain tricky limiting processes which might make G and G^{in} to be different from each other. It is an interesting question whether or not $G \neq G^{in}$ is possible in the case of spontaneous breakdown of Lie-group symmetries.

PART II. THE PATH-INTEGRAL METHOD

In the previous sections we have studied the dynamical rearrangement of symmetries by using the self-consistent formalism of quantum field theory. Although our study clarified various physical aspects of the spontaneous breakdown of symmetries, our arguments were based on the results obtained by using the pair approximation. In this part II, we shall present a simple and formal derivation of dynamical rearrangement of symmetries by making use of the path-integral method. In section 2.1 we shall show how to apply the path-integral method to problems of spontaneous breakdown of symmetries. In section 2.2, we shall illustrate the use of the path-integral method by studying the Goldstone-type model. Also a very simple derivation of the Goldstone theorem will be presented. In section 2.3 we present a simple derivation of the dynamical rearrangement of phase symmetry. Part II will be closed by several concluding remarks (see section 2.4).

2.1 How to Apply the Path-Integral Method to Problems of Spontaneous Breakdown of Symmetries?

The path-integral method has proved itself very useful for study of the problem of symmetries, because field transformations appear simply as changes of integration variables. However, special care is needed[32] when we look for asymmetric solutions.

As is well known, in the path-integral method, all the Green functions can be derived by repeated uses of functional derivatives acting on certain generating functional which will be denoted by W[J]. For example, when we consider a complex scalar field ψ and denote its Lagrangian density by L(x), the generating functional is given by

$$W[J] = \frac{1}{N}\int[d\psi][d\psi^*]\ \exp\{i\int d^4x[L(x) + J^*(x)\psi(x) + J(x)\psi^*(x)]\} \tag{116}$$

with

$$N = \int[d\psi][d\psi^*]\ \exp\{i\int d^4x\ L(x)\} \tag{117}$$

In the following the symbol N always means the one obtained from the numerator in W[J] by performing the limit $J \to 0$.

When we apply this method to problems of spontaneous breakdown of symmetries, we immediately meet a serious question. To explain how this question arises, we shall consider a case in which L(x) is invariant under the phase transformation $\psi \to e^{i\theta}\psi$.

When we apply the usual formulation of quantum field theory to such a model, our task is to solve the field equations under a condition such as $\langle 0|\psi|0\rangle = C$. It frequently happens that there appear both symmetric ($C = 0$) and asymmetric ($C \neq 0$) solutions. Furthermore, there appear infinite kinds of asymmetric solutions, which differ from each other by phase of C. The question to ask is how the single integration (116) can accommodate many such possibilities.

To settle this question, we first recall that the path-integral, W[J], is mostly contributed by those points (in the ψ-space) at which

$$\int d^4x\ [L + J\psi^* + J^*\psi]$$

is minimum. We shall call these points the stationary points. When there is only one stationary point, the path-integral picks up contribution only of this point. However, the situation is quite different when L(x) is invariant under a transformation. This is because[32] the stationary points move around in a certain domain (say D) of the ψ-space when the invariant transformations are performed. When the invariant transformations are continuous ones, the stationary points may form a line or a surface, etc. When such a situation happens, each of the points in D has a potentiality of being realized in nature, and in this way, we understand why there can appear many solutions from a given Lagrangian which is invariant under certain transformations. However, to realize each of these solutions, we need to modify W[J] in such a way that the domain D shrinks to a point.

As a simple example, consider a massless real scalar field:

$$W[J] = \frac{1}{N}\int [d\psi]\ \exp\{i\int d^4x\ [-\frac{1}{2}\partial_\mu\psi\ \partial_\mu\psi + J\psi] \tag{118}$$

Here $L(x) = -(1/2)\partial_\mu\psi\ \partial_\mu\psi$ is invariant under the translation $\psi \to \psi + \theta$, where θ is a constant. The stationary points are given by the solutions of the equation

$$\Box \psi = - J \ . \tag{119}$$

Since this equation is invariant under the translation, $\psi \to \psi + \theta$, the assembly of stationary points form a line in the ψ-space. This is an open line extending to both sides and the contribution of all the points on this line makes W[J] undetermined.[32] This may be seen as follows. Writing as

$$\psi(x) = \psi_o(x) + \psi'(x) \tag{120}$$

with

$$\psi_o(x) = - \frac{1}{\Box} J$$

which is a solution of (119), we obtain

$$W[J] = \exp \{\frac{i}{2} \int d^4x \int d^4y \ J(x) \ \langle x| \frac{1}{\Box} |y\rangle J(y)\} \tag{121}$$

which gives

$$\langle\psi\rangle = \frac{\delta}{i\delta J} W[J] \Big|_{J=0} = 0 \ . \tag{122}$$

On the other hand, if we use (120) after we change the integration variable $(\psi(x) \to \psi(x) + \theta)$ in W[J], we obtain

$$W[J] = \exp\{i\lambda \int J(x) d^4x\} \exp\{\frac{i}{2} \int d^4x \int d^4y \langle J(x) \langle x| \frac{1}{\Box} |y\rangle J(y)\} \tag{123}$$

which leads to

$$\langle\psi\rangle = \theta \ . \tag{124}$$

This shows that W[J] is ambiguous. To eliminate this ambiguity we shall modify W[J] by introducing a term[32] which breaks the translational symmetry:

$$W[J] = \frac{1}{N} \int [d\psi] \exp\{i \int d^4x [- \frac{1}{2} \partial_\mu \psi \ \partial_\mu \psi + J\psi$$
$$+ \frac{i\varepsilon}{2} [\psi(x) - v]^2] \tag{125}$$

Here v is a constant. The limiting process $\varepsilon \to 0$ should be performed at the end of calculations. It should be noted that N also contains the ε-term. Due to the existence of the ε-term, the domain D is not a line but a point satisfying

$$(\Box + i\varepsilon)\psi = -J + v$$

which is not invariant under the translation $\psi \to \psi + \theta$. The position of this point depends on the value of v. The use of

$$\psi_o = - \frac{J-v}{\Box+i\varepsilon}$$

in (120) leads to

$$W[J] = \exp[iv\int d^4x\ J]\exp[\frac{i}{2}\int d^4x\int d^4y\ J(x)<x|\frac{1}{\Box+i\varepsilon}|y>J(y)] \quad (126)$$

The change $\psi \to \psi + \theta$ in (125) induces the factor $\exp[i\theta\int d^4x\ J]$ and replaces v by $v-\theta$. We thus see that (126) is not modified by the above change $\psi \to \psi + \theta$: W[J] is now well-defined.(32) (126) gives

$$<\psi> = v\ . \quad (127)$$

It is worth noting that the Green function in (126) is now determined to be the Feynman function.

Let us now turn our attention to another example. We shall consider a phase invariance model of a complex scalar field ψ. In this case W[J] is given by (116) with L(x) being invariant under the phase transformation $\psi \to e^{i\theta}\psi$. In this case, since the phase transformation does not change the magnitude of ψ, the stationary points form a closed circle with its center at $\psi = 0$. Thus contributions to $<\psi>$ due to all the points on the circle compensate each other and we obtain(32)

$$<\psi(x)> = \frac{1}{N}\int [d\psi]\psi(x)\ \exp[i\int d^4x\ L] = 0\ . \quad (128)$$

This intuitive argument can be confirmed by the following mathematical consideration. When we change the variable as $\psi \to e^{i\theta}\psi$, then W[J] becomes $W[e^{-i\theta}J]$. This shows that $W[J] = W[e^{-i\theta}J]$ since

W[J] is well defined, implying that W depends on J and J^* only through the combination (J^*J). This gives $\langle\psi\rangle = 0$: in other words, the spontaneous breakdown of phase symmetry does not take place, although W[J] in (128) is well-defined.[32]

To obtain an asymmetric solution, we need an ε-term, which breaks the phase invariance. When such an ε-term is introduced, the domain of the stationary points is no more the circle but a point. Any ε-term which is not invariant under the phase transformation is acceptable for our purpose.[32]

As an example we can use the ε-term of the form $i\varepsilon|\psi(x)-v|^2$. The change $\psi \to e^{i\theta}\psi$ is equivalent to change J and v respectively by $e^{-i\theta}J$ and $e^{-i\theta}v$. Thus W depends on J through the combinations J^*J, J^*v, and Jv^*. This explains why $\langle\psi\rangle \neq 0$ is possible when $v \neq 0$. In the next section we shall show by an explicit computation how $\langle\psi\rangle \neq 0$ is established.

In general, when L(x) has a certain symmetry, we need the ε-term in order to obtain an asymmetric solution:

$$W[J] = \frac{1}{N}\int [d\psi][d\psi^*] \exp\{i\int d^4x[L + J^*\psi + J\psi^* + i\varepsilon\, f(\psi,\psi^*)]\} \tag{129}$$

where $f(\psi,\psi^*)$ is a term which does not have the symmetry under consideration.

2.2 The Ward-Takahashi Identity and the Goldstone Theorem

In this section we shall study the model (129) in which L is invariant under the phase transformation. Let us choose $f(\psi,\psi^*)$ as[32]

$$f(\psi,\psi^*) = |\psi(x)-v|^2 \qquad (v \neq 0) \tag{130}$$

This is not invariant under the phase transformation of ψ. We shall now change the integration variable in the numerator of W[J] as $\psi \to e^{i\theta}\psi$. Since the integration is not unaltered by the change of variables, we have

$$\frac{\partial W[J]}{\partial \theta} = 0 \tag{131}$$

which leads to[32]

$$\int d^4x \, \langle i(J^* \psi - J\psi^*) + \varepsilon \, (v^* \psi - v\psi^*) \rangle_{\varepsilon,J} = 0 \tag{132}$$

In the following we use the following notations:

$$\left.\begin{aligned} \langle F[\psi] \rangle &= \langle F[\psi] \rangle_{\varepsilon,J} \quad \text{with } \varepsilon = J = 0 \\ \langle F[\psi] \rangle_{\varepsilon} &= \langle F[\psi] \rangle_{\varepsilon,J} \quad \text{with } J = 0 \\ \langle F[\psi] \rangle_{J} &= \langle F[\psi] \rangle_{\varepsilon,J} \quad \text{with } \varepsilon = 0 \end{aligned}\right\} \tag{133}$$

The Ward-Takahashi identities can be derived by repeated uses of derivative $(\delta/\delta J)$ acting on (132).

Operating $\delta/\delta J^*(x)$ on (132), we obtain

$$\langle \psi(x) \rangle_{\varepsilon,J} + \varepsilon \int d^4y \, \langle \psi(x), \, v^* \psi(y) - v\psi^*(y) \rangle_{\varepsilon,J}$$
$$+ \int d^4y \, \langle \psi(x), \, J^*(y)\psi(y) - J(y)\psi^*(y) \rangle_{\varepsilon,J} = 0 \, . \tag{134}$$

Let us first note that the third term vanishes at $J = 0$, when the Green functions are well-defined. Indeed, if the third term was not zero at $J = 0$, then $\langle \psi(x), \psi(y) \rangle_{\varepsilon}$ and/or $\langle \psi(x), \psi^*(y) \rangle_{\varepsilon}$ would be singular at $J = 0$. (134) in the limit $J \to 0$ leads to

$$\langle \psi(x) \rangle_{\varepsilon} + \varepsilon \int d^4y \, \langle \psi(x), \, v^* \psi(y) - v\psi^*(y) \rangle_{\varepsilon} = 0 \, . \tag{135}$$

To simplify our considerations, we shall choose v to be real. Then (135) reads as

$$\langle \psi(x) \rangle_{\varepsilon} + \sqrt{2} \, i\varepsilon v \int d^4y \, \langle \psi_o(x), \, \chi(y) \rangle_{\varepsilon} = 0 \tag{136}$$

where χ is the imaginary part of ψ:

$$\psi = \frac{1}{\sqrt{2}} \, [\psi_R(x) + i\chi(x)] \, . \tag{137}$$

In the same way, operation $\delta/\delta J(x)$ on (132) leads to

$$<\psi^*(x)>_\varepsilon - \sqrt{2}\, i\varepsilon v\int d^4y <\psi^*(x),\ \chi(y)>_\varepsilon = 0$$

The above relations give

$$<\psi_R(x)>_\varepsilon - \sqrt{2}\, \varepsilon v\int d^4y <\chi(x),\ \chi(y)>_\varepsilon = 0 \quad (138)$$

$$<\chi(x)>_\varepsilon + \sqrt{2}\, i\varepsilon v\int d^4y <\psi_R(x),\ \chi(y)>_\varepsilon = 0 \quad (139)$$

When

$$<\psi_R(x)> = \tilde{v} \neq 0 , \quad (140)$$

(138) shows that $<\chi(x)\ \chi(y)>_\varepsilon$ has the $(1/\varepsilon)$-singularity at $\varepsilon = 0$. Let us introduction the Fourier transform of the two-point function:

$$<\chi(x),\ \chi(y)> = i(2\pi)^{-4}\int d^4p\ e^{-ip(x-y)}\Delta_\chi(p) \quad (141)$$

Here

$$\Delta_\chi(p) = -\lim_{\rho\to 0}\frac{Z_\chi}{p^2+m_\chi^2-i\rho} + \text{(continuum contribution)}. \quad (142)$$

The constant Z_χ is the wavefunction renormalization constant of the χ-field and m_χ is the mass. Feeding (141) and (140) into (138) with the limit $\varepsilon \to 0$, we obtain

$$\tilde{v} = -\, i\sqrt{2}\, vZ_\chi \lim_{\varepsilon\to 0}\frac{\varepsilon}{m_\chi^2-i\rho} \quad (143)$$

This shows that

$$m_\chi = 0 \quad (144)$$

$$\rho = a_\chi\varepsilon \quad (145)$$

where a_χ is a certain constant.

(144) leads us to the Goldstone theorem and the χ-field is

the Goldstone field. (143) shows that $\tilde{v} \neq 0$ is impossible when the ε-term is absent, reconfirming our argument in the previous section. (142), (143), (144), and (145) lead to

$$\left.\begin{aligned} \Delta_\chi(p) &= -\lim_{\varepsilon\to 0} \frac{Z_\chi}{p^2 - i\varepsilon a_\chi} + \ldots \\ \tilde{v} &= \sqrt{2}\, \frac{Z_\chi}{a_\chi} v \end{aligned}\right\} \tag{146}$$

This shows that $\tilde{v}$ is real when v is real.

All the Ward-Takahashi identities can be derived from (134). We leave the derivation of these relations to ref. (32).

We shall now prove[(33)] that the magnitude of $\tilde{v}$ is independent of the magnitude of v in the limit $\varepsilon \to 0$. For simplicity we choose v to be real. When $\varepsilon \neq 0$, $\tilde{v}$ is a function of v. We denote this function by $\tilde{v}(\varepsilon,v)$. Then we have

$$\begin{aligned} \frac{\partial}{\partial v} \tilde{v}(\varepsilon,v) &= \frac{\partial}{\partial v} \langle\psi(x)\rangle_\varepsilon \\ &= -\sqrt{2}\,\varepsilon\int d^4y\,[\langle\psi(x),\psi_R(y)\rangle_\varepsilon - \langle\psi(x)\rangle_\varepsilon\langle\psi(y)\rangle_\varepsilon] \\ &= -\sqrt{2}\,\varepsilon\int d^4y\,\langle\psi(x),\,\rho(y)\rangle_\varepsilon \end{aligned}$$

When ρ^{in} is not massless, this gives

$$\lim_{\varepsilon\to 0} \frac{\partial}{\partial v} \tilde{v}(\varepsilon,v) = 0\ .$$

(146) then shows that a_χ is proportional to v. When v is complex, it can be proved that $\tilde{v}/v$ is real, implying that $\tilde{v}$ and v have a same phase.

We shall close this section by a comment on expansions of asymmetric Green functions in terms of symmetric Green functions, each of which carries a power of asymmetric parameters. Suppose that such an expansion would exist. Then the expansion would be a power expansion in εv. Therefore, each term would approach zero at $\varepsilon = 0$, implying that the expansion would not make any sense.[32]

2.3 The Dynamical Rearrangement of Symmetries

By using as an example the phase invariant model presented in the previous section, we shall present a simple proof for the dynamical rearrangement of symmetry.

We have seen in the previous section that there are two incoming fields, χ^{in} and ρ^{in}, which correspond respectively to χ and ρ defined by

$$\psi = \tilde{v} + \rho + i\chi \qquad (147)$$

The field χ^{in} is massless. The mass of ρ^{in} will be denoted by m_ρ.

We now introduce the Fock space $\mathcal{H}$ associated with these incoming fields and define the S-matrix by

$$S = \sum_{n=0} \sum_{\ell=0} \frac{1}{n!\ell!} \int d^4x_1 \ldots d^4x_n \, d^4y_1 \ldots d^4y_\ell$$

$$s(x_1 \ldots x_n, y_1 \ldots y_\ell) : \rho^{in}(x_1) \ldots \rho^{in}(x_n) \chi^{in}(y_1) \ldots \chi^{in}(y_\ell) : \qquad (148)$$

in which

$$s(x_1 \ldots x_n, y_1 \ldots y_\ell) = (-i)^{n+\ell} K_\rho(x_1) \ldots K_\rho(x_n) K_\chi(y_1) \ldots K_\chi(y_\ell)$$

$$\langle \rho(x_1) \ldots \rho(x_n) \chi(y_1) \ldots \chi(y_\ell) \rangle \qquad (149)$$

Here notations are

$$\left. \begin{aligned} K_\rho(x) &= \Box - m_\rho^2 \\ K_\chi(x) &= \Box \end{aligned} \right\} \qquad (150)$$

We can rewrite (148) in the following simple form:

$$S = \langle : \exp A\,(\chi^{in}, \rho^{in}) : \rangle \qquad (151)$$

where

$$A(\chi^{in},\rho^{in}) = -i\int d^4x[\chi^{in}(x)K_\chi(x)\chi(x) + \rho^{in}(x)K_\rho(x)\rho(x)] \quad (152)$$

In a similar way we shall introduce a matrix ψ_H by

$$S\psi_H(x) = :<\psi(x) \exp A(\chi^{in}, \rho^{in})>: \quad (153)$$

The matrix ψ_H is called the Heisenberg operator.

Let us now introduce a real square-integrable function $g(x)$ satisfying

$$\Box\, g(x) = 0 \quad (154)$$

When we make the transformation

$$\chi^{in}(x) \to \chi^{in}(x) + \alpha g(x) , \quad (155)$$

with a constant α, S and $S\psi_H$ respectively become

$$S' = <: \exp A(\alpha, \chi^{in}, \rho^{in}):> \quad (156)$$

and

$$(S\psi_H)' = <\psi(x): \exp A(\alpha, \chi^{in}, \rho^{in}):> \quad (157)$$

where

$$A(\alpha, \chi^{in}, \rho^{in}) = -i\alpha\int d^4x\, g(x)K_\chi(x)\chi(x) + A(\chi^{in}, \rho^{in}). \quad (158)$$

It should be noted that the equation for χ^{in} is invariant under the transformation (155) because of (154).

We shall now derive the Ward-Takahashi identities by following the steps for derivation of the relation (132) with real v: we thus change the integration variable in S' as $\psi \to e^{i\theta}\psi$ and use the relation

$$\frac{\partial S'}{\partial\theta} = 0$$

We are then led to the following relation:

$$\lim_{\varepsilon\to 0} i\sqrt{2}\,\varepsilon v\int d^4x\, \langle\chi(x):\exp A(\alpha,\chi^{in},\rho^{in}):\rangle_\varepsilon$$

$$-\int d^4x: \langle[(\chi^{in}(x) + \alpha g(x))K_\chi(x)\rho(x)$$

$$- \rho^{in}(x)K_\rho(x)\chi(x)]\exp A(\alpha,\chi^{in},\rho^{in}) : = 0 \qquad (159)$$

When we make the same change of the integration variable in $(S\psi_H)'$, the relation

$$\frac{\partial}{\partial\theta}(S\psi_H)' = 0$$

leads to

$$\lim_{\varepsilon\to 0} i\sqrt{2}\,\varepsilon v\int d^4y\, \langle\chi(y)\psi(x):\exp A(\alpha,\chi^{in},\rho^{in}):\rangle_\varepsilon$$

$$-\int d^4y:\langle\psi(x)[(\chi^{in}(y)+\alpha g(y))K_\chi(y)\rho(y)$$

$$- \rho^{in}(y)K_\rho(y)\chi(y)]\exp A(\alpha,\chi^{in},\rho^{in})\rangle:$$

$$+ \langle\psi(x):\exp A(\alpha,\chi^{in},\rho^{in}):\rangle = 0 \qquad (160)$$

Considering the fact that $g(x)$ satisfies the massless equation (154) and the fact that the physical ρ-particle is massive, we can derive

$$\left.\begin{aligned}&\int d^4x\, \langle(\chi^{in} + \alpha g(x))K_\chi(x)\rho(x)Q\rangle = 0\\ &\int d^4x\, \langle\rho^{in}(x)K_\rho(x)\chi(x)\, Q\rangle = 0\end{aligned}\right\} \qquad (161)$$

where Q is either $\exp A(\alpha,\chi^{in},\rho^{in})$ or $\psi(x)\exp A(\alpha,\chi^{in},\rho^{in})$. The above relations are derived by studying the Feynman-diagrams in the momentum representation. The existence of function $g(x)$ in $A(\alpha,\chi^{in},\rho^{in})$ helps us in deriving relations in the momentum representation.*

Making use of (161), we obtain from (159) and (160) the following relations:

$$\lim_{\varepsilon\to 0} \varepsilon v\int d^4x <\chi(x): \exp A(\alpha,\chi^{in},\rho^{in}):> = 0 \tag{162}$$

$$<\psi(x): \exp A(\alpha,\chi^{in},\rho^{in}):>$$

$$= -\lim_{\varepsilon\to 0} i\sqrt{2}\ \varepsilon v\int d^4y <\psi(x)\chi(y):\exp A(\alpha,\chi^{in},\rho^{in}):> \tag{163}$$

These are the Ward-Takahashi identities which are useful for our purpose.

Let us now recall that $<\chi(x)Q>$ with any Q has the form

$$<\chi(x)Q> = \int d^4y <\chi(x)\chi(y)> q(y) \tag{164}$$

with certain q(y). Then (146) and (154) lead to

$$\lim_{\varepsilon\to 0} i\sqrt{2}\ \varepsilon v\int d^4x <\chi(x)Q>_\varepsilon$$

$$= i\tilde{v}\int d^4y\ q(y)$$

$$= \frac{\tilde{v}}{Z_\chi}\int d^4x\ K_\chi(x)\ <\chi(x)Q> \tag{165}$$

Thus (162) and (163) can be rewritten as

$$\int d^4x\ K_\chi(x)\ <\chi(x): \exp A(\alpha,\chi^{in},\rho^{in}):> = 0 \tag{166}$$

$$<\psi(x): \exp A(\alpha,\chi^{in},\rho^{in}):>$$

*If g(x) = 1, the first quantity in (161) contains terms such as

$$\frac{1}{m_\rho^2}\,\delta(p)\,p^2\left(\frac{\alpha}{p^2+i\varepsilon a}\right)^n, \qquad (n > 1)$$

which are not well-defined.

$$= -\frac{\tilde{v}}{Z_\chi}\int d^4y\ g(y)K_\chi(y)\langle\chi(y)\psi(x)\ :\ \exp A(\alpha,\chi^{in},\rho^{in}):\rangle \tag{167}$$

Let us now define

$$S_1 = \lim_{g(x)\to 1} S'$$

$$(S\psi_H)_1 = \lim_{g(x)\to 1} (S\psi_H)' .$$

The limit $g(x) \to 1$ should be taken after all the space-time integrations are performed. The existence of $g(x)$ is required by the same reason as the one for need of wave-packet states in calculation of matrix elements of the left hand side of (18) (cf. also footnote in preceeding page).

(155), (156) and (157) now give

$$\frac{\partial S_1}{\partial\alpha} = 0 \tag{168}$$

$$\frac{\partial}{\partial\alpha}(S\psi_H)_1 = i\frac{Z_\chi}{\tilde{v}}(S\psi_H)_1 . \tag{169}$$

(168) shows that S_1 is independent of α:

$$S_1 = S \tag{170}$$

(169) gives

$$(S\psi_H)_1 = e^{i\frac{Z_\chi}{\tilde{v}}\alpha} S\psi_H .$$

Using (170), we obtain

$$(\psi_H)_1 = e^{i\frac{Z_\chi}{\tilde{v}}\alpha} \psi_H . \tag{171}$$

We thus state that the transformation

$$\chi^{in} \to \chi^{in} + \eta\theta,\quad \rho^{in} \to \rho^{in} \tag{172}$$

with

$$\eta = \frac{\tilde{v}}{Z_\chi} \tag{173}$$

induces the phase transformation of the Heisenberg operator

$$\psi_H \to e^{i\theta}\psi_H \tag{174}$$

This is the content of dynamical rearrangement of the phase symmetry. Note that χ^{in} in this section corresponds to B^{in} in section 1.4.

We applied(33) a similar method to analysis of all the examples presented in section 1.4 and results showed that the forms of dynamical rearrangement of symmetries agree with those in section 1.4.

As we have seen, the precise meaning of (172) is

$$\chi^{in} \to \chi^{in} + \eta\theta g(x), \quad \rho^{in} \to \rho^{in} \tag{175}$$

with the limit $g(x) \to 1$. The limit should be performed only after the space-time integrations in the dynamical map (18) (with χ^{in} replaced by $\chi^{in} + \eta\theta g(x)$) are completed.

2.4 The Boson Transformation

We shall now present a proof for the statement that the boson transformations (cf. (118)) are invariant transformations. We shall explain this statement by using the phase invariant model studied in this section. The boson transformations are defined by

$$\left.\begin{aligned} \chi^{in}(x) &\to \chi^{in}(x) + \alpha(x) \\ \rho^{in}(x) &\to \rho^{in}(x) + \beta(x) \end{aligned}\right\} \tag{176}$$

in which $\alpha(x)$ and $\beta(x)$ stand for c-number functions satisfying

$$\Box\alpha = 0, \quad (\Box - m_\rho^2)\beta = 0 \tag{177}$$

The replacements in (176) change the Heisenberg operator ψ_H into a new operator which will be denoted by $\psi_H[\alpha,\beta]$. What we are going to prove is that[33] $\psi_H[\alpha,\beta]$ is a Heisenberg operator in the sence that $\psi_H[\alpha,\beta]$ has the form of ψ_H with suitable ε-terms.* To prove this, we note the relations

$$i\sqrt{2}\,\varepsilon v\int d^4x\ \alpha(x)\langle\chi(x)Q\rangle$$
$$= \frac{\tilde{v}}{Z_\chi}\int d^4x\ \alpha(x)K_\chi(x)\langle\chi(x)Q\rangle \tag{178}$$

and

$$i\varepsilon\int d^4x\ \beta(x)\langle\rho(x)Q\rangle = \frac{1}{a_\rho}\int d^4x\ \beta(x)K_\rho(x)\langle\rho(x)Q\rangle \tag{179}$$

Here Q is any quantity and a_ρ is the constant defined by

$$\Delta_\rho^{in}(p) = -\frac{Z_\rho}{p^2+m_\rho^2-i\varepsilon a_\rho} + (\text{continuum})$$

In derivation of (178) use was made of the relation (146). Since

$$\psi_H[\alpha,\beta] = \langle\psi(x){:}\ \exp A[\alpha,\beta,\chi^{in},\rho^{in}]\rangle \tag{180}$$

with

$$A[\alpha,\beta,\chi^{in},\rho^{in}] = -i\int d^4x[\alpha(x)K_\chi(x)\chi(x) + \beta(x)K_\rho(x)\rho(x)]$$
$$+ A(\chi^{in},\rho^{in})\ , \tag{181}$$

use of (177) and (178) leads to

$$\psi_H[\alpha,\beta] = \langle\psi(x){:}\ \exp A(\chi^{in},\rho^{in}){:}\ \exp\{\varepsilon\int d^4x[a_\chi\alpha(x)\chi(x)$$
$$+ a_\rho\beta(x)\rho(x)]\}\rangle \tag{182}$$

with $a_\chi = \sqrt{2}\,Z_\chi\,v/\tilde{v}$ (cf. (146)). Thus $\psi_H[\alpha,\beta]$ is the Heisenberg operator ψ_H in which v in the ε-term in the numerator is replaced

*Note that there is no need that the ε-terms in the numerator and denominator (N) in W[J] are the same.

by $v-(1/\sqrt{2})a_\rho\beta(x)-i(1/\sqrt{2})a_\chi\alpha(x)$. $\psi_H[\alpha,\beta]$ corresponds to the case in which $\langle\psi\rangle$ depends on x. As was pointed out in section 1.4, the boson transformation of the Goldstone field χ^{in} plays an important role in describing phenomena in superconductivity.

2.5 Concluding Remarks

We shall close part II with two remarks.

The first remark is concerned with the dynamical rearrangement of Lie group symmetries. In view of the fact that a careful analysis of the dynamical rearrangement of Abelian symmetries (such as the phase symmetry) were provided by the path-integral method, it is natural to apply the same method to the dynamical rearrangement of Lie group symmetries.[34] We are now busily working on this problem. The need of the limiting process $(g(x) \to 1)$ makes this problem quite complicated.

The second remark is concerned with application of the path-integral method to the Yang-Mills fields. Faddeev and Popov[35] invented an ingeneous method for establishing the gauge invariance of the generating functional. Our brief consideration has suggested that use of the ε-term presents another way of establishing the gauge invariance. It is an interesting problem to examine this possibility.

References

1. L. Leplae, R.N. Sen and H. Umezawa, Suppl. Prog. Theor. Phys. (Kyoto), Problems of Fundamental Physics (1965).

2. Y. Freundlich, Phys. Rev. D, Vol. 1, No. 12, 3290 (1970).

3. R. Hagedron, Introduction to Field Theory and Dispersion Relations (Pergamon Press, 1964).

4. H. Lehmann, K. Symanzik and W. Zimmermann, Nuovo Cimento, 11, 205 (1955).

5. H. Ezawa, K. Kikkawa and H. Umezawa, Nuovo Cimento 23, 751 (1962).

6. H. Ezawa, T. Muta and H. Umezawa, Prog. Theor. Phys. (Kyoto) 29, 877 (1963).

7. S. Weinberg, Phys. Rev. 137 (1965), B672.

8. L. van Hove, Physica 18, 145 (1952); A.S. Wightman and S.S. Schweber, Phys. Rev. 98, 312 (1955); K.O. Friedrichs, Mathematical Aspects of the Quantum Theory of Fields (Interscience Publishers, 1953).

9. H. Umezawa, Acta Phys. Hung. Tem. 19, 9 (1965).

10. J. Rest, V. Srinivasan and H. Umezawa, Phys. Rev. D, Vol. 3, No. 8, 1890 (1971).

11. Vaughan, Aaron and Amado, Phys. Rev. 124 (1961) 1258.

12. H. Umezawa, Nuovo Cimento 40, 450 (1965).

13. L. Leplae, F. Mancini and H. Umezawa, Derivation and Application of the Boson Method in Superconductivity (1973). This article contains a list of references on the subject.

14. F. Mancini, The Boson Method in Superconductivity, Thesis (1972), p. 110.

15. Benson, Phys. Rev. B, May (1973).

16. Madeleine Sirugne-Collin, Thesis, Universite D'Aix-Marseille (1967); Raoelina Andriambololona, Thesis, Universite D'Aix-Marseille (1967).

17. J. Goldstone, Nuovo Cimento 19, 154 (1961).

18. K. Nakagawa, R.N. Sen and H. Umezawa, Nuovo Cimento 42, 565 (1966).

19. H. Umezawa and R. Kawabe, Prog. Theor. Phys. (Kyoto) 4, 420 (1949).

20. Y. Nambu and Jona-Lasinio, Phys. Rev. 122, 345 (1961); ibid. 124, 246 (1961).

21. A. Aurilia, N. Papastamatiou, Y. Takahashi and H. Umezawa, Phys. Rev. D5, 3066 (1972).

22. A. Aurilia, Y. Takahashi and H. Umezawa, Phys. Rev. D5, 851 (1972).

23. C.G. Callen, S. Coleman and R. Jackiw, Ann. Phys. (N.Y.) 59, 42 (1970).

24. N. Papastamatiou and H. Umezawa, Phys. Rev. D7, No. 2, 571 (1973).

25. L. Leplae, F. Mancini and V. Srinivasan, UWM-4867-71-12 (preprint, 1971); L. Leplae and H. Umezawa, Jour. Math. Phys. 10, 2038 (1969).

26. M.P. Anderson, Phys. Rev. 110, 827, 1900 (1958).

27. Y. Nambu, Phys. Rev. 117, 648 (1960); S. Cremer, M. Sapir and D. Lurie, Nuovo Cimento 6B, 179 (1971).

28. P. W. Higgs, Phys. Rev. 145, 1156 (1966); T.W. Kibble, Phys. Rev. 155, 1554 (1967).

29. A. Aurilia, Y. Takahashi and H. Umezawa, Prog. Theor. Phys. 48, 290 (1972).

30. Y. Freundlich and D. Lurie, Nucl. Phys. B19, 557 (1970).

31. L. Leplae, R.N. Sen and H. Umezawa, Nuovo Cimento 49, 1 (1967); R.N. Sen and H. Umezawa, Nuovo Cimento 50, 53 (1967).

32. H. Matsumoto, N. Papastamatiou and H. Umezawa, The Formulation of Spontaneous Breakdown in the Path-Integral Method, Preprint, UWM-4867-73-3.

33. H. Matsumoto and H. Umezawa, UWM-preprint (1973).

34. H. Matsumoto, H. Umezawa, G. Vitiello and J. Wyly, UWM-preprint (1973).

35. L.D. Faddeev and V.N. Popov, Phys. Letters 25B, 29 (1967).

THE PRESENT STATUS OF THE COMPUTING METHODS IN QUANTUM ELECTRODYNAMICS

A. VISCONTI *

Université de Provence

Centre de Physique Théorique de Marseille (CNRS)

Contents

1. INTRODUCTION

My talk will deal with an application of the Renormalization Theory to Q.E.D., namely the numerical computation of radiative corrections to electromagnetic processes.

Although such a problem is an old one, characteristic advances have been realized these past years only through the introduction of computing techniques. I want to examine some fundamental aspects of these methods and since the majority of my audience does not appear to me as being completely familiar with computing techniques, let me, first of all, make some very general comments.

During one of his periodic visits to the Centre de Physique

---

* Contrat DRME n° 72 34 678 00 480 75 11

Théorique at Marseilles, A.C. Hearn was asking the question "Is the automated theoretical physicist a fact or a fancy ?" As a by-product of my talk, I want to show that in a very near future, there will be indeed automated theoretical physicists who will plan their work directly for the computer instead of using the computer as an auxiliary help to complete calculations prepared along the very traditionnal pattern of classical analysis.

Computers have been employed, first of all, as numerical or control devices and their use for algebraic or more generally symbolic and non numeric calculations has become more widespread only these last few years. As I already pointed out, the most recent advances in our understanding of Q.E.D. has been greatly helped by the ability of the computer to perform algebraic manipulations (like the tedious γ-matrices algebra) and to evaluate numerically Feynman multi-dimensional integrals. That rather slow expansion of computers programs dealing with non numeric expressions may be explained by the general structure of computers hardware which is designed chiefly for numerical calculations and not for the construction of algebraic or symbolic systems. As a matter of fact, the main difference between numeric programing languages and algebraic ones is due to the "value" of the variables which are algebraic expressions in the case of symbolic computation. But, because of its hardware build up, one has to teach the computer, through conveniently elaborated programs, the usual rules for all operations currently performed on algebraic expressions.

- Once such a program has been elaborated, the user will be then faced with practical problems, the main one being the "blow-up" problem : it happens often that the intermediate states of a program (which generally do not appear on the listing) even if the out-put has been reasonably condensed, may be of such a complexity and length that even the largest memories we can use at present may be easily saturated. Therefore, the manipulation of symbolic expressions requires not only powerful computers indeed but also very highly trained and specialized teams of programers, experts in the art of optimization, who will compile all the needed algebraic routines.

- However, even if we could afford the dream computer of the Physicist surrounded by its team of experts, we would have to face a second important problem : how should the results of algebraic calculations be presented to the user ? We are accustomed to present our mathematical expressions in a two-dimensional format using lavishly a large set of symbols belonging to several alphabets and types, but the number of characters which by now are available in the "printers" of our Computing Centers is really restricted. Furthermore, the meaning of a printed symbol in a book or that of an hand written sign on a paper depends drastically on the context :

we need, for instance, to distinguish between the k-th contravariant component x^k of a vector x (in a n-dimensional space) and the number x to the power k . Distinctions of that kind cannot be made at present in the printed output of a computer.

- The user is generally faced with a third problem of equal practical importance : many calculations result in a very profuse output. Indeed, this is due to the fact that a computer does not keep the same control over the size of the expressions that is possible in hand calculations. By hand, we do a lot of simple two-dimensional "pattern matching" in order to avoid the expansion of certain expressions unles, of course, such an expansion is necessary for a better comprehension or for further simplification. We may try to reproduce the same result using different tricks in our computer programs, but we still lack of a fully automatic method.

- Finally, one is often obliged to use two different languages with in the same program : LISP, for instance, for the symbolic part and FORTRAN for the numerical evaluations ; in that case one needs conversion programs which translate directly one language into another before any printing has been done. This is a feasible task.

- Besides many others, these are typical problems which arise when one tries to use a computer in a more sophisticated way than it is usually done, i.e. when the computer is considered mainly as a device and a tool for the manipulation of non numeric expressions 1).

2. THE PHYSICS OF ELECTRONS AND PHOTONS 2)

Q.E.D. is a wonder theory : it is the only one field theory whose predictions up to now have never been contradicted by experiments. Quantum electrodynamics is unchallenged in all its results for low or high energy and remains unchallenged even if one includes the electromagnetic interactions of hadrons.

That state of affairs is astonishing indeed since Q.E.D. is truly a very naive theory, it was built up using the formalism of Quantum Mechanics : the fields $\vec{E}$ and $\vec{H}$ obeying Maxwell equations are used as canonical operators which depend on a continuous set of variables $\vec{x}$ and t . Their interaction with matter requires the consideration of the Dirac spinor ψ , which in its turn is transformed into an operator satisfying a canonical hamiltonian formalism. From this relatively simple logical scheme came out a whole cascade of results which were always confirmed by experiments of the highest precision. But, despite of its simplicity, evaluation of radiative corrections are extremely combersome and very hard to perform because of the underlying algebra. Even if the set of rules for getting unambigous answers -through Renormalization Theory- is on the way of becoming a perfectly rigorous mathematical theory, still the numerical evaluation of the results appears in certain cases as a forbidding task.

Q.E.D. represents thus a remarkable playground where the capabilities and the power of the computational methods can be checked out through their application to at least two really important physical effects : the Lamb-Retherford shift and the anomalous magnetic moment of the electron. Indeed, contrarywise to the predictions resulting from the Dirac eq., radio frequency experiments show that the the $2S_{1/2}$ and $2P_{1/2}$ levels of the hydrogen atom have not the same energy. Their relative position is shown in the following figure :

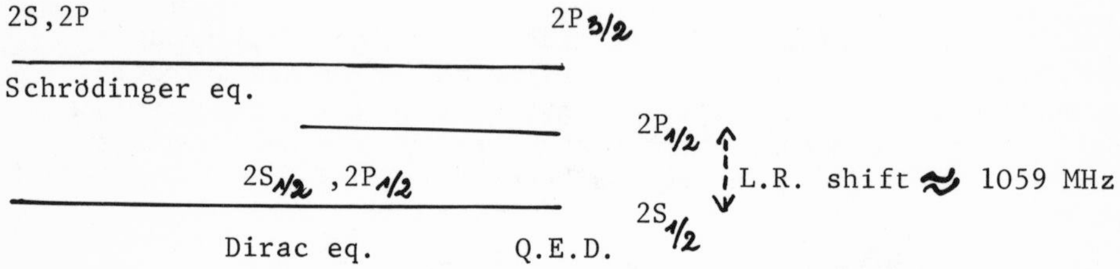

Using Q.E.D. one finds a shift of around 1059 MHz between $2S_{1/2}$ and $2P_{1/2}$ states : this shift is mainly due to the emission and absorption of virtual photons by the negatons and for a small amount (the S states are lowered by an amount of 27 MHz which constitutes the Uheling effect *) to the effective potential produced by the nucleus as seen by the negaton. More precisely, this last change results from the modification of the nucleus Coulomb potential by the vacuum polarization due to virtual negaton-positon pairs.

As far as the negaton magnetic moment is concerned, it is well known that the Gordon decomposition of the Dirac current shows that a Dirac electron has a magnetic moment :

$$\mu_0 = \frac{e}{2m}$$

which gives rise to an energy operator

$$\mu_0 \vec{\sigma} . \vec{H}$$

in a static magnetic field $\vec{H}$ ($\vec{\sigma}$ represents the Pauli matrices).

Q.E.D. shows that the magnetic moment of the electron in in-

* It may be noted by the way that the Uheling effect is greatly enhanced in muonic atoms.

teraction with photons is of the form :

$$\mu_0 + \mu_0 \delta\mu$$

where $\delta\mu$ is a dimensionless number which is expressed by a power series of the fine structure constant α

One may fit the former considerations in a more general framework by looking at the Q.E.D. vertex matrix element which can be written as

$$\bar{u}(p_2)\Gamma_\mu(p_2,p_1)u(p_1) = \bar{u}(p_2)\left[\gamma_\mu F_1(q^2) - i\frac{\sigma_{\mu\nu}}{2m}F_2(q^2)\right]u(p_1)$$

where q is the momentum transfer and $\sigma_{\mu\nu}$ is proportional to the commutator of γ_μ and γ_ν . This is the most general form of the vertex if one takes into account current and parity conservation*), the two scalar functions F_1 and F_2 of q^2 are respectively called the Dirac and Pauli form factor.
It is well-known that in the non-relativistic case, the charge density of the electron is the tridimensional Fourier transform of the function F_1

$$\rho(\vec{x}) = \frac{e}{(2\pi)^3}\int e^{i\vec{p}\vec{x}} F_1(-\vec{p}^{\,2})\, d^3p$$

With the auxiliary conditions $F_1(0) = 1$, the total charge of the electron will be e ; suppose now that $F_1(q^2)$ is expanded around the point $p^2/m^2 = 0$, then :

$$F_1(p^2) = 1 + \sigma\frac{p^2}{m^2} + \cdots$$

---

*) Le me make at this point a comment on gauge invariance : taking into account parity conservation only, the general matrix element of Γ_μ would have been obtained by adding to the former formula the supplementary term $\frac{q_\mu}{2m}\bar{u}(p_2)F_3(q^2)u(p_1)$. But if one introduces current conservation $q_\mu \bar{u}(p_2)\Gamma_\mu(p_2,p_1)u(p_1)$ the term F_3 has to be set equal to 0 , since the contribution of the term F_1 is 0 because of the Dirac eq. and that of F_2 is also 0 because of the antisymmetry property of $\sigma_{\mu\nu}$. In contradistinction to the internal gauge invariance, this may be called the external gauge invariance. There is a general belief, that classes of graphs which are gauge invariant in the former sense are also infrared divergencies free.

and the Taylor coefficient σ represents the "slope" of F_1 at $p^2=0$. In particular, for a negaton bound to a nucleus of charge Ze, physical considerations show that there is a contribution to the energy of the bound system

$$\Delta E = \frac{Ze^2}{m^2}\sigma\psi(0)$$

where $\psi(0)$ is the electron wave function taken at the origin. Let me finally quote the experimental value of the Lamb shift for the hydrogen atom as given by Robiscoe and Shyn 3) :

$$\Delta E_H\left(2S_{1/2}-2P_{1/2}\right) = \left(1057{,}90 \pm 0{,}06\right)\,MHz$$

Turning now our attention to the Pauli form factor, it is also well known that it contains the anomalous part of the magnetic moment of the electron. Indeed, from the vertex matrix element as previously written, one may deduce both Larmor term $\frac{e}{2m}\vec{L}.\vec{H}$ and the spin interaction term

$$\frac{e}{2mg_e}\vec{S}.\vec{H}$$

with

$$\frac{g_e-2}{2} = F_2(0) = a_e$$

thus the Pauli form factor $F_2(q^2)$ for the value 0 of its argument gives the anomalous part a_e of the magnetic moment of the electron.

This anomalous part has been measured by Wilkenson and Crane 4) and Wesley and Rich 4) by looking at the spin precession motion with respect to the momentum precession (Larmor precession), in other terms by looking at the Thomas precession. The best value of a_e seems to be :

$$(1159656{,}7 \pm 3{,}5)\ 10^{-9}$$

we cannot describe the experiments based on classical considerations, but we want to emphasize that an accurate value (taken for instance from the Josephson effect) of the fine structure constant $\alpha = e^2/\hbar c$ is of overhelming importance.

As far as muons are concerned, the most precise experiments have been made in CERN by Bailey et al 5).

$$a_\mu = (116616 \pm 31)\ 10^{-8}\ .$$

It has to be noted carefully that a_μ is not a purely electromagnetic quantity : in its theoretical evaluation, one has to take into account hadronic and weak-interaction effects which can be neglected in the case of electrons.

3. PRELIMINARY STEPS TOWARD AN AUTOMATED Q.E.D.

We want now to examine how far we can go in the automated evaluation of radiative corrections in Q.E.D. ; I am well aware that such a program is an extremely ambitious one. One considers indeed a given interaction (e.g. electron-photon interaction) and a given physical process involving electrons and photons and one wants to build up a program which will describe analytically and evaluate numerically the expected results. The pits one does encounter in the realization of such a vast program are really very many. Just to quote only one, the numerical evaluation of any Feynman integrals above its first threshold is a very hard task and, for the time being, an open question. Other problems have got working solutions for specific cases, but their general solution which is to be both general and effective with regard to its computational aspects, is not yet in a satisfactory form.

Anyway, the automatic handling of Q.E.D. requires several steps:

I. One has first of all to select, as a preliminary to any calculation of radiative corrections, the Feynman graphs which are associated with the process under consideration. This is indeed a trivial matter for low perturbative orders, but it is no more so for higher orders : we thus need to compile a program which will print for us all the needed graphs and only those.

II. Once the needed graphs have been drawn, we need to find out which ones are convergent or divergent. The second step will go through the automatic study of a given graph and the detection and enumeration of all its divergencies following the usual rules of Q.E.D.

III. For a given convergent graph in Q.E.D., we have then to perform

a) the γ -matrices algebras

b) the integration on the internal momenta following the Feynman α -parametrization method for instance

c) the final numerical integration which will lead to results we expect to agree with those obtained from experimental evidences.

IV. For a divergent Feynman graph we have to perform several renormalizations, i.e. to substract from the given graph several "shrinked" graphs following certain recipes.

The I , II , IIIa , IIIb and IV steps imply the use of non numeric methods, the IIIc involves either analytical or numerical analysis methods. Let us begin by studying the integration methods.

4. INTEGRATION METHODS 6)

It is a well known fact Feynman integrals -I mean the ones which are convergent or have been transformed into convergent one through the renormalization procedure- are particularly messy and that their evaluation is a very tedious task : these integrals, indeed, have rational functions as integrands but unhappily these functions of multidimensional variables have sharp spikes randomly distributed over the integration domain. As a consequence of such a behaviour, the usual Monte Carlo method fails in almost all the cases and for the time being, the most used and successfull numerical program (due to Sheppey) employs an adaptative Riemann sum method which has been tested indeed in several cases where independently an analytic procedure has been used. It seems that in a number of significant cases, the iterative process on which the method is based converges toward an accurate result but the foundation of the error analysis as performed by the program remains uncertain. We may also mention that the algebraic part of the total evaluation of a Feynman graph belonging to the class of graphs occuring in the Sixth order correction to the anomalous part of the magnetic moment of the electron takes a time which may counted in minutes, while its numerical evaluation requires hours. Numerical integration is thus a highly time consuming procedure.

In what follows, we shall give successively a quick survey of several numerical and analytic methods of integration : it is very clear indeed that analytic integration methods should represent the final answer to the evaluation of Feynman integrals.

a) The RIWIAD program 7)

This a FORTRAN IV subroutine designed to do N-dimensional integrations by a Riemann sum technic, it is based on an original idea by G. Sheppey (1963-1964), applied to Elementary Particles Physics first at SLAC by A.J. Dufner who improved certain points. Later one, B. Lautrup at CERN rewrote certain of its parts. This program is known under the names of SPCINT , LSD (Lautrup, Sheppey, Dufner) and RIWIAD, it is, by far, the most used program for the numerical evaluation of numerical convergent Feynman integrals whose integrand, as already been said, presents integrable singularities inside the integration domain and at its border.

Let me sketch in a few words some of the basic aspects of this program; suppose that we are dealing with a N-dimensional integral with a 0,1 integration interval

$$I = \int_0^1 dx_N \ldots \int_0^1 dx_1 \, f(x_1, \ldots, x_N)$$

where f , inside the limits 0,1 , is not a "too nasty" rational function of the variables $x_1, \ldots, x_N$. The p-th axis is divided into p_{max} infinitesimal intervals and let (i,j) be the j-th interval of the i-axis ; then to a set of N infinitesimal intervals

$$\{(1,j),(2,k),(3,\ell)\ldots\}$$

there corresponds a point $x \in \mathbb{R}^N$ such that

$$x = x(j,k,\ell,\ldots)$$

and an elementary volume around this point

$$d(1,j)d(2,k)d(3,\ell)\ldots$$

where $d(i,j)$ represents the length of the j-th interval of the i axis.

Therefore an elementary contribution to the integral will be

$$v = f(x(j,k,\ell,\ldots))d(1,j)d(2,k)d(3,\ell)\ldots$$

v is called a subvolume, and the total volume which will approximate the integral is

$$V = \sum_{\substack{1\leq j\leq j_{max}\\ 1\leq k\leq k_{max}\\ \ldots}} f(x(j,k,\ell,\ldots))d(1,j)d(2,k)d(3,\ell)\ldots$$

V is now expressed as a sum of subvolumes, the total number of which is

$$J_{max} = j_{max}.k_{max}.\ell_{max}. \ldots$$

The hypervolume of integration isnow divided in subvolumes v_J each having an index J and

$$V = \sum_{J=1}^{J_{max}} v_J .$$

Our main job will consist in the evaluation of the contribution of the subvolume v_J where the number J represents a given choice of intervals, say the j-th on the first axis, the k-th on the second axis,... with corresponding lengths $d(1,j),d(2,k),\ldots$ Such a set of intervals being chosen, we have to take into account that our intervals are not infinitesimal but have a finite small length ; we thus generate randomly n-systems of points, within the chosen J-th system of N intervals (a total of nN points for N-dimensional integral) and define an average value (v_J)

$$(v_J) = \frac{1}{n} \sum_{K=1}^{n} v_J^K$$

An estimate error for the evaluation of the subvolume v_J will be given by the variance

$$\sigma(v_J) = \frac{1}{n-1} \sum_{K=1}^{n} (v_J^K - (v_J))$$

and the total error is taken as

$$\sigma(V) = \sum_{J=1}^{J_{max}} \sigma(v_J) .$$

These are the basic operations which will be repeated a sufficient number of times, says s times, in order to get the approximation of the given integral with an error fixed a priori by the user. If V_q is the result of the q-th operation, we shall define an average (V) by the formula of weighted means

$$(V) = \frac{\sum_{q=1}^{s} V_q w(q)}{\sum_{q=1}^{s} w(q)}$$

where $w(q)$ is the relative weight assigned to the V_q measurement value.

The choiceof the weight functions defined as being proportional to the inverse of the variance seems reasonable enough, we then introduce

$$w(q) = \frac{V_q^2}{\sigma(V_q)}$$

and the accumulated standart deviation will be

$$\sigma_{tot.} = \frac{(V)}{\sqrt{\sum_{q=1}^{s} w(q)}}$$

The s operations we just mentioned represent an iterative procedure designed to improve the original estimate of the integral ; our fundamental problem lies therefore in the discovery of the regions in which the previous estimation was the poorest and the progressive improvement in these regions without an undue sacrifice in the rest of the space of integration. Our task consists first in the determination of the new size d(i,p)' of a previously used d(i,p) interval and second in the decision to change or not the number of intervals per axis.

For the first task, one uses a formula of the type

$$d(i,p) \rightarrow d(i,p)' = d(i,p)\left\{\log\left(\frac{\sigma(V)}{\sigma(i,p)}\right)(1-d(i,p))\right\}^{\delta}$$

where $\sigma(i,p)$ is the sum of the variances found in all subvolumes to which d(i,p) contributed and δ is an adjustable constant which varies from 0 to 1 . There is an analogous formula which fixes a new number of intervals per axis as a function of the old one.

If we now come into practical details, we do include in a typical input the number of dimensions over which the integration is to be performed ; an initial distribution of intervals over each of the axis ; the desired accuracy, such that when the standart deviation reaches this value, the program returns automatically the expected answer ; the total number of subvolumes into which the hyperspace is to be divided ; the maximum number of iterations to be performed (in case the desired accuracy is not reached). As an output, one gets the value of the calculated integral, the error for each iteration separately and also the accumulated error. The time needed by the RIWIAD program depends essentially on the complexity of the graph ; for the sixth order correction of the anomalous magnetic moment of the electron the time varies from 30 to 45 minutes on a Univac 1108 (or 1110) to 160 minutes on the very powerful CDC 7600 at CERN (Calmet and Peterman). This is the longest numerical calculation known up to day, it deals with the insertion of the light by light scattering graph into the sixth order radiative correction of the electron anomaly.

b) An application of the Gauss method : the Levine and Wright method 8)

It has been noted by Levine and Wright that the worst integrable singularities are those which lie on the faces of hypercubes, they are more troublesome than the ones inside of the integration domain since such singularities can be indeed completely surrounded by a sufficient number of chosen points. Thus, the recipe these authors are using consists in performing transformations which remove or at least weaken the singularities on the border, they use afterwards a Gauss method in order to calculate the transformed integral.

We cannot go into the details but we may, as an example, consider the integral

$$\int_D \frac{dx\ dy}{x+y}$$

where D is the square : $0 \leq x \leq 1$; $0 \leq y \leq 1$. Although, the integral is well defined, the origin is a singular point for the integrand. Perform now the change of variables

$$x = u^2 \qquad y = v^2$$

introducing the transformation jacobian, the integrand becomes $4uv(u^2 + v^2)^{-1}$ and the integral is to be taken over the same domain $D : 0 \leq u \leq 1 \; ; \; 0 \leq v \leq 1$. But, one observes immediately that such an integral is no longer singular along the lines $v/u = cst$; it is then easy to verify that a Gauss evaluation of the integral give a better result than a Gauss evaluation of the untransformed integral. Application of this method to Feynman integrals is of course much more complicated, but the numerical results fit very nicely indeed with the analytical ones ; the only drawback being that the error evaluation is mainly an educated guess in that technique.

c) The semi analytic method of Levine and Roskies 9)

Another method which, for the time being, is specialized to vertex graphs, has been recently given by Levine and Roskies. The idea of the method stems from the observation that the form factors are scalar functions of a single four-vector q and thus do not depend on the orientation of q .

Since no angle is intrinsically defined by the problem, it might be possible to perform analytically the four dimensional integrals for each Feynman graph. There will remain only ℓ radial integrals, where ℓ is the number of loops momenta. Generally the dimensionality of these remaining integrals is lower than that which would have been obtained by parametrizing the given integral by the (conventional) α-method due to Feynman. Therefore, this technique could be very useful since we are left with simpler integrals which can be performed numerically.

Furthermore, the authors have shown that some graphs may be calculated completely analytically, in particular the method has been applied to the six diagrams

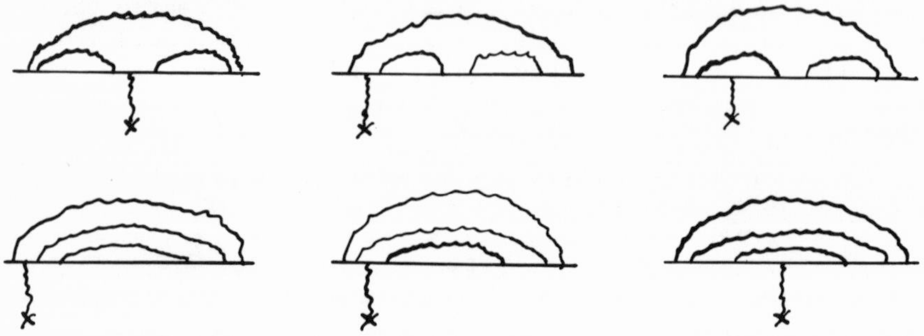

it involves the introduction of spherical harmonics, Gegenbauer polynomials and their algebra.

Up to now, the Levine and Roskies method has been applied only to certain classes of vertices (the so-called two colors graphs), but the work in this domain is still continuing.

d) Programs for the analytic evaluation of Feynman graphs

The analytic evaluation of Feynman integrals, when possible, gives unambigous and definite results. The first calculations of the anomalous magnetic moment of the electron (Karplus and Kroll, Petermann, Soto,...) were performed analytically along Feynman classical technics.

More recently, using conventional and dispersive technics (introduced by G. Källen and later on by E. Terentiev), E. Remiddi and his collaborators obtained extremely important results. Such a work needs indeed patience, ingenuity, ressourcefulness ; it is very valuable since such results are the only ones which are exact and may be used as tests for the numerical integration methods 10).

In particular, the work by A. Petermann on the evaluation of some sixth-order graphs, the evaluation of the discontinuities of electromagnetic form factor (vertex) at fourth order by Barbieri, Mignacco and Remiddi show that the main tool for analytic evaluation is the theory of Nielsen generalized poly-logarithms functions

$$S_{n,p}(x) = \frac{(-1)^{n+p-1}}{(n-1)!\ p!} \int_0^1 \frac{\log^{n-1} t \ \log^t (1-xt)}{t} dt$$

which seem to constitute the natural family of functions for the study of radiative corrections in Q.E.D.

The Spence functions and the Euler di- or tri-logarithm are examples of Nielsen functions, e.g.

$$Li_2(x) = -\int_0^x \frac{\log(1-t)}{t} dt \quad ; \quad Li_3(x) = \int_0^1 \frac{Li_2(t)}{t} dt$$

which are examples of the identity :

$$Li_n(x) = S_{n-1,1}(x) .$$

The SINAC program (Standart Integral Analytic Calculation) has been established by Petermann using the CDC 6600 SCHOONSHIP program by Veltman at CERN and allows to express analytically several Feynman integrals through the Nielsen functions ; as a matter of fact the introduction of Renormalization technics led Petermann to define the m-subtracted $S^{mm}_{1p}(x)$ Nielsen function as the series obtained when one neglects in the Taylor expansion around the point $x = 0$ of a Nielsen function the $m-p-1$ first terms 11). Another program the RSIN has been written by J.A.Fox and A.C. Hearn in the REDUCE language and is therefore available in computers other than CDC. That program is inspired by Petermann's program 12).

e) Symbolic integration

The SINAC and RSIN programs represent an attempt to solve in a very special case the general problem of the indefinite integration. It should be nevertheless made clear that, for the time being, nobody has been able to apply the general symbolic methods we are going to sketch to integrals arising in Q.E.D., but my own feeling is that, sooner or later, the solution of the numerical evaluation of radiative corrections lies in that direction.

Let me, first of all, point out that around 1960 the integrations technics were the same as 200 years before : mathematicians considered such a problem as requiring heuristic solutions and a good deal of ressourcefulness and intelligence. Between 1960 and 1970, it became evident that computer programs were faster and more powerful than humans, while using technics similar to theirs.

The first program which appeared in this field of research was Slagle's 13) SAINT program (Symbolic Automatic INTegrator) in 1961, it was followed by Moses 14) SIN program (Symbolic INtegrator) in 1967. Such programs have two goals in view : to choose among a wide variety of methods leading to the indefinite integral of a well defined function, the method which is the most efficient and as a next task to express the final answer in a way which is familiar to the user. For instance, integrals of trigonometric functions can be written in several forms, in terms of sinus and cosinus or of tangents of half angles or even through complex exponentials. The user will prefer the answer in a form which he is able to comprehend easily.

The strategy of the SIN program is developed in three stages : in the first stage, one tries to solve the problem at hand by using a table of derivatives of known functions ; in the second stage, one attempts to solve the same problem by using one of the eleven methods which are specific to certain classes of integrals : trigonometric functions, exponentials, radicals, rational functions. When the first two stages fail, one uses a general method (based on Liouville theory of integration) which generates a guess for the form of the integral based on the form of the integrand. The first steps in that direction are due to Laplace and Abel who showed that the integral of an algebraic function $y(x)$ (i.e. solutions of the polynomial $P(x,y) = 0$ with integer coefficients) contains only those algebraic functions which are present in the integrand. The next step was undertaken by Liouville and was used in Risch's decision procedure (1969). This is a method which concerns the class of "elementary functions" which are obtained by extending the field D of the rational functions by an algebraic extension and a transcental one*). Such an extended field $\mathcal{F}$ is called the field of elementary functions, the theorem states that if the integral of f belongs to such an extension, then

$$\int f(x)dx = V_o + \sum C_i \log V_i$$

where $V_o, V_i \in \mathcal{F}$ and the C_i are constants.

We thus have a way of obtaining the indefinite integral of most elementary functions (in the ordinary sense of the word) provided that the indefinite integral exists as a member of $\mathcal{F}$, correlatively we are also able to decide if any given integral belongs or not to such a class.

One may hope that these ideas will be successfully extended to multidimensional integrals and that in a near future, a program applying such methods to polylogarithmic functions will allows to perform all the symbolic integrations appearing in the high radiative corrections. This is an open field to the investigations of applied mathematicians, as well as theoretical physicists.

5. SYMBOLIC PROGRAMS AND THE LISP LANGUAGE

When we try to held a conversation with a computer, we have to use a language which can be understood by our interlocutor who, like any human being, has his own collection of signs (which constitute the "machine code language") to which he is able to answer. But, to converse with a computer in its own language is a very highly impractical method : a program written in machine code language can be used only with a given type of computer ; one therefore should try to avoid that kind of communication with the computer.

Luckily enough, there are universal language which can be understood by a variety of computers : this is the case of the most used language, the FORTRAN and of another more sophisticated one, the LISP language (LISt Processing) which has imbedded in its syntax an extremely important property, namely its recursive character.

---

* For the precise meaning of these extensions, the reader should refer to Ref. 14. The transcendental extension introduces into the field logarithmic and exponential functions and it should be noted that Risch's algorithm depends heavily on the properties of these functions.

Among all the programs which have been devised to perform all the symbolic manipulations related to the γ -matrices algebra, reduction and α -parametrization of Feynman integrals, etc... four systems programs survived as given in the following table

Machine code language ; CDC 6000 series	FORTRAN	LISP	REDUCE (based on LISP)
SCHOONSHIP (author : M. Veltman[15]))	ASHMEDAI (M.J.LEVINE 16))	(Campbell and A.C. Hearn) (J. Calmet and M. Perrottet[17]))	(author : A.C. Hearn 18))

SCHOONSHIP is a really fast working program, it is well adapted to calculations with γ -matrices and to all analytic transformations needed for the parametrization of Feynman integrals. It has only one drawback : it is specific to the 6000 series CDC machines. The group working in Marseilles has at his disposal a terminal connected to Univac 1110 ; unhappily, we were not able to use that Language. We could have used ASHMEDAI whose performances are comparable to SCHOONSHIP ; there is an historical reason if we did not, since at the time we performed our calculations, ASHMEDAI was very confidential and used by its author only.

Our group in Marseille could have used REDUCE (with some adaptations due to compilation problems), but after several discussions with its Author, it appeared that it was worthwhile to write down a special program for the evaluation of the renormalized radiative corrections, a program based on the source language of REDUCE , i.e. LISP . I shall, thus, sketch very briefly what is the meaning of LISP language 19).

Let me therefore start from a very specific example : consider any graph in Q.E.D., three sorts of fields are involved : the field ψ denoted by the symbol A , the field $\bar{\psi}$ denoted by the symbol B , and the photon field denoted by C . We shall represent a vertex by an integer s ; an external line of type A , for instance, attached to the vertex s will be defined by such a symbol as

(s A) .

An internal photon line connecting the k and j vertices will be noted by

(k C j C)

the two "atoms" C reminding the reader that it corresponds to the vacuum expectation value $\langle 0|P\{A(x)A(y)\}|0\rangle$. An electron line joining the vertex k to the vertex j will correspond to either of the symbols

(k A j B) or (k B j A)

following its orientation.

Each of the three expressions we just introduced is a "list" and a Feynman graph is represented by the list of the lists of its lines: the external lines are to be put at the beginning and the lines will appear in the listing in the order A , B , C . For instance, the graph corresponding to the first radiative correction of the simple vertex is represented by the set of following symbols

((2 A)(3 B)(1 C)(1 A 2 B)(1 B 3 A)(2 C 3 C))

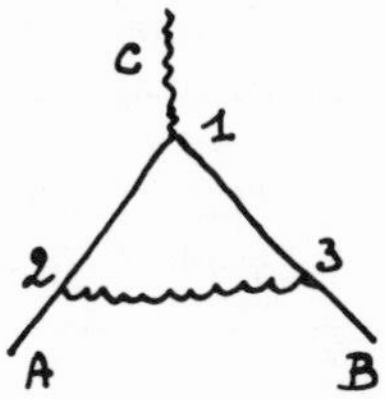

a Feynman diagram, in computing language, is a list of lists and it appears that the concept of list is a basic ingredient in the study of diagrams, it is also a basic ingredient of the LISP language.

LISP is a formal and mathematical language which differs from FORTRAN or ALGOL in that it is designed mainly for symbol manipulations (and not essentially for numerical ones) : its simplest element is the atomic symbol (or atom) which is any chain of alphanumerical characters, for instance

1 , K , Trace , ...

while a list is formed by a left-hand bracket several atoms or lists separated by blanks and a right hand bracket*), as examples

(A B C) , (A (B C)) , ((A B) C) , ...

---

*) In a more complete study of the LISP language, the lists are introduced as special cases of the so-called S -expressions (symbolic expressions).

NIL corresponds to an empty list

```
NIL = ( )
```

and represents an atom as well as a list.

The LISP language uses a few basic functions : the name of the function appears as the first atomic symbol in a list where the other elements are the arguments of the function. We define three functions which will act on lists : the CONS function has two lists as arguments (although its first argument may be an atom), it adds its first argument to the list which constitutes its second argument

```
(CONS A (BC))  =  (A B C)
(CONS (AB) (CD))  =  ((AB) C D)  ≠  (A B C D)
```

the CAR or CDR functions either isolate or suppress the first argument of a list. Although the output may be an atom, these functions are not defined for atomic arguments, let us quote as examples

```
(CAR (A B C))  =  A          (CDR (A B C))  =  (B C)
(CAR ((A B)C)) =  (A B)      (CDR ((A B)C))  =  C
                             (CDR (A))  =  ( )
```

Besides these fundamental functions, there are several others which are of a very common use : for instance SETQ is a function with two arguments : its first argument is an atom which takes the value of the second argument. LIST is another function which builds up the list of its arguments, for instance :

```
(LIST A)  = (A)     ,     (LIST A B)  =  (A B)
(LIST (A B)C)  =  ((A B)C) , ...
```

APPEND is a function which has two lists as arguments : it makes a single list out of its two arguments by dropping the inner right and left-hand brackets, for instance

```
(APPEND (A B)(C))  =  (A B C)
(APPEND((A))(B C)) =  ((A)B C)  .
```

The function CADR chooses the second argument of a list, it is defined as follows :

```
(CADR X)  ≡  (CAR (CDR X))
```

as an example, one has :

```
(CADR (A B C))  ≡  (CAR (CDR(A B C)))  =  (CAR(B C))  =  B .
```

The LISP language uses also <u>predicats</u>, i.e. functions whose value is true (T) or false (F) : for instance, ATOM is a function whose value is T only if its argument is atomic.

(ATOM A) = T (ATOM (A B)) = F

EQ is a function with two atomic arguments, its value is T if both arguments are identical. NOT is another predicat which has the value F if the list which is its argument is different from NIL .

We finally want to define the very important function COND , it has the following form

$$(\mathrm{COND}\ (p_1\ e_1)(p_2\ e_2)\ldots(p_n\ e_n))$$

each of the p_j is a predicat whose value is T or F , each of the e_j is the value one wants to assign to the whole expression. If p_1 is T , the function takes the value e_1 , if p_1 is F we go over to the second bracket : if p_2 is T , the whole expression takes the value e_2 ... and so on until we reach the first of the p_k which is T . If none of the p_k is T , the COND function returns the value FALSE . It must be noted that T and F must be considered as predicats with T and F as a value.

As an example, consider the definitions of the absolute value $|x|$ of the number x

$$|x| = (\mathrm{COND}\ (x<0\ \ -x)\ T\ x))$$

we may also, as another example, define $n! \equiv$ FACT n for the positive integer n :

FACT n = (COND (n=0 1)(T nFACT n-1)) .

Remark that in the last case, the function we want to define appears in its own definition, we thus have a recursive definition. As a last example we consider the predicat (NULL X), where X is an atom, it takes the value T if X is NIL (it plays more or less the same rule as NOT)

NULL X = (COND ((ATOM X)(EQ x NIL)) (T F))

which means, if X is an atom then the value of the function is (EQ X NIL), if not the value of the function is F .

The binding of an expression to a variable or to a function is performed through the functions SETQ and DEFINE . Furthermore, a function is defined by the LAMBDA statement which introduces unambigously both the form and the value of the function under consideration. For instance, the last example has to be written as follows

```
DEFINE(((NULL(LAMBDA(X)(COND((ATOM X)(EQ X NIL))(T F))))))
```

The former comments may give a very rough idea indeed about the way how LISP works ; its main property, i.e. its recursive character, is exhibited by the examples on the COND function.

6. PROGRAMS FOR THE ENUMERATION OF FEYNMAN GRAPHS

We now come back to step I of paragraph 3 , say the enumeration and the construction of all diagrams corresponding to a given electromagnetic process. Three types of programs dealing with that problem have been published until now :

a) one may elaborate the Wick algorithm for ordered products, such a program has been worked out by J.A. Campbell and A.C. Hearn in REDUCE language 20).

b) one may use combinatorial methods : M. Perrottet chooses this method using FORTRAN 21).

c) finally, J.Calmet using LISP language worked a program in the generating functional formalism 22).

Let me say a few words about the method c) : this program belongs to a wider mathematical framework which was elaborated by the group of the Centre de Physique Théorique de Marseille for dealing with Renormalization problems in Q.E.D. 23). The functional formalism through the introduction of external sources permits to generate all vacuum expectation values of time ordered products of field operators (propagators) from the so-called generating functional of sources. A special exponential form of this generating functional allows its expansion in the fine structure constant and construction of the corresponding graphs. As far as Renormalization is concerned : one first expresses the generating functional in terms of the physical parameters ; one then goes over to the bare parameters and introduces also the so-called wave functions renormalization by a change of scale (the Z factors) of the sources. Once, all these transformations have been performed, one writes down the renormalized generating functional in an exponential form and performing its expansion in the physical fine structure constant, one obtains, in this way, the Feynman graphs describing a given Q.E.D. process and the corresponding counterterms which have to be subtracted from the diagram under consideration in order to get a physical, finite contribution. All operations, all needed numerical coefficients are obtained unambiguously in this way. My time being short, I cannot go into technical details which may be found in the set of papers given in the references.

The only comments I want to present about the application of this method to the enumeration of Feynman graphs contributing to a given radiative correction is the following : a program called DIAG I enumerates all the graphs up to a given number of vertices. Among the graphs one obtains, several are identical (corresponding to the numerical coefficients of the Wick formula). Another program DIAG II chooses the graphs which are topologically inequivalent, also the elemination of tadpoles graphs or the application of Furry's theorem can be executed by two specific functions. Finally, another program called DIFFUSION permits to choose the graphs describing the physical electromagnetic process one has in mind.

These programs work very efficiently and it is worthwile to note, by the way, that the technic of generating functionals may be very usefull for problems in the mathematical theory of graphs.

7. AUTOMATIC RECOGNITION OF THE CONVERGENT OR DIVERGENT GRAPHS AND THE PROBLEM OF COUNTERTERMS

The expansion of the renormalized generating functional of propagators in the physical fine structure constant connects unambigously any given graph to its counterterms, except if the diagram is convergent, since in the case of Q.E.D., there are no subtractions to be made.

Thus, the first step of any calculation is to recognize whether or not a graph is divergent : such a program was recently compiled by J. Calmet. It determines also the kind of the divergences : vacuum polarization, electrons self-energies or vertices. One may remark that it does not consider infra-red divergences. The scanning of a graph begins by looking at loops with 2,3,... sides, when either a loop or a union of loops is found to correspond to a divergent subgraph, then the graph under study is reduced to a lower order graph. The list of the divergent subgraphs is then examined in order to handle properly the overlapping divergences and to put them in an order suitable for Renormalization.

The complete description of the program can be found in the paper quoted in the references, the reader will find two examples which are reproduced below which show the efficiency of the method 24).

AUTOMATIC RECOGNITION OF THE DIVERGENCIES
(J. Calmet)

```
NIL
*TEST  '(EX10);
```

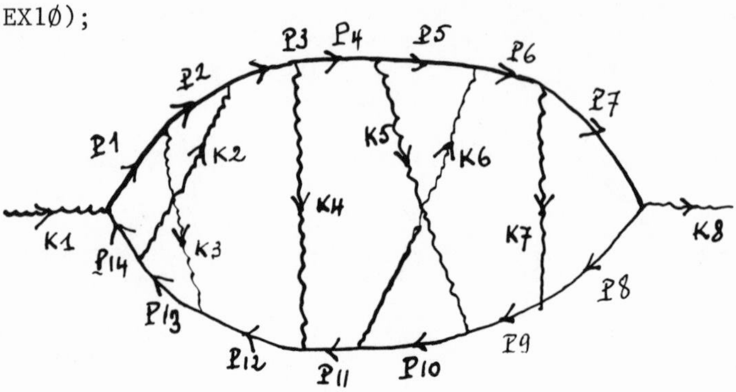

This diagram includes the
following divergences :

1 OVERLAPPING DIVERGENCE INCLUDING :

VERTEX
INTERNAL LINES : K7 P8 P7
EXTERNAL LINES : K8 P6 P9

VERTEX
INTERNAL LINES : K5 P5 P9 P10 K6 P6
EXTERNAL LINES : P4 P11 K8

VERTEX
INTERNAL LINES : K4 P4 P11
EXTERNAL LINES : K8 P12 P3

VERTEX
INTERNAL LINES : K3 P13 P2 K2 P3 P12
EXTERNAL LINES : K8 P1 P14

PHOTON SELF-ENERGY
INTERNAL LINES : P1 P14
EXTERNAL LINES : K1 K8

AUTOMATIC RECOGNITION OF THE DIVERGENCIES
(J. Calmet)

TEST '(EX5) ;

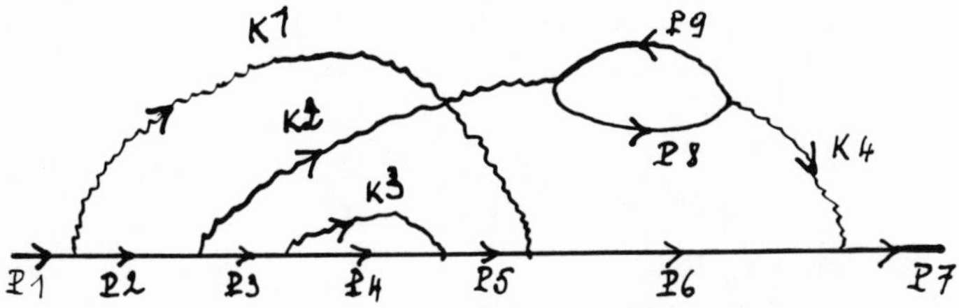

This diagram includes the
following divergences :

1 FERMION SELF-ENERGY

INTERNAL LINES : K3 P4
EXTERNAL LINES : P3 P5

2 PHOTON SELF-ENERGY

INTERNAL LINES : P8 P9
EXTERNAL LINES : K2 K4

3 OVERLAPPING DIVERGENCE INCLUDING :

VERTEX

INTERNAL LINES : K1 P3 P2
EXTERNAL LINES : K2 P1 P6

FERMION SELF-ENERGY

INTERNAL LINES : P6 K2
EXTERNAL LINES : P1 P7

8. ALGEBRAIC REDUCTION OF FEYNMAN GRAPHS

I shall mention very briefly different methods dealing with the algebra involved in the evaluation of Feynman integrals and elaborate more carefully about the method used by Calmet and Perrottet which follows very closely the original method of Feynman.

We note first of all that because of the recursive properties of the LISP language, when a program is set up for a given low order, it works also for higher orders. Consider for simplicity sake a scalar field φ interacting with itself via a cubic interaction term $g\varphi^3$. This is the simplest example one can imagine since it is well known that the integrals one encounters in Q.E.D. include (contrarywise to the $g\varphi^3$ interaction) numerators with γ-matrices and need a fair amount of supplementary algebra. In both cases, the Feynman method leads to the parametrization of the integral, the reduction of the quadratic form to its principal axis, the integration over the loop-momenta : for each of these steps, Calmet and Perrottet listed LISP functions and the whole system builds up the ACOFIS program (Algebraic COmputation of Feynman Integrals Scalar 25)).

We shall describe any given graph by several lists which constitute partly the input of our program :

a) We shall give the list of the propagators, a propagator of the form $\left[(p-k)^2 + m^2\right]^{-1}$ is described by the list

((M E)(-1 K)(1 P))

where M E represents the square of the mass m of the particle and the external and internal momenta are respectively denoted by p and K . For a more compex expression as

$$\frac{1}{\left[(p_2-k_1)^2+m^2\right]\left[(p_2-k_1-k_2)^2+m^2\right]\left[(p_1-k_1)^2+m^2\right]}$$

we shall have the list :

(((M E)(-1 K1)(1 P2))((M E)(-1 K2)(-1 K1)(1 P2))((M E)(-1 K1)(1 P1)))

More generally, to any Feynman integral

$$\int \frac{d^4k_1 \ldots d^4k}{a_1 a_2 \ldots a_n}$$

where the a_j^{-1} are of the form

$$a_j^{-1} = \left(\sum \xi_r p_r + \eta_r k_r\right)^2 + m^2$$

we shall associate a list of this type.

b) The list of the α Feynman parameters in the form

LISTALFA = (AF1 AF2 AF3 ...)

c) The list of all the products k_i^2 , p_i^2 , $k_i k_j$, $k_i p_j$, $p_j p_\ell$,.. ...

We parametrize the integral under consideration by the well-known formula

$$\frac{1}{a_1 \ldots a_n} = (n-1)! \int_0^1 d\alpha_n \ldots \int_0^1 d\alpha_1 \, \frac{\delta\left(\sum \alpha_i - 1\right)}{\left(\sum \alpha_i a_i\right)^n}$$

and a special LISP function called PARAMETRIZATION brings this step into the program.

One looks then to the terms to be shifted in order to get a quadratic form in the integration variables ; several functions (as POLYKP which brings the denominator into the form of a polynomial in k_1^2 , k_2^2 , $k_1 k_2$, ... p_1^2 , p_2^2 , as TRANSLATION which translates the integration variables and several other functions which may be found in the quoted references) build up that part of the program. One finally integrates over the internal variables, using the formula

$$\int \frac{d^4k}{(k^2+a^2)^n} = \frac{i\pi^2}{(a^2)^{n-2}} \, \frac{1}{(n-1)(n-2)} \, .$$

This is one of the last operations before the numerical integration ; it is performed by two functions GLOBAL and INTEGRATION ; GLOBAL goes through the list of the internal variables $k_1^2, k_2^2, \ldots$ and checks if their degree n in the quadratic form in the denominator is smaller or greater than 2 : if $n > 2$ then the INTEGRATION function performs indeed the required operation. In the special case of the $g\varphi^3$ theory which is superrenormalizable, if $n < 2$ one has to make a subtraction following the formula

$$\frac{1}{a^2} - \frac{1}{b^2} = -2 \int_0^1 \frac{a-b}{\left[(a-b)\beta + b\right]^3} \, d\beta$$

such a procedure transforms the divergent integrals of the $g\varphi^3$-theory into convergent ones, but does not constitute a renormalization, for instance the analyticity properties of the scattering amplitudes which can be calculated following that procedure are lacking. We shall see below that Q.E.D. requires the handling of divergences in a more refined way.

This last step constitutes the end of the program and as a final result we get an integral over the α -parameters which is a function of the external momenta p and where the α_j satisfy both conditions :

$$0 \leq \alpha_j \leq 1 \qquad\qquad \sum \alpha_i = 1$$

Before studying the realistic case of Q.E.D., let me illustrate the previous considerations by looking at one of the simplest function of the program, namely the DENOMINATEUR function which performs the decomposition in monomials of the denominator of a typical scalar Feynman graph.
Such a function is as follows :

```
DEFINE ((
(DENOMINATEUR (LAMBDA (X) (PROG (M DA DB MASS L U S R RESULT
V W)
EO    (SETQ M (CAR X))
      (SETQ DA NIL)
      (SETQ DB NIL)
      (SETQ MASS (LIST 1 (CAR M))
      (SETQ M (CDR M))
E3    (SETQ L (CAR M))
      (SETQ U (CDR L))
      (SETQ V (CAR U))
      (SETQ W (CAR L))
      (SETQ DA (APPEND (LIST 1 (CONS (QUOTE *)(CONS V U)))DA))
      (COND ((NOT (SETQ S (CDR M)))(GO E1)))
E2    (SETQ R (CAR S))
      (SETQ DB (APPEND (LIST (TIMES W (CAR R))(LIST (QUOTE *)
      2 V (CADR))) DB))
      (COND ((SETQ S (CDR S))(GO E2)))
      (COND ((SETQ M (CDR M))(GO E3)))
E1    (SETQ RESULT (CONS (APPEND MASS (APPEND DA DB)) RESULT))
      (COND ((SETQ X (CDR X))(GO EO)))
      (RETURN RESULT)    )))
            ))
```

We first note that there are in the former listing a few new functions which are very simple indeed but have not been yet defined. The first new function we come accross is the PROG function which executes sequentially all the instructions which are given by its argument.

SETQ is a function with two arguments : its first argument is a variable which takes the value of the second argument of the function. The sign (*) is not a variable, the function (QUOTE

*) means that * should not be evaluated.

There are arithmetic functions in LISP , in the listing we considered above we met the function TIMES which has for value the numerical product of its arguments.

Let us now see how the function DENOMINATEUR works by applying it to the decomposition of a single Feynman denominator of the form

$$(p-k)^2 + m^2 = p^2 + k^2 - 2pk + m^2 .$$

Such a propagator is represented by the list

((ME)(-1 K)(1 P))

and the list denoted by X in the denominator function is reduced to a single element, namely the previous one

X = (((ME)(-1 K)(1 P)))

it is clear that the first bracket in X means that we consider the list of all propagators whose product constitutes the Feynman integral under study. The X we are considering is the simplest example where there is only one propagator, it is very unrealistic indeed.

We now apply the DENOMINATEUR function to X :

X = ((ME)(-1 K)(1 P)))

```
E0  (SETQ M (CAR X))            M = ((ME)(-1 K)(1 P))
    (SETQ DA NIL)               DA = ( )
    (SETQ DB NIL)               DB = ( )
    (SETQ MASS (LIST 1          MASS = (1 (ME))  (one gets the
     (CAR M)))                                    term  + ME)
    (SETQ M (CDR M))            M = ((-1 K)(1 P))
E3  (SETQ L (CAR M))            L = (-1 K)
    (SETQ U (CDR L))            U = (K)
    (SETQ V (CAR U))            V = K
    (SETQ W (CAR L))            W = -1
    (SETQ DA (APPEND (LIST 1    DA = (1 (* K K))  (one gets the
     (CONS (QUOTE *)(CONS V                     term  + K^2)
     U))) DA))
    (COND((NOT(SETQ S (CDR      S = ((1 P)
     M)))(GO E1)))              Since S is not empty, one goes
                                over to the next instruction
E2  (SETQ R (CAR S))            R = (1 P)
    (SETQ DB (APPEND (LIST      DB = (-1 (* 2 K P)) (one gets the
      (TIMES W (CAR R)(LIST                       term  -2 K P )
      (QUOTE *) 2 V (CADR R)))DB))
```

```
    (COND ((SETQ S (CDR S))      S = ( )
     (GO E2)))                   S  is empty, one goes over to the
                                 next instruction.
    (COND ((SETQ M (CDR M))      M = ((1 P))
      (GO E3)))                  We therefore go back to  E3
E3  (SETQ L (CAR M))             L = (1 P)
    (SETQ U (CDR L))             U = (P)
    (SETQ V (CAR U))             V = P
    (SETQ W (CAR L))             W = 1
    (SETQ DA (APPEND (LIST 1     DA = (1 (* P P) 1 (* K K))
     (CONS (QUOTE *)(CONS V U)
     ))DA))
    (COND ((NOT (SETQ S (CDR     S = (  )
     M)))(GO E1)))               the argument of  NOT  being  NIL,
                                 one goes over to  E1
E1  (SETQ RESULT (CONS (APPEND   RESULT = ((1 (ME) 1 (* P P) 1
     MASS (APPEND DA DB))                 (* K K) -1 (* 2 K P)))
     RESULT))
    (COND ((SETQ X (CDR X))      X = (  )
     (GO EO )))                  X  being  NIL , one goes over to
                                 the next instruction
    (RETURN RESULT)              the value of the function
                                 DENOMINATEUR  is now the value of
                                 RESULT.
```

This is the simplest example of the different functions one needs for the evaluation of the radiative corrections in the scalar theory with a cubic interaction, the total ACOFIS program contains 14 functions and the interested reader should go back to Perrottet's original thesis.

Q.E.D. differs from a scalar theory by several features which make the symbolic part of the calculation more cumbersome and the analytical one more refined. The denominators of the propagators are slightly different from the ones of the scalar theory since we have now two masses to consider : the mass m_e of the electron and the mass 0 of the photon. But we have also to characterize the numerator by giving in addition to the lists of internal and external momenta and masses the list of the γ-matrices. We shall see later that to eliminate ultraviolet divergencies, we shall need to introduce in the input some more informations concerning the counterterms.

I cannot even think of describing all the needed LISP functions in Q.E.D., such a task would require several lectures. Because of the complexity of the algorithm, interested listeners have to look at the listings which are included in Calmet thesis 22). As a matter of fact, one catches every opportunity provided by the symmetries of the integral or the physics of the problem in order to simplify parts of the program. For instance, in the solution of one of the more important problems which have been solved in the past few years, namely the sixth-order radiative correction to the anomalous magnetic moment of the electron, there are important simplifications since we are interested in the contribution of the vertex function to the $\sigma_{\mu\nu}$ term.

Let us now consider divergent graphs : through the generating functional formalism, we know how to associate to a definite graph on unambigous number of counterterms which are given by divergent integrals of perfectly defined rational functions as integrands. The main idea of the method which has been used is to consider the sum of the integrand of the unrenormalized integral and the integrands of the integrals arising from the counterterms. Since such a sum has a well-defined integral, we may use a numerical integration method (such as RIWIAD) in order to evaluate the renormalized integral.

As far as the anomalous magnetic moment is concerned, there are some further simplifications. We note first of all that in the evaluation of a vertex graph, the skeleton divergence may be dropped since its tensorial dependence is only in γ_μ and we are interested only in the $\sigma_{\mu\nu}$ coefficient ; this remark means also that the contribution of any graph without internal divergences to the anomalous magnetic moment a_e can be computed in a straightforward way. Then, considering the way these counterterms are obtained, we can always associate the corresponding lines of the original graph and of its counterterms with the same α parameter ; this is an important remark, since we already saw that the points where the integrands are numerically calculated are chosen at random. We may finally remark that one can get the contributions to a_e from the counterterms of a diagram by performing only slight modifications in the program used for the original graph. In fact the main modification is that some internal momenta have to be placed on their mass shell : these are the momenta which are external for each divergent subdiagram. A simple example is given by the following vertex and its counterterm

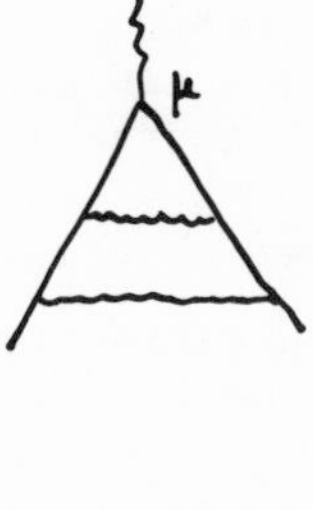

the bubble in the counterterm stays for the divergent integral over the momentum k of the well-defined rational function :

$$\frac{\gamma_\nu (i\hat{m}_s - i\hat{k} - m) \gamma_\mu (i\hat{m}_s - i\hat{k} - m) \gamma_\nu}{\left[(m_s - k)^2 + m^2 \right]^2 \left[k^2 + \lambda^2 \right]}$$

where $\hat{m}_s$ is a four-momentum on the mass-shell. By that we mean that because of the anticommuting properties of the γ -matrices, the former expression is decomposed into a sum of terms whose first and last factors are the factors $\hat{m}_s$: then one replaces $i\hat{m}_s$ by the electron mass -m .

One may proceed in the same way for self-energy subdiagrams, nevertheless in that case one may use another feature of the program, i.e. the introduction of the finite parts of the self-energies or vacuum polarization. One is led to simpler expressions and one may save much computing time.

Programs with the same scope than the one we have been describing have been compiled by other physicists : Levine and Wright 26) have been using the ASHMEDAI language which is a specialized powerful language working on computers of several makes, Kinoshita and Cvitanoviç 28) performed all the needed algebra in an extremely economical way using a formulation due to Brodsky and Kinoshita 27) (which was improved by the last Author) and a program on a PDP-10 in the TECO editing language.

It should be remarked that all the output of the programs written in LISP or REDUCE need to be translated into FORTRAN when one has to perform the numerical integration : such a translation is assured automatically by means of several conversion functions.

Let me add a few words about another problem which arises in Q.E.D., namely the problem of infrared divergences (IRD) whose removal is straightforward when the integrals are analytically evaluated, in which case one needs only to assign a mass to the photon and drop factors in either terms in $\log \lambda$ or $\log^2 \lambda$ where λ is the ratio of the photon and electron masses ; one then goes over to the limit $\lambda = 0$. Unfortunately, it is not possible to computerize this procedure at present. The removal of IRD is a troublesome task and requires some computing times. One may, for instance, employ least square fitting techniques and in order to improve the accuracy of the result one computes in the same run a sum of diagrams which are known to be free of IRD . More specifically, one performs the calculations for a set of values of λ between 10^{-1} and 10^{-3} , the results are then fitted by a curve with the equation

$$G(\lambda) = A + B\lambda + C\lambda\log\lambda + D\lambda^2\log\lambda + E\lambda^2\log^2\lambda + \dots$$

and one looks for the intersection of that curve with the λ axis. The coefficients A , B , C , ... which are referred as background terms are generally smooth enough for a good determination of the term A .

One may use also the method of Intermediate Renormalization since there are no infrared divergences for the vertex Γ_μ at the points $p_2 = p_1 = 0$ and one computes :

$$\Gamma_\mu^{ren}(p_2,p_1) = \left[\Gamma_\mu(p_2,p_1) - \Gamma_\mu(0,0)\right] + \left[\Gamma_\mu(0,0) - \Gamma_\mu(p_0,p_0)\right]$$

where p_o represents the momenta p_2, p_1 on the mass-shell.

9. THEORETICAL AND EXPERIMENTAL RESULTS : COMPARISON AND COMMENTS [28)]

I intend to give in this last paragraph a brief survey of the calculated estimates of the electrons and muons anomaly and of the Lambshift in hydrogen atom, I will then compare theory and experimental data.

a) Anomalous magnetic moment of leptons.

Let us consider first the electron anomaly ; only one well-known graph contributes to the second order radiative correction and gives the famous Schwinger contribution $\alpha/2\pi$. There are the seven well-known graphs which contribute to the fourth-order correction in e

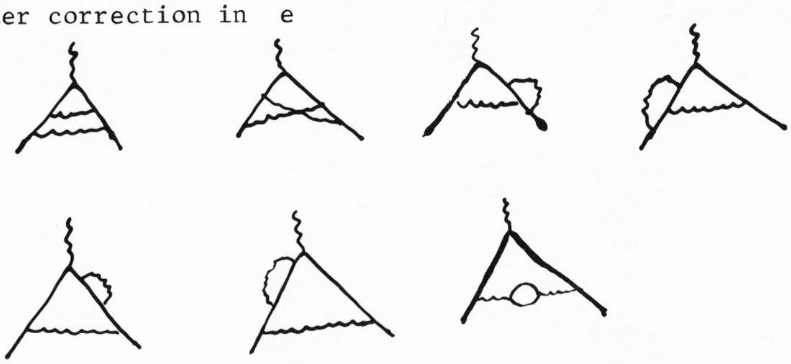

the analytic result : $-0.328(\frac{\alpha}{\pi})^2$ (which includes di- and tri-logarithmic functions) is in a very good agreement with the one obtained through a computer 29) .

The increase of difficulty between the fourth and the sixth order is really striking : first of all there are 72 graphs and their complexity is such that an analytic evaluation in closed form can be performed for only a few diagrams. Among these 72 graphs only 40 have to be calculated if one takes into account all existing symmetries and due to gauge invariance considerations, all those graphs may be classified within 6 classes :

I) The 6 graphs including the fourth-order light by light scattering graph (graph 1,...6)

II) The 12 graphs with second order photon self-energies insertions (graph 7,...18)

III) The 4 graphs with fourth-order photon self-energies insertions (graphs 19,...22)

IV) The 6 graphs including three photons exchanged (graphs 23,... 28)

V) The 20 graphs including two photons exchanged (graphs 29,... 48)

VI) The last 24 graphs with one photon exchange (graph 49,... 72)

From B. E. Lautrup, A. Petermann and E. De Rafael (2)

The graphs belonging to the first three classes have been at least calculated twice, generally the results obtained using computers or for few of them an analytic method are in rather good argument 30, 28).

The remaining three classes include 50 graphs and have been evaluated separately by Kinoshita, Cvitanoviç and by Levine, Wright the known results -since not everything has been published- do not agree, in units of $(\alpha/\pi)^3$ they obtain respectively : 1,02 ± 0,04 and 0,883 ± 0.06 . Because of the fact that each group handles differently the ultra-violet and infrared divergences, it appears as impossible to compare the values of individual graphs. Some of the graphs among the 50 we just mentioned have been calculated by several authors using either classical Feynman techniques or the new method of Levine-Roskies 31, 28 , 9).

If we now come to the comparison with experimented data, it turns out that we have two figures, one resulting from the Kinoshita-Cvitanoviç group : $(1,29 \pm 0.07)(\frac{\alpha}{\pi})^3 = (1159652,9 \pm 2,4)10^{-9}$ and the second from the Levine-Wright group : $(1.15 \pm 0.07)(\frac{\alpha}{\pi})^3 = (1159651,1 \pm 2.5)10^{-9}$ to be compared with the experimental value obtained by Wesley and Rich : $(1159656,7 \pm 3.5)10^{-9}$.

Let us now consider the muon anomalous magnetic moment and begin with a remark valid for the electron anomaly. It has to be noted that in our former considerations we did not include graphs with muons (or heavier particles) as virtual particles. For instance looking at the fourth-order corrections to electron anomaly, there are insertions of vacuum polarization graphs into the vertex part where the virtual photon emits and absorbs a virtual pair of muons. It turns out that, as far as electrons are concerned, the values of such graphs depend on the ratio $(m_e/m_\mu)^2$, thus such corrections terms are small indeed. But rigorously speaking, we should have decomposed the electron anomaly as follows :

$$a_e = \frac{g_e - 2}{2} = a_1 + a_2(\frac{m_e}{m_\mu}) .$$

Let us now turn our attention to the muon anomaly, the same line of reasoning as before leads us the decomposition of the muon anomaly a_μ :

$$a_\mu = a_1 + a_2(\frac{m_\mu}{m_e})$$

and finally :

$$a_\mu - a_e = a_2(\frac{m_\mu}{m_e}) - a_2(\frac{m_e}{m_\mu})$$

We just need to calculate only graphs contributing to $a_2(\frac{m_\mu}{m_e})$ if we suppose known an accurate value of a_e .

There is also another effect which contributes to a precise measurement of a_μ : the resonances in the $\pi - \pi$ system do indeed contribute in the evaluation of the vacuum polarization effects. This effect has been calculated by dispersive techniques. Due to a lack of time, I cannot enter into the details of such calculations and the interested reader should go to the quoted reference, but let me quote the overall result for the muon anomaly.

$$a_\mu - a_e = (624.5 \pm 6)\ 10^{-9}$$

from which one gets the theoretical value

$$a_\mu = (116589.7 \pm 8)\ 10^{-9}$$

to be compared with the most recent experimental value 28)

$$a_\mu^{exp} = (116616 \pm 31)\ 10^{-8} \quad .$$

b) Lamb shift

As already said, the lamb shift depends on the value of the slope σ of the Dirac form factor $F_1(p_2^2)$. The experimental accuracy requires the fourth-order contributions to that slope, one has therefore to calculate :

$$\sigma^{(4)} = m^2 \left. \frac{dF_1(p^2)}{dp^2} \right|_{p^2=0}$$

The seven fourth-order graphs, (see §9, a), contribute to the value of that slope from which one has to extract, following the formula in §2 giving the canonical form of the electromagnetic vertex, the factor of γ_μ .

Historically, after the pioneering paper by H. Bethe, the first calculations were made by Weneser, Bersohn, and Kroll 32) in 1953 and were subsequently followed by several others which may be found in Ref. 28). It turns out that

$$\sigma^{(4)} = 0.4699 \times \left(\frac{\alpha}{\pi}\right)^2$$

which leads to the following value of the Lamb shift in hydrogen atom

$$\left(1057{,}911 \pm 0.012\right) \text{ MHz}$$

in excellent agreement with the experimental value.

There is a last class of problems which were not mentioned in my survey, namely the bound states problems. They are important indeed and interesting corrections have been calculated, but a systematic method dealing with the Bethe-Salpeter equation through computing techniques is unfortunately lacking.

ACKNOWLEDGMENTS

I enjoyed several discussions with Dr. J. Calmet who provided me with his not yet published review paper at the Third Colloquium on Advanced Computing Methods in Theoretical Physics.

REFERENCES

1. General review papers on Computing Science :
D. BARTON and J.P. FITCH - Applications of Algebraic Manipulative Programs in Physics. Reports on Progress in Physics, 1972 35 253 .
J.A. CAMPBELL - Comparative Survey of Programing Languages in Computing as a Language of Physics (CLP) International Atomic Energy Agency - Vienna 1972 .
A.C. HEARN - Computer Solution of Symbolic Problems in Theoretical Physics CLP Atomic Energy Agency - Vienna 1972 .
See also several reports in :
Colloquium on Computational Methods In Theoretical Physics, Centre de Physique Théorique de Marseille 1970 , 1971 , 1973 . (The 1973 Proceedings are on the way to be published).

2. General review papers on Physics of electrons and photons.
J. CALMET - A Review of Computational Q.E.D. to be published in the Proceedings of the "Third Colloquium on Computational Methods in Theoretical Physics", Centre de Physique Théorique de Marseille 1973 .
B.E. LAUTRUP, A. PETERMANN and E. DE RAFAEL - Physics Reports 4 3 1972 .
N.M. KROLL - Proceedings of the Third International Conference on Atomic Physics - Boulder 1972 (Plenum N.Y.).
A. RICH , J.C. WESLEY - Rev. Mod. Phys. 250 44 1972 .
All the above mentioned reports contain very detailed bibliography on Q.E.D. problems.

3. R. ROBISCOE and T. SHYN - Phys. Rev. Lett. 24 559 1970 .

4. D.T. WILKINSON and H.R. CRANE - Phys. Rev. 130 852 1963 .
J.C. WESLEY and A. RICH - Phys. Rev. A4 1341 1971 .

5. J. BAILEY et al. - Phys. Lett. 28 B , 287 1968 .

6. For general comments :
D. BARTON and J.P. FITCH - see ref. 1.
F. JAMES : in Proc. Third Colloquium on Advanced Computational Methods in Theoretical Physics - Marseille 1971 (to be published).

7. W. CZYZ , G.G. SHEPPEY and J.D. WALECKA - Nuov. Cim. 34 420 1964 .
See also Reports by A.J. DUFNER , B.E. LAUTRUP in Proc. 1st and 2nd Coll. on Adv. Comp. Meth. in Theor. Phys. - Marseille 1970 and 1971 .

8. J. LEVINE and J. WRIGHT - Proc. 2nd Coll. on Adv. Comp. Methods in Theor. Phys. - Marseille 1971 .

9. J. LEVINE and R. ROSKIES - Phys. Rev. Lett. 30 772 1973 .
See also R. ROSKIES in Proc. "Third Coll. on Adv. Comp. Meth. in Theor. Phys." - Marseille 1973 .

10. R. BARBIERI, J. MIGNACO and E. REMIDDI - Nuov. Cim. 6A 21 1971 , 11A 826 1972 .

11. A. PETERMANN - Proc. 2nd Coll. on Adv. Comp. Meth. in Theor. Phys. - Marseille 1971 .

12. J.A. FOX and A.C. HEARN - To appear in J. Comp. Physics.

13 J. SLADGLE - Ph.D. Thesis MIT 1961 .

14. J. MOSES - Commun. Ass. Comp. Mach. 14 548 1971 where a bibliography may be found.

15. M. VELTMAN - CERN Preprint 1967 .
H. STRUBBE - Proc. 3d Coll. on Adv. Comp. Meth. in Theor. Phys. - Marseille 1973 .

16. M.J. LEVINE - Proc. 3d Coll. on Adv. Comp. Meth. in Theor. Phys. - Marseille 1973 .

17. J.A. CAMPBELL and A.C. HEARN - J. Comp. Phys. 5 280 1970 .
J. CALMET and M. PERROTTET - J. Comp. Phys. 7 191 1971 .

18. A.C. HEARN - REDUCE users manual preprint UCP-19 (March 1973) Univ. of Utah.
A.C. HEARN - Proc. 3d Coll. on Adv. Comp. Methods in Theor. Phys. - Marseille 1973 .

19. J. MAC CARTHY - LISP 1.5 Programers Manual MIT Press.

20. J.A. CAMPBELL and A.C. HEARN - J. Comp. Phys. 5 280 1970 .

21. M. PERROTTET - Computing as a Language of Physics. Int. Atomic Energy Agency - Vienna 1972 .

22. J.CALMET - PhD Thesis - Centre de Physique Théorique - Marseille 1970 .

23. Y. LE GAILLARD and A. VISCONTI - J.Math.Phys. 6 1774 1965.
J. SOFFER and A. VISCONTI - Phys. Rev. 162 1386 1967 .
J. CALMET, R. SENEOR, J. SOFFER and A. VISCONTI - Phys. Rev. 162 1390 1967 .

24. J. CALMET - Preprint Centre de Physique Théorique Marseille 73/P.53 .

25. M. PERROTTET - Thèse de 3ème Cycle - Centre de Physique Théorique 1970 .

26. M.J. LEVINE - Proc. 2nd Coll. on Adv. Comp. Meth. in Theor. Phys. Marseille 1971 .

27. T. KINOSHITA - Proc. 3d Coll. on Adv. Comp. Meth. in Theor. Phys. Marseille 1973 .
P. CVITANOVIC - Proc. 3d Coll. on Adv. Comp. Meth. in Theor. Phys. Marseille 1973 .
S. BRODSKY and T. KINOSHITA - Phys. Rev. D3 356 1971 .

28. General bibliography may be found in either of the first three papers quoted in Ref. 2 .

29. R. KARPLUS and N.M. KROLL - Phys.Rev. 77 536 1950 .
A. PETERMANN - Nucl. Phys. 5 677 1958 .
C.M. SUMMERFIELD - Ann. Phys. (N.Y.) 5 26 1958 .

30. J.A. MIGNACO and E. REMIDDI - Nuov. Cim. 60A 519 1969 .
S. BRODSKY and T. KINOSHITA - Phys. Rev. D3 356 1971 .
J. CALMET and M. PERROTTET - Phys. Rev. D3 2101 1971 .
S. ALDINS, S.J. BRODSKY , A.J. DUFNER and T. KINOSHITA - Phys. Rev. D1 2378 1970 .
J. CALMET and A. PETERMANN - CERN Preprint 1755 to be published.

31. T. KINOSHITA and P. CVITANOVIC - Phys.Rev.Let. 29 1534 1972 .
M.J. LEVINE and J. WRIGHT - Phys. Rev. Let. 26 1351 1971 and University of Illinois Preprint (July 1973) .

32. J. WENESER , R. BERSOHN and N.M. KROLL - Phys. Rev. 91 1257 1953 .

RELATIVISTIC STRINGS AND SUPERGAUGES

Bruno ZUMINO

CERN

Geneva, Switzerland

1. - INTRODUCTION

In these lectures I shall try to describe some work which has been done by a number of people over the past few years to give a description of the hadronic world [1], and some of their ramifications. The main point is that a field theory, such as a quark field theory with a gluon, even if it could be in principle correct, is hopelessly remote from the properties of hadrons one is interested in. This is so because in the strong interaction perturbation theory methods are unreliable (although recent ideas on asymptotic freedom may modify this situation) and also because what one wants is to calculate the entire spectrum of excitations and their interactions, a very difficult mathematical problem. It seems natural, therefore, to consider models which contain from the very beginning the entire spectrum of excitations as well as the main part of their interaction, so that higher order corrections may be expected to introduce only relatively small modifications. A possible approach of this type is the use of infinite component wave equations. Another possibility is to consider extended systems with internal excitations, such as the relativistic string described in the next section. Other extended systems could be the two-dimensional relativistic membrane or possibly a three-dimensional system with some kind of internal structure.

2. - THE NAMBU-GOTO ACTION

The simplest extended system is one-dimensional. Its action can be obtained by analogy with that of the relativistic point particle, which is

$$S = -m \int \sqrt{-\left(\frac{dx}{d\tau}\right)^2}\, d\tau$$

where the path of integration is varied between two fixed end points. One takes the action proportional to the (Minkowski) area of a two-dimensional surface spanned between the initial and the final position of the string. If the surface is described parametrically by the equations

$$x_\mu = \phi_\mu(\xi^0, \xi^1) \qquad \mu = 0,1,2,3$$

the action is then [2]

$$S = -\frac{1}{2\pi\alpha'} \int d^2\xi \sqrt{-g}$$

where

$$g = \det g_{ij}$$

is the determinant of the metric tensor

$$g_{ij} = \partial_i \phi \cdot \partial_j \phi \qquad i,j = 0,1$$

on the surface. The constant α' turns out to be the slope of the Regge trajectory formed by the excitations of the system. We shall set it equal to 1. Because of its geometrical meaning, the action is invariant with respect to a reparametrization of the surface. This invariance gives rise to identities. This is analogous to the situation with the point particle. There the Lagrangian is

$$L = -m\sqrt{-\dot{x}^2} \qquad \dot{x} = \frac{dx}{d\tau}$$

Under an infinitesimal change in the parameter τ

$$\delta x = u(\tau)\frac{dx}{d\tau}$$

the Lagrangian transforms like a density

$$\delta L = \frac{d}{d\tau}(uL) .$$

On the other hand

$$\delta L = \frac{\partial L}{\partial \dot{x}}\delta\dot{x} + \frac{\partial L}{\partial x}\delta x = \frac{\partial L}{\partial \dot{x}}\frac{d}{d\tau}(u\dot{x}) + \frac{\partial L}{\partial x}u\dot{x} =$$

$$= \frac{d}{d\tau}\left(\frac{\partial L}{\partial \dot{x}}\dot{x}u\right) + \left(\frac{\partial L}{\partial x} - \frac{d}{d\tau}\frac{\partial L}{\partial \dot{x}}\right)\dot{x}u .$$

Therefore, since $u(\tau)$ is arbitrary,

$$\frac{\partial L}{\partial \dot{x}}\dot{x} - L \equiv 0 \quad , \quad \left(\frac{\partial L}{\partial x} - \frac{d}{d\tau}\frac{\partial L}{\partial \dot{x}}\right)\cdot\dot{x} = 0 .$$

The Hamiltonian

$$H = p\dot{x} - L \equiv 0 , \qquad p = \frac{\partial L}{\partial \dot{x}} = \frac{m\dot{x}}{\sqrt{-\dot{x}^2}}$$

is identically zero. Now

$$\dot{x} = -\frac{pL}{m^2} ,$$

so one can write

$$H = -(p^2 + m^2)\frac{L}{m^2} .$$

To quantize the system one can take

$$[x_\mu, p_\nu] = i\eta_{\mu\nu}$$

and

$$H = f(\tau)(p^2 + m^2)$$

where $f(\tau)$ is an arbitrary function, corresponding to the arbitrariness in the choice of the parameter τ. It is, of course, convenient to pick a special gauge in which f is a constant, $\dot{x}^2 = -1$, and the parameter is the proper time. The equations of motion which are in general

$$\frac{d}{d\tau}\left(\frac{m}{\sqrt{-\dot{x}^2}}\dot{x}_\mu\right) = 0 \quad ,$$

become then the simple linear equations

$$\ddot{x}_\mu = 0 \; .$$

The situation with the string is very similar. Now

$$L = -\frac{1}{2\pi}\sqrt{-g}$$

$$P_\mu{}^0 = \frac{\partial L}{\partial(\partial_0\phi_\mu)} = \frac{1}{2\pi}\,\frac{\partial_0\phi_\mu\,(\partial_1\phi)^2 - \partial_1\phi_\mu\;\partial_0\phi\cdot\partial_1\phi}{\sqrt{-g}}$$

$$P_\mu{}' = \frac{\partial L}{\partial(\partial_1\phi_\mu)} = \frac{1}{2\pi}\,\frac{\partial_1\phi_\mu\,(\partial_0\phi)^2 - \partial_0\phi_\mu\;\partial_0\phi\cdot\partial_1\phi}{\sqrt{-g}}$$

and the equations of motion are

$$\partial_0 P_\mu{}^0 + \partial_1 P_\mu{}' = 0 \; .$$

Furthermore, for an open string one allows free variations of the boundary of the surface. This gives the boundary conditions

$$P_\mu{}' = 0$$

at the end points of the string. Under a reparametrization

$$\delta\phi_\mu = u^i(\xi)\,\partial_i\phi_\mu$$

the Lagrangian transforms like a density. This gives rise to the identities

$$\partial_1\phi\cdot P^0 \equiv 0 \qquad (P^0)^2 + \frac{1}{4\pi^2}(\partial_1\phi)^2 \equiv 0$$

$$H = \partial_0\phi\cdot P^0 - L \equiv 0 \; .$$

One can choose special gauges in which the co-ordinate lines on the surface satisfy the orthonormality conditions

$$\partial_1\phi \cdot \partial_0\phi = 0$$

$$(\partial_1\phi)^2 = -(\partial_0\phi)^2$$

(the minus sign because of the Minkowski metric) so that the metric $g_{ij} = \lambda(\xi)\eta_{ij}$ is proportional to the flat metric (-1,1). Then

$$P_\mu^0 = -\frac{1}{2\pi}\partial_0\phi_\mu \qquad P_\mu^1 = \frac{1}{2\pi}\partial_1\phi_\mu$$

and the differential equations become

$$(\partial_0^2 - \partial_1^2)\phi_\mu = 0,$$

while the boundary condition is now

$$\partial_1\phi_\mu = 0$$

at the end points, which can be chosen to be $\xi_1 = 0, \pi$. A special solution which satisfies everything is

$$\phi^0 = \xi^0, \quad \phi^1 = \cos\xi^1 \cos\xi^0, \quad \phi^2 = \cos\xi^1 \sin\xi^0, \quad \phi^3 = 0.$$

Other solutions are obtained by a normal mode expansion. This way one can find the spectrum of excitations.

The special gauge chosen does not fix completely the parametrization. Restricted gauge transformations are still possible, which are conformal transformations of the parameters ξ^0, ξ^1,

$$\partial_0 u^1 = \partial_1 u^0 \quad , \quad \partial_0 u^0 = \partial_1 u^1,$$

so that

$$(\partial_0^2 - \partial_1^2)u^i = 0.$$

Observe, however, that the full set of equations of the system, i.e., the differential equations, the boundary conditions and the quadratic gauge conditions, could not be derived from a Lagrangian invariant only under two-dimensional conformal transformations. The Nambu-Goto Lagrangian, which is invariant under general co-ordinate transformations, has the great merit of giving the entire set of equations, some as differential equations, some as identities. The non-linear gauge conditions are essential to show that the system has no ghost states when quantized.

I shall not describe here in detail the quantization procedure. I only wish to mention a very remarkable result [3]. It turns out that a consistent relativistic quantization is possible only in a space-time of 25+1 dimensions. This fact does not discourage people working in this field, because the number of dimensions necessary for consistent quantization depends on the number of degrees of freedom of the system. For instance in the Neveu-Schwarz model (Section 4) which contains also a spinor it is 10 instead of 26, and it is possible to make models with internal symmetries where it goes down to 4. So one hopes that when a realistic model describing the true spectrum of particles and resonances is found, it will be consistent in four dimensions.

Instead of taking an open string one could have taken a closed string. One would have obtained the same differential equations but, instead of the boundary conditions given above, one would have a periodicity condition. The spectrum of excitations would be different. While the open string gives rise to the spectrum of the Veneziano model [4], the closed string gives rise to that of the Shapiro-Virasoro model [5].

How should interactions be introduced ? Contrary to the procedure customary in quantum field theory, it does not seem necessary to add interaction terms to the Lagrangian of the system (such interaction terms would modify the spectrum of the free string). Instead, due to the non-linear nature of the equations (gauge conditions), the same Lagrangian can describe the process of scattering of two or more strings just by extending the integral over the surface having the given strings as initial and final configuration. The quantization of such a topologically more complicated situation is not easy, and even classical solutions are hard to find.

It is easier to solve the problem of a string in interaction with external fields [6]. The interaction must and can be introduced in such a way as to be invariant under co-ordinate transformations on the surface. This is necessary so that ghost states not be generated. It turns out that for the open string the appropriate external field is a vector field, while for the closed string it is a tensor field of spin two. From the resulting, e.g., vector n point functions one can obtain by factorization all the amplitudes of the Veneziano model. Similarly for the Shapiro-Virasoro model.

3. - THE RELATIVISTIC STRING AS A VORTEX LINE

One may ask what is the relation between the theory of the relativistic string and the usual local quantum field theory. There are two ways of looking at this. The first is to consider the analogy, mentioned in the previous section, between the point particle and the string. Just like the first quantization of the point particle introduces a wave function $\psi(x)$ which then becomes a local field operator in second quantization, one can formulate the first quantization of the relativistic string in terms of a functional $\Psi[\ell]$ of the string. In second quantization this would become a multilocal field operator satisfying equations involving variational derivatives, analogues of the differential equations satisfied by the local operator $\Psi(x)$. I shall not pursue here this very interesting approach, which could develop into a complete quantum theory of multilocal fields.

A second point of view tries to find something like a relativistic string among the solutions of local quantum field theory. This is exemplified by some work of Nielsen and Olesen [7] to which I made a small contribution. One starts with the observation that a (closed or infinite) string resembles very much a vortex line. In non-relativistic physics vortex lines occur in superfluids (liquid helium or superconductors). A relativistic vortex line can be expected to occur in a relativistic field theory which resembles mathematically the equations of a superfluid. This is the case for the Higgs model, which has equations which in the static limit become identical with the Landau-Ginzburg equations for superconductors [8], which are well known to have vortex line solutions.

The Lagrangian of the Higgs model is

$$L = -\frac{1}{4}F_{\mu\nu}^2 - |(\partial_\mu - ieA_\mu)\varphi|^2 - \mu^2|\varphi|^2 - h|\varphi|^4$$

$$F_{\mu\nu} = \partial_\mu A_\nu - \partial_\nu A_\mu .$$

In the case $\mu^2 < 0$ (h is always taken > 0 for stability) one has "spontaneous symmetry breaking". The differential equations have a solution such that for large space-like distances $|\varphi| = \sqrt{-\mu^2/2h} = \lambda/\sqrt{2}$. Observe that the vector current is

$$j_\mu = 2e^2 A_\mu \varphi^*\varphi + ie(\varphi^*\partial_\mu\varphi - \varphi\,\partial_\mu\varphi^*)$$

or

$$eA_\mu - \frac{j_\mu}{2e\varphi^*\varphi} = \partial_\mu \chi$$

where χ is the phase of φ

$$\varphi = |\varphi| e^{i\chi} .$$

Since the wave function must be uniform, the phase must change by a multiple of 2π along any closed curve. Therefore

$$\oint \left(eA_\mu - \frac{j_\mu}{2e|\varphi|^2} \right) dx^\mu = 2\pi n .$$

This is the quantization of the "fluxion". Where the current vanishes the line integral becomes just the magnetic flux. In the usual treatment of the Higgs model one considers solutions for which $n = 0$. The phase χ can then be transformed to zero by means of a gauge transformation. Vortex lines occur for solutions where $n \neq 0$ for some line integrals.

One can see things more explicitly by exhibiting a solution in the static, cylindrically symmetric case. Let ρ, z, ω be the cylindrical co-ordinates and take $A_0 = A_z = A_\rho = 0$ $A_\omega(\rho) \equiv A(\rho) \neq 0$. One can easily see that the equations of motion become

$$-\frac{1}{\rho}\frac{d}{d\rho}\left(\rho \frac{d}{d\rho}|\varphi|\right) + \left[\left(\frac{n}{\rho} - eA(\rho)\right)^2 + \mu^2 + 2h|\varphi|^2\right]|\varphi| = 0$$

$$-\frac{d}{d\rho}\left(\frac{1}{\rho}\frac{d}{d\rho}(\rho A)\right) + e|\varphi|^2\left(eA - \frac{n}{\rho}\right) = 0$$

where

$$\varphi = |\varphi| e^{in\omega} .$$

Let us consider the case $n = 1$. Asymptotically for large ρ one has $|\varphi| = \lambda/\sqrt{2}$. Putting this asymptotic value into the equation for A one finds the approximate solution

$$eA = \frac{1}{\rho} + C K_1\left(\frac{e\lambda}{\sqrt{2}}\rho\right) \sim \frac{1}{\rho} + C\sqrt{\frac{\pi}{e\sqrt{2}\lambda\rho}}\; e^{-\frac{e\lambda\rho}{\sqrt{2}}}$$

and the magnetic field is

$$H(\rho) = \frac{1}{\rho}\frac{d}{d\rho}\left(\rho A(\rho)\right) = C\,\frac{\lambda}{\sqrt{2}}\,K_0\left(\frac{e\lambda\rho}{\sqrt{2}}\right)$$

$$\sim \frac{C}{e}\sqrt{\frac{\pi\lambda}{2\sqrt{2}\,e\rho}}\;e^{-\frac{e\lambda\rho}{\sqrt{2}}} \,.$$

Here C is a constant which could be determined by numerical integration. The approach of $|\varphi|$ to its constant value is described by

$$|\varphi| = \frac{\lambda}{\sqrt{2}} + \beta(\rho) \qquad \beta(\rho) \sim e^{-\sqrt{-\mu^2}\,\rho} \,.$$

Of course, the above solutions are valid only for large ρ. One can study the behaviour of the solutions for small ρ. One finds that the magnetic field is regular and that φ vanishes. For a solution with quantum number n, it behaves like

$$\varphi \sim \rho^n e^{in\omega} \,.$$

The solution described above corresponds to an infinitely long vortex line coinciding with the z axis. The vortex extends over a region characterized by the two masses $e\lambda$ (which determines how far the magnetic field goes) and $|\mu|$ (which determines the region of variation of φ). If both $e\lambda$ and $|\mu|$ are very large, the vortex resembles a string. The magnetic energy per unit length is given by

$$\frac{1}{2}\int_0^\infty H^2\, 2\pi\rho\, d\rho \sim \lambda^4 e^2 \int_0^\infty \left[K_0\left(\frac{e\lambda}{\sqrt{2}}\rho\right)\right]^2 2\pi\rho\, d\rho =$$

$$= 2\pi\lambda^2 \int_0^\infty \left[K_0(y)\right]^2 y\, dy \sim \lambda^2$$

while the energy carried by the field φ is approximately

$$\mu^2\,\frac{1}{\mu^2}\,\lambda^2 = \lambda^2 \,.$$

So the total energy per unit length $\sim \lambda^2$ and should be identified with the constant $1/(2\pi\alpha')$ of the Nambu-Goto action. This means that the characteristic length of the hadronic excitations is $1/\lambda$ or the characteristic mass λ. To be able to talk about a string one must assume

$$e\lambda \gg \lambda \quad \text{and} \quad |\mu| \gg \lambda$$

In particular this means $e \gg 1$, a strong coupling limit. Observe that $e\lambda$ and $|\mu|$ are essentially the masses of the vector and of the scalar particle which one finds in the usual treatment of the Higgs model.

Clearly the Higgs model admits vortex type solutions in which the vortex line is not infinite but closed. To have an open vortex line presents difficulties because the quantized magnetic flux would have to be generated at one end and absorbed at the other. It may be possible to make such a model by putting magnetic monopoles of opposite magnetic charge at the end points. Parisi [9] has observed that the presence of the quantized vortex line would change the nature of the potential between the two monopoles. Instead of a Coulomb potential which is what one would expect when the field φ vanishes at infinity, in the case of "spontaneous symmetry breaking" one expects a potential increasing linearly with the distance between the monopoles, as soon as this distance is sufficiently large to use the above estimate of the energy per unit length. This is interesting because it could explain why quark-monopoles cannot escape.

Finally, let us ask how the "string" will move, as a consequence of the time-dependent field equations. It would be interesting to make a detailed mathematical analysis, but in the case of a sufficiently thin string of not too large curvature it is very plausible that the string will move approximately according to the equations following from the Nambu-Goto action. The reason is simply that the Nambu-Goto action is the only invariant action one can write without using higher derivatives.

4. - FERMIONS AND SPIN

A very interesting model has been studied by Neveu and Schwarz [10]. It is related to a model for spinors studied by Ramond [11]. Written as a two-dimensional field theory the equations of the model are

$$\Box \phi = 0 \qquad \gamma^i \partial_i \psi = 0$$

$$4\,\partial_i \phi \cdot \partial_j \phi + i\,\bar{\psi} \cdot (\gamma_i \partial_j + \gamma_j \partial_i)\psi = \lambda\,\eta_{ij}$$

$$\gamma^i \gamma^j \partial_i \phi \cdot \psi = 0$$

Here ϕ and ψ are respectively a two-dimensional scalar and spinor (two components) which are also four-vectors in space-time. The dot denotes the product of two four-vectors and η_{ij} is the two-dimensional metric $(-1,1)$. The two gamma matrices are

$$\gamma^0 = \begin{pmatrix} 0 & 1 \\ -1 & 0 \end{pmatrix} \qquad \gamma^1 = \begin{pmatrix} 0 & 1 \\ 1 & 0 \end{pmatrix}.$$

The spinor can be taken to be Hermitian. To the above differential equations one must add boundary conditions at the end points of the "string". There are two possible choices. Writing

$$\psi = \begin{pmatrix} \psi_1 \\ \psi_2 \end{pmatrix} ,$$

one may impose that $\psi_1 = \psi_2$ at both end points or that $\psi_1 = \psi_2$ at one end point and $\psi_1 = -\psi_2$ at the other end point. When one goes to a normal mode expansion, in the first case one obtains the Ramond model, in the other the Neveu-Schwarz model.

Going back to the above equations one sees that in addition to the linear Klein-Gordon and Dirac equation one has two non-linear conditions. The first is a generalization of the gauge condition described in Section 2. The second is a side condition of spinor character and is called the supergauge condition.

The above model has been extensively studied. The spectrum and the amplitudes possess very interesting properties and the model is ghost-free for the critical dimension 10. We do not have the time to describe all these interesting developments. Here we would like to look in some detail at a particular aspect, which comes from the occurrence of the supergauge conditions.

Just like the equations of the Veneziano model are invariant under conformal transformations, those of the Neveu-Schwarz model are invariant under a larger group, that of supergauge and conformal transformations. The supergauge transformations are defined in infinitesimal form by [12]

$$\delta\phi = i\bar{\alpha}\psi$$

$$\delta\psi = \partial_i\phi\,\gamma^i\alpha + F\alpha$$

$$\delta F = i\bar{\alpha}\gamma^i\partial_i\psi$$

where $\alpha(\xi)$ is the infinitesimal parameter of the transformation ; it is a (two-component) Hermitian spinor totally anticommuting with itself and with ψ, commuting with ϕ and F, and satisfying the equations

$$(\gamma_i\partial_j + \gamma_j\partial_i + \eta_{ij}\gamma^\ell\partial_\ell)\alpha = 0 .$$

F is an auxiliary field, which is necessary to ensure the group structure, although it is easily eliminated from the Lagrangian

$$L = -\tfrac{1}{2}\partial_i\phi\partial^i\phi - \tfrac{i}{2}\bar{\psi}\gamma^i\partial_i\psi + \tfrac{1}{2}F^2$$

since its equation of motion is simply F = 0. The commutator of two supergauge transformations δ_1 and δ_2 of parameters α_1 and α_2 is easily evaluated. For instance

$$\delta_2\delta_1\phi = i\bar{\alpha}_1\delta_2\phi = i\bar{\alpha}_1\gamma^i\alpha_2\partial_i\phi + i\bar{\alpha}_1\alpha_2 F.$$

Since α_1 and α_2 anticommute, one has

$$\bar{\alpha}_1\alpha_2 = \bar{\alpha}_2\alpha_1$$

$$\bar{\alpha}_1\gamma^i\alpha_2 = -\bar{\alpha}_2\gamma^i\alpha_1 .$$

Therefore

$$(\delta_2\delta_1 - \delta_1\delta_2)\phi = a^i\partial_i\phi$$

where

$$a^i = 2i\bar{\alpha}_1\gamma^i\alpha_2 .$$

Observe that $a^i(x)$ satisfies the equations for an infinitesimal conformal transformation

$$\partial_i a_j + \partial_j a_i + \eta_{ij} \partial^\ell a_\ell = 0$$

as a consequence of the equation satisfied by α_1 and α_2. The same result could be obtained for ψ and F. The commutator of two supergauge transformations is a conformal transformation. On the other hand, two conformal transformations commuted give a conformal transformation while a conformal and a supergauge transformation give a supergauge transformation. We are in presence of a Lie algebra. Actually, it is not a Lie algebra of the usual kind, since some of the parameters are commuting quantities and some anticommuting quantities. It is a Lie algebra with parameters belonging to a Grassmann algebra. Under a supergauge transformation or a conformal transformation the above Lagrangian changes by a total derivative. The action integral is invariant.

The above Lagrangian gives the Klein-Gordon equation and the Dirac equation, together with the equation $F = 0$. One may ask whether one can find a Lagrangian from which follows also the gauge condition and the supergauge condition of the Neveu-Schwarz model. This would be the analogue of the Nambu-Goto Lagrangian, a "string picture" of the Neveu-Schwarz model. Clearly, one must at least write the theory in a form invariant under general (rather than conformal) co-ordinate transformations in two dimensions. This was done by Kikkawa and Iwasaki [13]. However, in order to obtain not only the gauge condition but also the supergauge condition, this is not enough. To achieve this aim Wess and I [14] have introduced generalized supergauge transformations which have the property that their commutator gives a general co-ordinate transformation. We have then an enlarged algebra consisting of general co-ordinate transformations and generalized supergauge transformations rather than the algebra discussed above consisting of conformal and supergauge transformations. A Lagrangian theory invariant under this enlarged algebra can be constructed and provides a string picture of the Neveu-Schwarz model.

From the point of view of four-dimensional field theory, one may ask the question whether supergauge transformation can be defined in four space-time dimensions. The answer is positive, but here their commutator gives a four-dimensional conformal transformation accompanied by a γ_5 transformation. It is also possible to construct invariant Lagrangians with interaction. Since the supergauge transformations link tensors (scalars, etc.) to spinors, the coupling constants for spinor and tensor interactions are related, as are their masses. One finds here a situation similar to that of SU(6) invariance (or its relativistic generalizations) with two important differences. SU(6) linked different kinds of tensors (scalars, pseudoscalars, vectors) to each other and different spinors (spin $\frac{1}{2}$ and spin $\frac{3}{2}$) to each other ; supergauge transformations link tensors with spinors. SU(6) was not an invariance of the Lagrangian (the kinetic terms were not invariant) ; supergauge transformations truly leave the action invariant. The possibility of constructing truly invariant actions giving rise to a non-trivial S matrix seems to violate some of the anti SU(6) no go theorems. The reason is probably that those theorems were proved under the assumption that the group in question was an ordinary Lie group and not one of the extended Lie groups with anticommuting as well as commuting parameters which we have considered above. It is well possible that supergauge transformations, originally studied in two-dimensions in connection with dual models, will play their most natural rôle as invariances of four-dimensional field theory. In the simple renormalizable four-dimensional models studied until now one finds that the invariance under supergauge transformations, with the relations among masses and coupling constants it gives rise to, implies cancellation among divergent diagrams. As a consequence the field theory becomes less divergent than it would have been if those relations had not been satisfied.

REFERENCES

1) For an excellent description of the theory of dual models, see the 1970 Brandeis lectures of S. Mandelstam, edited by Deser, Grisaru and Pendleton, MIT Press, Vol. 1, and the more recent article by J.H. Schwarz in Physics Reports 8C, No 4 (September 1973).

2) Y. Nambu - Lectures at the Copenhagen Summer Symposium (1970) ;
O. Hara - Progr.Theor.Phys. 46, 1549 (1971) ;
T. Goto - Progr.Theor.Phys. 46, 1560 (1971) ;
L.N. Chang and T. Mansouri - Phys.Rev. D5, 2535 (1972) ;
T. Mansouri and Y. Nambu - Phys.Letters 39B, 375 (1972).

3) P. Goddard, J. Goldstone, C. Rebbi and C.B. Thorn - Nuclear Phys. B56, 109 (1973).

4) G. Veneziano - Nuovo Cimento 57A, 190 (1968).

5) M.A. Virasoro - Phys.Rev. 177, 2309 (1969) ;
A. Shapiro - Phys.Letters 33B, 361 (1970).

6) R. Ademollo, A. d'Adda, R. d'Auria, E. Napolitano, S. Sciuto, P. Di Vecchia, F. Gliozzi, R. Musto and F. Nicodemi - CERN Preprint TH. 1702 (1973).
See also T. Mansouri and Y. Nambu - Ref. 2).

7) H.B. Nielsen and P. Olesen - Preprint Niels Bohr Institute, Copenhagen.

8) See :
J.B. Keller and B. Zumino - Phys.Rev.Letters 7, 164 (1961) ; earlier relevant references can be found there.

9) G. Parisi - Private communication to the author.

10) A. Neveu and J.H. Schwarz - Phys.Rev. D4, 1109 (1971).

11) P. Ramond - Phys.Rev. D3, 2415 (1971).

12) This way to introduce supergauge transformations and the use of the auxiliary field F is due to J. Wess and the author. For an earlier related investigation, see :
J.L. Gervais and B. Sakita - Nuclear Phys. B34, 632 (1971).

13) Y. Iwasaki and K. Kikkawa - Preprint City College of New York.

14) J. Wess and B. Zumino - in preparation.

APPLICATION OF GAUGE THEORIES TO WEAK AND ELECTROMAGNETIC INTERACTIONS

B. Zumino

CERN - GENEVA

INTRODUCTION

Quantum electrodynamics is the typical example of a successful theory. Its accuracy is quoted to be of a few parts in a million, from comparison with experimental data on the g-2 of the muon or on hyperfine structure in hydrogen. This remarkable success is due to the smallness of the coupling constant

$\frac{e^2}{\hbar c} \approx \frac{1}{137}$ and to the possibility of renormalizing the pertubation theory divergences. The usual theory of weak interactions, on the other hand, is not renormalizable. Therefore, in spite of the smallness of the coupling constant ($Gm_P^2 \approx 10^{-5}$, where G is the Fermi constant) higher order calculations are meaningless. Even in its less divergent form (that with an intermediate boson) the theory has quadratic divergences. These present a problem not only because the theory is not finite. Worse, it is not possible to understand how the higher order corrections can be smaller than the lowest order. For instance, if a process goes like G in lowest order, the next order will go like $G^2\Lambda^2$, where Λ is a cut-off and it will not be smaller unless $G\Lambda^2$ is appreciably smaller than unity. For some processes, this kind of interpretation requires rather small values of Λ, of a few GeV. This corresponds to admitting that the theory fails already at uncomfortably small energies.

It would clearly be desirable to have a renormalizable theory of weak interactions. Now the renormalizability of quantum electrodynamics depends to a large extent on gauge invariance. On the other hand, electromagnetic and weak interactions share the property of being universal. This suggests the idea of a unified theory of both interactions, in which gauge invariance is extended to a kind of Yang-Mills invariance which, hopefully, has the effect of reducing the degree divergence and of rendering the theory renormalizable.

The theory of massless Yang-Mills fields is renormalizable at least in the sense of ultraviolet divergences, the infra-red problem being extremely severe but the theory of Yang-Mills fields with mass is not, as one can see by considering diagrams with more than one loop. A renormalizable theory can be obtained if the vector mesons are given a mass by the Higgs mechanism.

2. THE HIGGS MODEL

The Lagrangian of the Higgs model is

$$L = -\frac{1}{4} F^2_{\mu\nu} - |(\partial_\mu - ieA_\mu)\varphi|^2 - \mu^2 |\varphi|^2 - h|\varphi|^4 ,$$

where φ is a complex scalar field and $F_{\mu\nu} = \partial_\nu A_\mu - \partial_\nu A_\mu$. Except for the self-interaction of the scalar field, it looks like the electrodynamics of a scalar. It is invariant under the gauge transformation

$$A_\mu \rightarrow A_\mu + \partial_\mu \Lambda , \qquad \varphi \rightarrow \varphi e^{ie\Lambda} ,$$

and, for $\mu^2 > 0$, $h > 0$ corresponds to a massive complex scalar (bare mass μ) and a massless vector.

On the other hand, let $\mu^2 < 0$, $h > 0$. In this case the potential

$$V(|\varphi|) = \mu^2 |\varphi|^2 + h |\varphi|^4$$

will have its minimum not for $|\varphi| = 0$ but for $|\varphi| = \frac{\lambda}{\sqrt{2}}$ where

$\lambda = \sqrt{-\frac{\mu^2}{h}}$. This indicates that in the quantized theory the field will have a non-vanishing vacuum expectation value, which to lowest order (tree approximation) is given b- $\lambda/\sqrt{2}$. Using the invariance of the Lagrangian this vacuum expectation value can always be chosen real, so that if one writes

$$\varphi = \frac{1}{\sqrt{2}}(\lambda + \varphi_1 + i\,\varphi_2)$$

the two real fields φ_1 and φ_2 have vanishing vacuum expectation values.

Let us now introduce the two new real fields χ and Θ by the transformation

$$\varphi = \frac{1}{\sqrt{2}}(\lambda + \chi)e^{i\frac{\Theta}{\lambda}}$$

Since

$$\varphi_1 = \chi + \ldots$$

$$\varphi_2 = \Theta + \ldots$$

where the dots denote terms at least bilinear in the fields, the transformation between φ_1, φ_2 and χ, Θ is invertible and one expects that eihter set of fields could be used with the same result for all on shell amplitudes.

The above gauge transformation leaves χ unchanged while Θ transform as $\Theta \to \Theta + e^{\lambda}\,\Lambda$. It is clear that, if one chooses $\Lambda = -\frac{1}{e^{\lambda}}\Theta$, one transforms Θ to zero. Therefore, taking

$$A_\mu = B_\mu + \frac{1}{e^{\lambda}}\partial_\mu\Theta$$

the Lagrangian becomes independent of Θ and contains only the gauge invariant fields χ and B_μ. Expressed in terms of these fields it takes the forms

$$L = -\frac{1}{4}F_{\mu\nu}^2 - \frac{1}{2}(\partial_\mu\chi)^2 - \frac{1}{2}e^2\lambda^2B_\mu^2 - \frac{1}{2}e^2B_\mu^2(2\lambda\chi + \chi^2)$$

$$-\frac{1}{2}(\mu^2 + 3h\lambda^2)\chi^2 - \frac{h}{4}(4\lambda\chi^3 + \chi^4) - (\mu^2 + h\lambda^2)\lambda\chi - \frac{\mu^2}{2}\lambda^2 - \frac{h}{4}\lambda^4.$$

The coefficient $\mu^2 + h\lambda^2$ of the term linear in χ vanishes and the constant $-\frac{\mu^2\lambda^2}{2} - \frac{h}{4}\lambda^4$ can be dropped.

Now $F_{\mu\nu} = \partial_\mu B_\nu - \partial_\nu B_\mu$. This Lagrangian describes the interaction of a vector field B_μ, of mass $e\lambda$, with a real scalar field χ, of mass $\mu^2 + 3h\lambda^2 = -2\mu^2 > 0$. No gauge group is left, since it has been possible to express the Lagrangian completely in terms of gauge invariant fields. However, the various tri- and quadri-linear couplings are related in a particular way, which is the remnant of the original gauge invariance. Observe that the possibility of eliminating completely the phase Θ depends on the non-vanishing of λ. In the case $\mu^2 > 0$, $\lambda = 0$, there is no invertible field transformation like the above which permits to eliminate Θ.

3. A YANG-MILLS THEORY FOR LEPTONS (Weinberg, Salam)

Let us consider the case of the electron e and its neutrino ν_e; the muon and its neutrino can be treated similarly. We know that the electromagnetic current $\bar{e}\gamma_\mu e$ and the weak current $\bar{e}\gamma_\mu\frac{1+\gamma_5}{2}\nu_e$ (together with its hermitian conjugate) play a dynamical role. We wish to find the minimal Yang-Mills type group which involves these currents.

Introducing the leptonic left-handed doublet

$$L = \frac{1+\gamma_5}{2}\begin{pmatrix}\nu_e \\ e\end{pmatrix}$$

one can write the weak current as

$$\bar{e}\gamma_\mu \frac{1+\gamma_5}{2}\nu_e = \bar{L}\gamma_\mu \tau_- L$$

and its hermitian conjugate as $\bar{L}\lambda_\mu \tau_+ L$. Since $[\tau_+, \tau_-] = \tau_3$, we see that commuting these two currents one obtains the neutral current

$$-\bar{e}\gamma_\mu \frac{1+\gamma_5}{2} e + \bar{\nu}_e \gamma_\mu \frac{1+\gamma_5}{2}\nu_e = \bar{L}\gamma_\mu \tau_3 L.$$

These three currents form an SU(2) structure. In order to obtain the electromagnetic current one must introduce another neutral current. Observe that the current

$$\bar{e}\gamma_\mu \frac{1+\gamma_5}{2} e + \bar{\nu}_e \gamma_\mu \frac{1+\gamma_5}{2}\nu_e = \bar{L}\gamma_\mu L$$

commutes with the above SU(2) group. Furthermore, introducing the right-handed singlet

$$R = \frac{1-\gamma_5}{2} e,$$

we notice that the current

$$\bar{e}\gamma_\mu \frac{1-\gamma_5}{2} e = \bar{R}\gamma_\mu R$$

also commutes with the SU(2) group. The electromagnetic current can be written as

$$\bar{e}\gamma_\mu e = \frac{1}{2}(-\bar{L}\gamma_\mu \tau_3 L + \bar{L}\gamma_\mu L) + \bar{R}\gamma_\mu R.$$

Therefore it is sufficient to adjoin to the currents of the SU(2) group the commuting current

$$\frac{1}{2}\bar{L}\gamma_\mu L + \bar{R}\gamma_\mu R.$$

We see that a leptonic SU(2) x U(1) is the minimal Yang-Mills type group which contains the electromagnetic and weak currents. We see also that it is necessary to introduce a neutral current other than the electromagnetic one, and therefore a neutral intermediate vector boson besides the photon.

The Yang-Mills fields A_μ and B_μ of the SU(2) x U(1) group will be coupled to the corresponding currents giving the interaction

$$i\,\frac{g}{2}\,\bar{L}\gamma_\mu \vec{\tau} L\,\vec{A}_\mu - i\,g'\left(\frac{1}{2}\,\bar{L}\gamma_\mu L + \bar{R}\gamma_\mu R\right)B_\mu .$$

The coupling constants g and g' are independent. It is clear that the charged intermediate boson is given by

$$W_\mu^- = \frac{A_{\mu 1} + iA_{\mu 2}}{\sqrt{2}} ,$$

and its hermitian conjugate, and must be given a mass. A certain linear combination Z_μ of $A_{\mu 3}$ and B_μ must also be massive, while the orthogonal combination A_μ, which describes the photon must remain massless. The photon field is determined by the fact that it couples to the electromagnetic current.

For the electric charge e one finds

$$\frac{1}{e^2} = \frac{1}{g^2} + \frac{1}{g'^2} ,$$

Furthermore,

$$Z_\mu = \frac{g A_{\mu 3} - g' B_\mu}{\sqrt{g^2 + g'^2}} \qquad A_\mu = \frac{g' A_{\mu 3} + g B_\mu}{\sqrt{g^2 + g'^2}}$$

Observe that $g > e$ and $g' > e$.

If initially all vector fields are masssless, we must now introduce masses so that the photon field remains massless. This could be done by introducing the masses for the fields W_μ and Z_μ by hand. One knows, however, that the massive Yang-Mills theory is not renormalizable, although less divergent than a theory with arbitrary couplings. Therefore, one prefers to introduce the vector masses by means of the Higgs mechanism, which breaks Yang-Mills invariance in a less brutal way and can be hoped not to spoil the renormalizability of the zero mass theory.

One introduces a scalar

$$\varphi = \begin{pmatrix} \varphi^+ \\ \varphi^\circ \end{pmatrix}$$

which is an insodoublet with respect to the leptonic SU(2) group. Its coupling to the vector fields must be such that the electromagnetic field remains massless when the field φ° acquires a non-vanishing vacuum expectation value λ as a consequence of a suitable invariant self-interaction such as

$$-\mu^2 (|\varphi^+|^2 + |\varphi^\circ|^2) - h(\varphi^+|^2 + |\varphi^\circ|^2)^2$$

This can be achieved by adding to the Lagrangian a term

$$-|(\partial_\mu - i\frac{g}{2}\vec{\tau}\cdot\vec{A}_\mu - i\frac{g'}{2} B_\mu)\varphi|^2$$

This expression is invariant under the SU(2) x U(1) group (by transforming appropriately φ under the U(1) group) and the couplings are chosen so that the electromagnetic field does not couple to φ^0. In order to give the argument in the same form as we have done for the Higgs model, let us introduce, instead of the four real fields entering in φ the four new real fields $\vec{\Theta}$ and χ by the invertible transformation

$$\begin{pmatrix} \varphi^+ \\ \\ \varphi^0 \end{pmatrix} = e^{\frac{i}{\lambda}\vec{\Theta}.\vec{\tau}} \begin{pmatrix} 0 \\ \\ \frac{\lambda + \chi}{\sqrt{2}} \end{pmatrix} = \begin{pmatrix} \frac{i\Theta_1 + \Theta_2}{\sqrt{2}} + \ldots\ldots \\ \\ \frac{\lambda + \chi - \Theta_3}{\sqrt{2}} + \ldots\ldots \end{pmatrix}$$

Since the Lagrangian is gauge invariant, Θ can be transformed away and the net effect is equivalent to the replacement

$$\begin{pmatrix} \varphi^+ \\ \\ \varphi^0 \end{pmatrix} \longrightarrow \begin{pmatrix} 0 \\ \\ \frac{\lambda + \chi}{\sqrt{2}} \end{pmatrix}$$

together with a reinterpretation of the vector and spinor fields. There results a mass $m_W = \frac{\lambda g}{2}$ for the charged field W_μ and a mass $m_Z = \frac{\lambda}{2}\sqrt{g^2 + g'^2}$ for the neutral field Z_μ, with $m_Z > m_W$. The Fermi constant is given by

$$\frac{G}{\sqrt{2}} = \frac{g^2}{8m_W^2} = \frac{1}{2\lambda^2} .$$

Because of the chiral nature of the Yang-Mills group of our model the spinor fields L and R must also start by being massless. Their masses can also be generated by means of the interaction with the field φ.

The interaction

$$-\frac{f}{\sqrt{2}}(\bar{R}\,\varphi^{*}L + \bar{L}\varphi R)$$

which is invariant under the SU(2) x U(1) gauge group gives rise, by the above described replacement for φ, to an electron mass $m_e = f\lambda$, while the neutrino remains massless. Since the coupling constant f is arbitrary one can obtain the mass of the muon in the same way by choosing a different value for f in that case. The couplings of the χ field to the leptons turn out then to be proportional to the lepton mass.

4. EXTENSION OF THE THEORY TO HADRONS

In trying to extend the model to hadrons one problem immediately presents itself. As we have seen, the Yang-Mills structure, desirable because it gives a more convergent theory, requires the introduction of a neutral weak current. In the case of hadrons, it is known that $\Delta S = 1$ neutral currents, if they are at all present, are enormously suppressed with respect to the charged currents. Hadronic $\Delta S = 0$ neutral currents are acceptable, with a strength perhaps up to one tenth of the charged currents. An extension of the theory to hadrons must therefore be formulated so as to avoid $\Delta S = 1$ neutral currents. At the same time it is crucial to test experimentally the existence of neutral leptonic and $\Delta S = 0$ hadronic currents. The present experimental situation is unclear.

It is easy to see that, in the usual quark picture with three quarks, the $\Delta S = 1$ neutral currents are unavoidable. Let us denote the three quarks as p, n, λ. Omitting gamma matrices and space - time indices, the SU(3) structure of the charged Cabibbo current can be written as

$$\bar{p}\,(cn + s\lambda) = (\bar{p}\;\bar{n}\;\bar{\lambda})C_{+}\begin{pmatrix} p \\ n \\ \lambda \end{pmatrix}$$

where c and s are the cosine and sine of the Cabibbo angle and the matrix

$$C_+ = \begin{pmatrix} 0 & c & s \\ 0 & 0 & 0 \\ 0 & 0 & 0 \end{pmatrix}$$

The hermitian conjugate current has the same form with the hermitian conjugate matrix

$$C_- = \begin{pmatrix} 0 & 0 & 0 \\ c & 0 & 0 \\ s & 0 & 0 \end{pmatrix}$$

The commutator of the two charged currents involves the commutator of the two matrices

$$\left[C_+, C_-\right] = \begin{pmatrix} 1 & 0 & 0 \\ 0 & -c^2 & -sc \\ 0 & -sc & -s^2 \end{pmatrix}$$

A Yang-Mills theory will necessarily involve the neutral current

$$(\bar{p}\ \bar{n}\ \bar{\lambda}) \left[C_+, C_-\right] \begin{pmatrix} p \\ n \\ \lambda \end{pmatrix} = \bar{p}p - c^2\,\bar{n}n - s^2\,\bar{\lambda}\lambda - sc(\bar{n}\lambda + \bar{\lambda}n)$$

which clearly induces $\Delta S = 1$ transitions.

Glashows, Iliopoulos and Maiani have found a way out of this difficulty. They introduce a fourth quark p' and write the charged current as

$$\bar{p}'(-sn + c\lambda) + \bar{p}(cn + s\lambda) = (\bar{p}' \; \bar{p} \; \bar{n} \; \bar{\lambda})C_+ \begin{pmatrix} p' \\ p \\ n \\ \lambda \end{pmatrix}$$

where c and s are as before but now

$$C_+ = \begin{pmatrix} 0 & 0 & -s & c \\ 0 & 0 & c & s \\ 0 & 0 & 0 & 0 \\ 0 & 0 & 0 & 0 \end{pmatrix}$$

The commutator of C_+ with its hermitian adjoint C_- is now purely diagonal

$$\left[C_+, C_-\right] = \begin{pmatrix} 1 & & & 0 \\ & 1 & & \\ & & -1 & \\ 0 & & & -1 \end{pmatrix}$$

and the corresponding neutral weak current $\bar{p}'p' + \bar{p}p - \bar{n}n - \bar{\lambda}\lambda$ conserves strangeness. It is clear that the same result can be achieved in other models with more than three quarks, although this seems the simplest.

From the point of view of the strong interactions, the fourth quark indicates an approximate SU(4) invariance and the existence of a new class of particles ("charmed" particles) characterizes by a new quantum number (the "charm"). This new quantum number could be rigorously conserved by the strong interactions. Observe that the neutral weak current conserves charm, but the charmed particles can decay very rapidly via the charged currents. On the other hand, the vanishing of $\Delta S = 1$ matrix elements of the neutral weak current is true even in presence of SU(4) and SU(3) breaking. Therefore, to lowest order in the weak interactions,

$K^o \rightarrow \mu^+ + \mu^-$ is absent. What is the effect of higher order weak and electromagnetic interactions? Even if their contribution is finite, one may fear that they may give a result numerically too large. This would indeed be true if one found that the effective cut-off, corresponding to the quadratic divergences of the usual theory without neutral intermediate boson, is now the mass of the intermediate boson. Fortunately, this is not the case and one can prove that if SU(4) were valid, the amplitude for K_1 or K_2 into $\mu^+ + \mu^-$ would vanish to all orders. Therefore, the amplitude is expected to be proportional to the strenght of SU(4) breaking and, being of higher order, may turn out to be sufficiently small.

5. RENORMALIZATION AND THE PROBLEM OF ANOMALIES

When the Yang-Mills group involves γ_5 transformations, the Ward identities can be expected to be spoiled by the presence of anomalies and unitarity becomes problematic. Indeed, if one considers diagrams such as

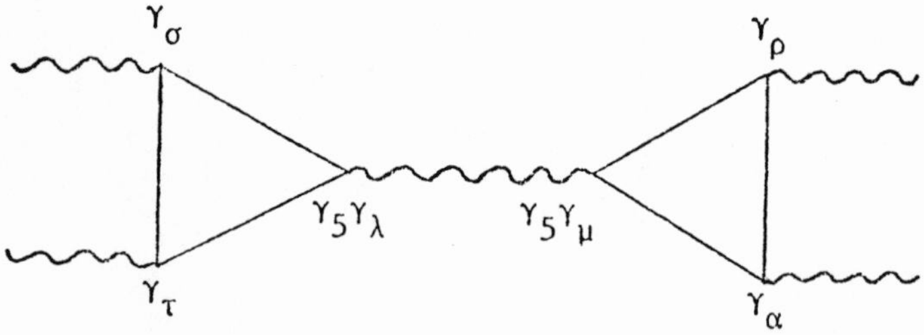

and

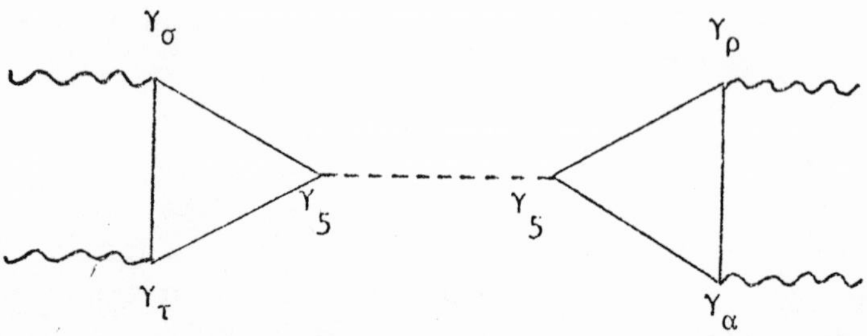

where the wavy lines represent vectors and the dotted line the scalar ghost, one sees immediately that the cancellation of the singularity for $k^2 = 0$ in the vector and in the scalar propagator requires the Ward identities associated with the λ_5 gauge transformations. If Adler type anomalies are present in the Ward identities the cancellation no longer occurs and one is left with an unphysical singolarity at $k^2 = 0$ (in the manifestly unitary formulation the anomalies complicate the divergence problem). Fortunately, it is possible to give models in which the anomalies due to various spinor fields cancel (at least in the lowest order loops such as the triangular loops above). The condition of cancellation gives rise to interesting restrictions for the couplings of the elementary spinor fields to the vectors.

Let us put all elementary spin one-half fields into a single column ψ, which contains therefore all leptons and all quarks, and let us denote by

$$L = \frac{1+\gamma_5}{2}\psi \quad \text{and} \quad R = \frac{1-\gamma_5}{2}\psi$$

its left and right-handed parts. In analogy with the case of ν_e and e we write the interaction of ψ with the vector fields as

$$i\bar{L}\gamma_\mu\left(\frac{g}{2}\vec{C}\cdot\vec{A}_\mu - \frac{g'}{2}C_o B_\mu\right)L - i\bar{R}\gamma_\mu g' D B_\mu R.$$

For instance, in the case of the doublet ν_e, e the matrices $\vec{C}$, C_o and D were respectively $\vec{\tau}$,1 and $\frac{1-\tau_5}{2}$. In general, the Yang-Mills properties of the vector fields require that the matrices $\vec{C}$ satisfy the algebra of SU(2) and that C_o commute with them. These matrices as well as D will in general be in reduced form, with submatrices corresponding to various groups of spinor fields. The electromagnetic potential, which is a known linear combination of $A_{\mu 3}$ and B_μ should have no axial vector coupling, and its vector coupling identifies the electromagnetic current. From this it follows that

$$D = -Q, \qquad C_o = -2Q + C_3$$

Where Q is the (diagonal) charge matrix.

Separating the vector and the axial vector parts, the above interaction can be written as

$$i\bar{\psi}\gamma_\mu \mathcal{V}_\mu \psi + i\bar{\psi}\gamma_\mu \gamma_5 \mathcal{A}_\mu \psi$$

where the vector and axial vector matrices are

$$\mathcal{V}_\mu = \frac{g}{4}\vec{C}\cdot\vec{A}_\mu + \frac{g'}{4} C_3 B_\mu - g'(-q + \frac{C_3}{2}) B_\mu$$

$$\mathcal{A}_\mu = \frac{g}{4}\vec{C}\cdot\vec{A}_\mu - \frac{g'}{4} C_3 B_\mu$$

In general the anomaly can be expressed completely in terms of the structure constants of the group under consideration and of the totally symmetric d_{abc} symbol, defined as usual as being proportional to the $\mathrm{Tr}\,\lambda_a \{\lambda_b , \lambda_c\}$ where λ_a are the matrices representing the generators in the representation realized by the fundamental spinor fields of the theory. For any group involving vector and axial vector currents and for any representation, the condition for the cancellation of the anomaly is just that the d_{abc} symbol vanish. In our case the matrices representing the generators are C_i $(i = 1,2,3)$ and $C_o = -2Q + C_3$. Since anyway

$$\mathrm{Tr}\, C_i \{C_j , C_k\} = 0$$

(there is no anomaly in SU(2)), and there is no axial vector coupling to C_o, we are left with

$$\mathrm{Tr}\, C_i C_j C_o \equiv \mathrm{Tr}\, C_i C_j(-2Q + C_3) = 0$$

The trace of the left-hand side is proportional to δ_{ij} and $\mathrm{Tr}\, C_i C_j C_k$ to ε_{ijk}. Therefore, it is sufficient to require

$$\mathrm{Tr}\, C_3^{\,2}\, Q = 0,$$

which is the basic restriction for absence of anomalies. In practice, this equation can be further simplified, since in our model $C_3^2 = 1$. This is true for the leptons as well as for the four quarks of the previous section. We are then left with

$$\mathrm{Tr}\ Q = 0,$$

the sum of the charges of all elementary spin one-half fields (leptons and quarks) must vanish. If we assume that the only leptons in nature are the usual electron and muon with their neutrinos, their charges add up to -2. Then

$$\mathrm{Tr}\ Q_h = 2,$$

where Q_h is the purely hadronic part of the charge matrix.

To the above restrictions on the charges of the hadronic fields one may add that which comes from requiring that the purely hadronic part of the anomaly give the correct value and sign of the $\pi^o \to 2\gamma$ decay amplitude

$$2\ \mathrm{Tr}\ T_3\ Q_h^2 \quad 1$$

where T_3 is the hadronic isospin matrix. Finally, of course, one must be able to construct the usual baryons with their correct charges. All these requirements, together with that discussed in the previous section of absence of $\Delta S = 1$ neutral currents, restrict considerably the possible models of hadrons. A model which satisfies all these requirements consists of three quadruplets of quarks, the quarks p', p, n, λ in each quadruplet having charges $\frac{2}{3}, \frac{2}{3}, -\frac{1}{3}, -\frac{1}{3}$ and $T_3 = 0, \frac{1}{2}, -\frac{1}{2}, 0$ respectively. The construction of the baryon octet is the same as in the usual SU(3) quark model (p' does not enter) except that a baryon is composed of three quarks taken one from each of the three quartets. As a side bonus, in this model there is no problem in understanding the symmetric baryon ground states since the tree fermions in it are different. Another possible model having the same advantage is the generalization to SU(4) of a model suggested for SU(3) by Han and Nambu [16]. It is a tree quartet model with charges (1, 1, 0, 0) (1, 1, 0, 0) and (0, 0, -1, -1). In spite of the existence of these possibilities one must admit that the hadronic part of all models proposed until now is still far from satisfactory.

INDEX

Date Due